Microemulsions: Structure and Dynamics

Editors

Stig E. Friberg, Ph.D., D.Sci.

Curators' Distinguished Professor
Department of Chemistry
University of Missouri
Rolla, Missouri

Pierre Bothorel

Professor
Centre Recherches Paul Pascal
Domaine Universitaire
Talence, France

CRC Press, Inc.
Boca Raton, Florida

7346-7625

CHEMISTRY

Library of Congress Cataloging-in-Publication Data

Microemulsions : structure and dynamics.

Includes bibliographies and index.
1. Emulsions.   I. Friberg, Stig, 1930-
II. Bothorel, Pierre.
TP156.E6M518   1987   660.2'94514   86-6168
ISBN 0-8493-6598-8

Direct all inquiries to CRC Press, Inc., 2000 Corporate Blvd., N.W., Boca Raton, Florida, 33431.

© 1987 by CRC Press, Inc.

International Standard Book Number 0-8493-6598-8

Library of Congress Card Number 86-6168
Printed in the United States

FOREWORD

The progress of microemulsion research has taken place in well-defined stages. The introduction period was founded on Schulman's original discovery and was, as expected, focused on the interfacial free energy, because Schulman obtained his microemulsions from a macroemulsion by the addition of a cosurfactant.

In the second stage the microemulsion systems were derived from micellar solutions by the Scandinavian groups. This development was also as expected against the strong research on surfactant association structures by Ekwall and his collaborations. In this period, the value of a phase diagram representation was realized and the relation between long range order anisotropic structures and the microemulsions clarified.

The present stage is characterized by an extensively enhanced knowledge about structure and dynamics in these systems. This has led to the realization that the structure of the microemulsions is related both to solutions with critical behavior and long range order structures, the lyotropic liquid crystals. These two aspects have been elucidated independently by the French groups and by the Lund spectroscopy group.

With these recent developments at a mature state, we felt that a book describing them in a systematical manner would be well justified. We have been very fortunate to attract outstanding authorities in these different aspects of the microemulsion phenomenon and hope this book will not only provide an up-to-date introduction to microemulsions but also serve as a basis for further development in the area.

Stig E. Friberg
Pierre Bothorel

THE EDITORS

Dr. Stig E. Friberg is Curators' Distinguished Professor of Chemistry at the University of Missouri-Rolla.

He obtained his undergraduate education at University of Stockholm, continuing with his graduate investigations at the Swedish Research Institute for National Defense. After his D.Sc. in 1966 and Ph.D. in 1967, he accepted the Directorship when the Swedish Institute for Surface Chemistry was founded in 1969. This institute developed into an internationally renowned research center, with more than 60 researchers employed, in the 7 years of Dr. Friberg's directorship.

In 1976, Dr. Friberg became Chairman of the Chemistry Department at the University of Missouri-Rolla. As a reward for his contributions to the Department, Dr. Friberg was named the first Curators' Professor at the Rolla campus in 1979.

In parallel with these heavy administrative responsibilities, Dr. Friberg has been active in research, with more than 300 publications and four books to his name, as well as countless seminars, lectures, and continuing education courses. His research has, among other phenomena, resulted in the following scientific discoveries: the influence of liquid crystals on emulsion stability (1970), a mechanism for foam stability in nonaqueous systems (1971), an explanation for hydrotropic action (1972), proof of the catalysis by inverse micelles (1972), nonaqueous lyotropic liquid crystals (1979), nonaqueous microemulsions (1983), and the liquid crystalline model for the lipid part of stratum corneum (1985).

Dr. Friberg has received many honors, of which the following may be mentioned. He is a member of the Swedish Royal (National) Academy for Engineering Sciences (since 1974); he is one of four Americans to have received the Award by the Japanese Chemical Society for Excellence in Colloid Research; he was awarded the 1979 Literature Award from the Society of Cosmetic Chemists; and he received the ACS National Award in Colloid and Surface Chemistry (The Kendall Award) in 1985.

Pierre Bothorel is Professor of Chemical Physics at Domaine University, Talence, France, and Director of the Paul Pascal Research Centre, a laboratory of the Centre National de la Recherche Scientifique (CNRS).

He received his doctorate (thése d'Etat) in 1957 at Bordeaux University. He was a research fellow appointed by CNRS from 1953 to 1960 and worked in the Chemical Physics Department of the university.

Dr. Bothorel is a member of numerous scientific committees in France. He started the French Group of Microemulsions (Groupe de Recherches Coordonnées sur les Microemulsions), which includes six laboratories of chemistry and physics localized in different cities in France. The Centre Paul Pascal, of which he has been the Deputy Director since 1969 and Director since the beginning of 1986, is one of the main laboratories of chemical physics in France.

Dr. Bothorel has authored or coauthored more than 100 scientific publications in the field of chemical physics. His major research interests include the study of molecular conformations of aliphatic chains, the biophysics of phospholipidic membrane models, microemulsions, and emulsions.

CONTRIBUTORS

Anne-Marie Bellocq, D.Sc.
Maitre de Recherche CNRS
Centre de Recherche Paul Pascal
CNRS
Domaine Universitaire
Talence, France

Jacques Biais
Centre de Recherche Paul Pascal
CNRS
Domaine Universitaire
Talence, France

Bernard Clin
Centre de Recherche Paul Pascal
CNRS
Domaine Universitaire
Talence, France

Pierre Lalanne
Centre de Recherche Paul Pascal
CNRS
Domaine Universitaire
Talence, France

Jacques Lang, D.Sc.
Institut Charles Sadron
CNRS-CRM
Strasbourg, France

Dominique Langevin
Laboratoire de Spectroscopie Hertzienne
 de l'Ecole Normale Supérieure
Paris, France

Yuh-Chirn Liang, B.S.
Research Assistant
Department of Chemistry
University of Missouri
Rolla, Missouri

Björn Lindman, Ph.D.
Professor
Department of Physical Chemistry 1
Lund University
Lund, Sweden

P. Neogi, Ph.D.
Assistant Professor
Department of Chemical Engineering
University of Missouri
Rolla, Missouri

J. C. Ravey, D.Sc.
Directeur de Recherche CNRS
Laboratoire de Physico-Chimie des
 Colloides
Université de Nancy I
Faculte des Sciences
Vandoeuvre les Nancy, France

Didier Roux, D.Sc.
Chargé de Recherche CNRS
Centre de Recherche Paul Pascal
CNRS
Domaine Universitaire
Talence, France

Peter Stilbs, Ph.D.
Professor
Royal Institute of Technology
Stockholm, Sweden

Raoul Zana, D.Sc.
Institut Charles Sadron
CNRS-CRM
Strasbourg, France

TABLE OF CONTENTS

Chapter 1

PHASE DIAGRAMS AND PSEUDOPHASE ASSUMPTION

Jacques Biais, Bernard Clin, and Pierre Lalanne

TABLE OF CONTENTS

I. INTRODUCTION

The aim of this chapter is not to present a detailed study of phase diagrams for systems giving rise to microemulsions. We shall be concerned mainly with systems made of water or brine (noted W), oil (O), surfactant (S), and cosurfactant (A for alcohol). Such systems have been largely studied in Chapter 2 or in reviews and recent papers.[1] These diagrams are fairly often complex.[2] They clearly exhibit "simple" phase isotropic or oriented liquids/ or solids.[3] The different systems can be single-, two-, three-, even four-phase systems and they can be more or less close to critical and even tricritical points.[3]

In addition to that complexity (a large part of the state space for such systems) is relevant to the Winsor nomenclature,[4] two and three phase (microemulsions) states are distinguished as:

1.   Winsor I (WI) two-phase states for a microemulsion in equilibrium with an organic phase or Winsor II (WII) states for a microemulsion in equilibrium with a water phase.
2.   Winsor III (WIII) three-phase states when a microemulsion is in equilibrium both with an organic phase and an aqueous phase.

Such systems are presented in Figure 1. The ideal Winsor diagrams of Figure 2 will be labeled Winsor I, II, and III, respectively. They are usually considered relevant to water (W), oil (O), and surfactant (S) ternary systems.[5] Meanwhile, one should note that they

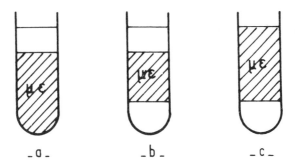

FIGURE 1. Winsor I, III, and II systems. (a) Winsor I: a microemulsion in equilibrium with an organic phase; (b) Winsor III: a microemulsion in equilibrium with an organic and an aqueous phase; (c)
Winsor II: a microemulsion in equilibrium with an aqueous phase.

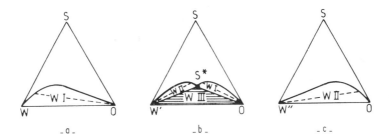

FIGURE 2. Ideal Winsor's diagrams. (a) Winsor I (WI); (b) Winsor III (WIII); (c)
Winsor II (WII).

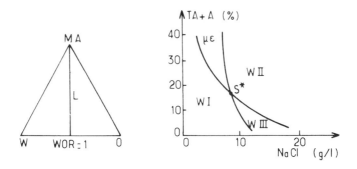

FIGURE 3. Different states (microemulsion = με, WI, WIII, WII) on a
given line (WOR = 1) in a given plane a/s.

have been only seldomly observed,[6,7] particularly Winsor III states. Nevertheless, a transition
can readily be observed, for a given system, from one Winsor state to another, e.g., WI →
WIII → WII. This can be obtained: (1) in the case of nonionic surfactant systems, by
modifying the temperature, (2) in the case of ionic surfactant systems, by increasing the
salinity of brine, (3) by modifying the surfactant-cosurfactant (active blend) concentration
ratio.

The diagram on Figure 3 corresponds to the second case and represents the different states
of a quinary system (brine, A, O, S) in a two-dimensional space: active blend (a + s) wt%
vs. salinity. In this diagram, the water/oil ratio (WOR) has been kept constant as has the
surfactant/cosurfactant ratio (S/A).

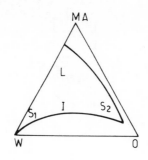

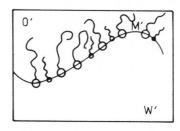

FIGURE 4.   Differentiating the three microdomains: oil (O'), water
(W'), and interface (M').

The practical interest of three-phase states need no more be emphasized. It is well known indeed that these states give rise to very low interfacial tensions between microemulsion and organic or aqueous phase.[1] Creating such a situation in a petroleum field by introducing a microemulsion, giving rise to an equilibrium between water and oil would allow an optimal recovery. This is the reason why S* systems of Figures 2 and 3 are usually named optimized systems. The best optimization, judiciously defining the physicochemical properties of active blend, should, in fact, lead to such an S* system with the minimal amount of active blend.

Hence, it appears fundamental, from a practical point of view as well as from a fundamental one, to dispose of concepts or of models allowing one to deal with properties of quaternary systems (W, A, O, S) that are clearly related to thermodynamical stability conditions. Numerous studies deal with stability[1] but difficulties encountered are too numerous and inherent to: (1) system complexity (quaternary or quinary in case of salted ones), and (2) the fact that microemulsions are structurated media, with different structures, the kind which depend, among other things, upon the composition.

One should note here that for any kind of structure, the structurated state gives rise to three kinds of domains:

1.   Aqueous microdomains, mainly constituted of water and of some alcohol. We shall call them the water pseudophase (W');
2.   Organic microdomains, constituted by oil, a part of cosurfactant, and a few water. We shall call them the oil pseudophase (O');
3.   Interface or membrane microdomains constituted by nearly the whole surfactant (especially in case of ionic ones) and the remaining cosurfactant. We shall call them the membrane pseudophase (M').

Such a structurated domain is shown on Figure 4.

Though the pseudophase model we proposed[9] cannot simultaneously deal with microemulsion stability, its development allows an interesting approach to this problem.

This model allows, from a restricted number of experimentally accessible parameters,

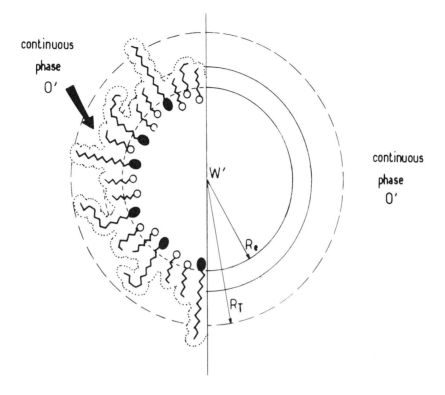

FIGURE 5.   Schematic representation of an O/W microemulsion (micellar system).

1.   An *a priori* calculation of pseudophase (W', O', M') compositions
2.   A determination in the phase states of the quaternary system (WAOS) of so-called pseudo three-phase planes, wherein the system can be treated as a ternary system made up of three pseudocompounds that are the three pseudophases W', O', and M'
3.   To preface that, when the system is multiphase, (WI, WIII, or WII), the excess phases have compositions identical to those of the microemulsion pseudophases. Therefore, in the pseudophase plane, the phase diagrams should be similar to the ideal Winsor diagrams of Figure 2
4   An easy study of phenomena like salt partitioning in Winsor II systems
5.   A theoretical interpretation of vapor pressure measurements for microemulsion components (mainly oil and alcohol). These measurements can also provide very interesting information on chemical potentials within the microemulsion phase

Hence, this chapter provides an experimental illustration of the points (1 to 5) above.

II. THE PSEUDOPHASE MODEL[8]

A. Representing Microemulsions in Terms of Pseudophases and Main Hypotheses
1. Pseudophases

In case of S_2 micelles (water-in-oil microemulsion), a microemulsion can be represented as in Figure 5.[9-11,13]

On a microscopic scale, such systems exhibit three kinds of domains or "pseudophases" — the core of the micelle, membrane or interface, and the continuous phase. These pseudophases are noted W', M', and O', respectively. Each consists of at least two components and the whole is in thermodynamical equilibrium. The volume composition notations are given in Table 1.

Table 1
VOLUME COMPOSITION
NOTATIONS FOR O′, M′, AND
W′ PSEUDOPHASES

	O′	M′	W′
Alcohol	V_A^O	V_A^M	V_A^W
Oil	V_O		
Surfactant		V_S^M	
Water	V_W^O		V_W^W

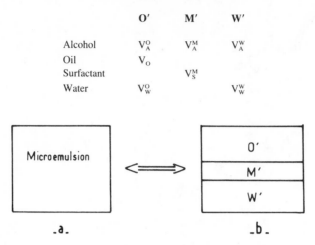

FIGURE 6. Modeling of a microemulsion in terms of three pseudophases in equilibrium (oil [O′], membrane [M′], water [W′]).

2. Hypotheses

The thermodynamical stability of microemulsions requires[11] very low interfacial tensions (approximately 10^{-3} mN/m). In such conditions, discrepancy between inner and outer pressure for the micelle is very low. In case of liquids it can be neglected and this leads to the following hypotheses:

First hypothesis: The chemical potentials of every compound only depend, within a pseudophase, upon the composition of this pseudophase and in no way upon the geometrical structure.

Second hypothesis: Though the aggregate size can be rather low (some tenths of angströms as the radius of micellar microemulsions), and although the membrane psuedophase is likely a more or less complex monolayer, we shall employ the hypothesis that pseudophase composition can be considered within the classical frame of interphase equilibria thermodynamics.

The equilibrium, for a given compound i, between the three pseudophases O′, M′, and W′, will be governed by

$$\mu_i^{O'} = \mu_i^{M'} = \mu_i^{W'} \tag{1}$$

where μ_i^j is the chemical potential of species i in pseudophase j.

Hence, any geometrical feature is removed from the model and for our purpose, the system can be stylized as in Figure 6. An experimental vindication for this hypothesis has been obtained[15] through partial vapor pressure measurements for oil and alcohol; the results of microemulsion systems have been compared to those of two-phase water-alcohol-oil ternary systems.

B. Models of Solutions — Chemical Potentials

To quantitatively treat the equilibria between pseudophases, i.e., of their own compositions, one has to know the explicit expression of chemical potential for every compound. Therefore, one has to choose a model of solution: this problem will be solved by successively considering each of the three pseudophases — O′, W′, and M′.

1. O' Pseudophase

The O' pseudophase is made of three components: oil (O), alcohol (A), and some water (W). We shall simplify, postulating that all the oil is in this very pseudophase. What makes this pseudophase peculiar is that the alcohol is self-associated. This leads to complexes (dimers, trimers, etc.).[16,17]

$$2 \text{ R-OH} \rightarrow \text{R-O} \overset{H}{\underset{\cdot\cdot H}{\diagup\diagdown}} \text{O-R}$$

Hence, one has to choose not only a model of solution but a self-association model for the alcohol: we shall call it the association mode.

Model of solution — The model we chose is the one of "ideal associated solutions" of Prigogine[16] with the entropic correction of Flory.[19] Expressing chemical potentials, this model led us to use volume (ϕ_i) in place of molar fractions.

$$\phi_i \text{ is defined as } \phi_i = \frac{n_i v_i}{n_s v_s + \Sigma n_i v_i} \tag{2}$$

where v_s is the molar volume of solvent and v_i the molar volume of ith order associated species. Excess molar volumes are supposed to be negligible.

$$\mu_i = \mu_i^* + RT \text{ Ln } \phi_i \tag{3}$$

Association mode — If $A_1, A_2 \ldots, A_i \ldots$ are the complexes formed, the equilibria can be written

$$A_1 + A_i \rightleftarrows A_{i+1} \qquad K_i$$

where A_i is the ith order associated species for the alcohol and K_i the relevant self-association constant. Between numerous proposed modes,[18,20,21] we chose the one leading to the simplest mathematical form: the Mecke's hypothesis.[22] This author postulates that all the self-association constants are identical.

$$K_1 = K_2 \ldots = K_i = \ldots\ldots = K \tag{4}$$

The knowledge of K and of the volume fraction of alcohol ϕ_A allows one to calculate volume fractions for each species and particularly for monomers (ϕ_{A_1})

$$\phi_{A_1} = \frac{\phi_A}{K \cdot \phi_A + 1} \tag{5}$$

Prigogine's theorem[17] — Prigogine has shown that for any association mode the self-associated alcohol chemical potential (in a given phase) is, in fact, identical to the chemical potential of the monomer.

$$\mu_A^o = \mu_{A_1}^o \tag{6}$$

where μ_A^o holds for potential of alcohol in oil and $\mu_{A_1}^o$ for monomers in oil.

$$\mu_A^o = \mu_{A_1}^o = \mu_{A_1}^{o*} + RT \text{ Ln } \phi_{A_1}^o \tag{7}$$

Chemical potentials for water and oil can be written for their part

$$\mu_W^O = \mu_W^{O*} + RT \, Ln \, \phi_W^O$$

$$\mu_O^O = \mu_O^{O*} + RT \, Ln \, \phi_O^O \tag{8}$$

where ϕ_W^O and ϕ_O^O hold for volume fractions of O/W in oil pseudophase.

2. Water Pseudophase W'

The water pseudophase is mainly constituted of water and some alcohol. Commonly used alcohol usually has very poor solubility in water and hence it can be postulated that no aggregates are formed. Therefore, the alcohol is looked at in this pseudophase as free solvated monomers:

$$\mu_A^W = \mu_A^{W*} + RT \, Ln \, \phi_A^W$$

$$\mu_W^W = \mu_W^{W*} + RT \, Ln \, \phi_W^W \tag{9}$$

where ϕ_A^W and ϕ_W^W are alcohol and water volume fractions in water pseudophase.

3. M' Pseudophase

The interface is made of alcohol and surfactant. The study of this very pseudophase is undoubtedly the most difficult. We postulate that the membrane is a homogeneous medium wherein the alcohol chemical potential depends only upon the concentration. This hypothesis is realistic in view of experimental evidence for weak variations of membrane composition vs. global system composition variations.[8,23]

Therefore, the activity coefficients can be looked at as nearly constant. Then, to not make explicit use of them is equivalent to including them in the standard reference potential.

Chemical potentials in membrane pseudophase then can be written

$$\mu_s^M = \mu_s^{M*} + RT \, Ln \, \phi_s^M$$

$$\mu_A^M = \mu_A^{M*} + RT \, Ln \, \phi_A^M \tag{10}$$

where ϕ_s^M and ϕ_A^M are surface fractions of surfactant and alcohol defined as

$$\phi_s^M = \frac{\alpha \, V_S^M}{V_A^M + \alpha \, V_S^M} \qquad \phi_A^M = \frac{V_A^M}{V_A^M + \alpha \, V_S^M}$$

where α is the ratio of mean lengths of alcohol and surfactant: $\alpha = l_A/l_S$.

C. Equilibria Constants — Pseudophase Compositions Calculation

1. Equilibria Constants

The system is defined by the knowledge of volumes of the four components: V_w for water, V_O for oil, V_A for alcohol, and V_S for surfactant. At equilibrium, chemical potentials have to satisfy the following relations:

1. Partition of alcohol between O', M', and W':

$$\mu_{A_1}^O = \mu_A^M = \mu_A^W \tag{11}$$

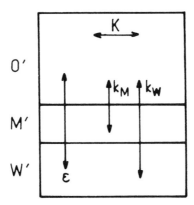

FIGURE 7. The different constants associated to the equilibria between the three pseudophases (O′, M′, W′).

Table 2
PARTITION AND ASSOCIATION CONSTANTS OF EQUILIBRIA BETWEEN PSEUDOPHASES, AS FUNCTIONS OF VOLUME FRACTIONS

K	K_w	K_m	ϵ'
$K = \dfrac{\phi^o_{A2}}{(\phi^o_{A1})^2} = \cdots = \dfrac{\phi^o_{A(i+1)}}{\phi^o_{A1}\,\phi^o_{Ai}}$	$K_w = \dfrac{\phi^w_A}{\phi^o_{A1}}$	$K_m = \dfrac{\phi^M_A}{\phi^o_{A1}}$	$\epsilon' = \dfrac{\phi^o_w}{\phi^w_w}$
$K = \dfrac{\phi^o_A - \phi^o_{A1}}{\phi^o_{A1}\,\phi^o_A}$			

2. Partition of water between O′ and W′:

$$\mu^W_W = \mu^O_W \tag{12}$$

Partition constants are associated to each one of these relations as schematized in Figure 7. In Table 2, these constants are expressed as functions of volume and surface fractions.

In view of the very low solubility of water in oil, we shall suppose as a crude approximation the water solubility in O′ to be related to that of alcohol. Hence, we shall use:

$$V^O_W = \epsilon(V^O_A)^2 \tag{13}$$

This formula accounts satisfactorily for experimental values of this solubility.[8]

Remarks — When dealing with the compositions of the pseudophases O′, M′ and W′ we shall make use of the following parameters:

$$\gamma = \frac{V^O_A}{V_O} \qquad \sigma = \frac{V^M_A}{V^M_S} \qquad \lambda = \frac{V^W_A}{V^W_W} \tag{14}$$

2. Solving the Equations

A convenient way to solve the equations of Table 2 is to use the parameter γ as variable (with Approximation 13). This parameter can easily be expressed as a function of alcohol monomeric concentration (ϕ^o_{A1}), the others being an explicit function of it: for a given system (V_A, V_O, V_S, V_w) one gets the whole set of values (V^O_A, V^M_A, V^W_A, V^W_W, V^O_W) defining the pseudophases.

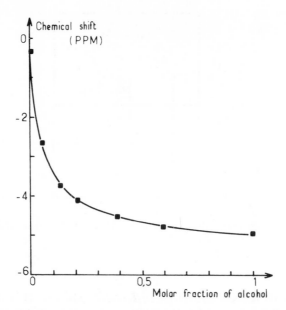

FIGURE 8.   Chemical shift of butanol hydroxylic proton (system water-toluene-butanol) vs. alcohol molar fractions: (■) = experimental points; (——) = calculated curve.

D. Making Use of the Model — Equilibria Constants Determination

1. Principle

To make use of the model, one first has to note that the equilibrium shown in Figure 7 is not modified (displaced) if one of the pseudophases is removed, for example, M'. This particularly means that K, K_w, and ϵ constants can be obtained studying the simple W, A, O system.

- K is deduced from a NMR study of the hydroxylic proton of alcohol
- K_w and ϵ fall from a study of two-phase systems (tie-lines) in the alcohol-water-oil ternary study
- K_M is the only constant that has not been studied in terms of microemulsions systems.

The experiment we ran is named the "dilution technique" and is suited to saturated water in oil systems.[23,24] It allows one to measure γ.

2. Some Examples of Constant Determinations

a. Self-Association Constant of Alcohol in O', K

The experiment gives the evolution of chemical shifts for the hydroxylic proton of alcohol when the molar fraction of alcohol is varied within the organic phase of a W, A, O system. The results obtained for butanol-toluene-water are shown in Figure 8. The agreement between theoretical curve corresponding to K = 55 and experimental points is clearly observed.

b. Water Partition Between O' and W', K_w and ϵ

From the equations of Table 2 one can calculate the compositions of water and organic phases in equilibrium in 2 W, O, A diagram, i.e., the two-phase domain tie-lines. We shall not develop these equations hereafter; we only want to point out a very simple case when the tie-line slope is reversed. Let us assume that the partition is water favorable for low alcohol concentration and becomes oil favorable for high alcohol concentration; then, the zero slope tie-line corresponds to

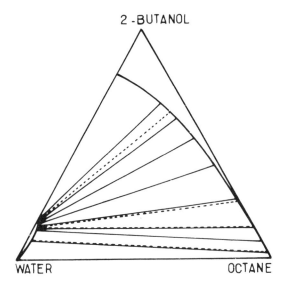

FIGURE 9. Comparison between experimental (---) tie-lines (Bavière and Coll.) and calculated (——) tie-lines for the system water-octane-2 butanol.

Table 3
NUMERICAL VALUES OF MODEL CONSTANTS
FOR DIFFERENT EXPERIMENTAL SYSTEMS

Constants systems	K	K_w	ϵ	K_m
n-Butanol octane SDS water	190	13	0.06	158
n-Pentanol dodecane OBS water	130	2.5	0.06	110
n-Hexanol dodecane SDS water	120	0.5	0.06	83

$$\frac{V_A}{V_A + V_O + V_W} = \frac{K_W - 1}{K} \tag{15}$$

where V_A, V_O, and V_W define the composition of the whole system.

In Figure 9 we present the comparison between experimental results obtained by Bavière et al.[25] on water-octane, 2-butanol (this is a system with a clear evidence for tie-line slope inversion), and the results of our calculation.

Remarks — It is important to note that this agreement can be obtained only by making use of the self-association of alcohol in oil: to get K_W one obviously has to know K as a preliminary. ϵ is obtained directly from the solubility curve of water in oil-alcohol mixtures.

c. Alcohol Partition Between M' and O', K_M

As already stated, this determination (from K, K_W, and ϵ values) needs the knowledge of at least one continuous phase (O') studied by dilution of saturated microemulsions.

Table 3 is a summary of results obtained (K, K_W, ϵ, and K_M) for different systems.

Numerous diluting experiments have been performed on these systems. We present in Figure 10 the agreement between calculated compositions for different O' (γ_{calc}) and experimental pseudophases (γ_{exp}).

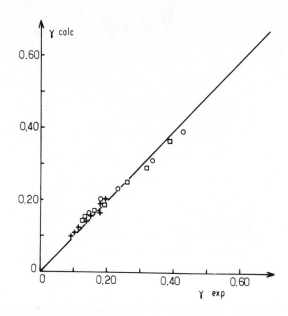

FIGURE 10.   Comparison between experimental and calculated γ for three systems: ($\square$) pentanol-dodecane-OBS-water; (+) butanol-octane-SDS-water; (O) hexanol-dodecane-SDS-water.

Remarks — One should note, regarding the values of Table 3, that:

1.   Self-association constant K and water-oil partition constant K_W decrease as alcohol chain length increases.
2.   Self-association constant K depends on the solvent, too (K has been found to be equal to 55 for *n*-butanol in toluene).
3.   A rigorous discussion of K_M and K_W variations is made quite difficult because of their dependency upon the self-association constant value (see, for instance, Equation 15).

E. The Pseudophase Model and Some of its Implications
1. Pseudotrinodal Triangles Reality

To any point P (composition defined by V_A, V_O, V_W, V_S) from the space of states (Figure 11), we can associate three pseudophases, O′, M′, and W′, the compositions of which, defined by γ, σ, and λ, can be calculated.

The points W′, M′, and O′ are the vertice of a triangle we call pseudotrinodal (pseudo three-phase triangle). Any point within such a triangle is associated to the same pseudophases (namely O′, M′, and W′). Two points within a given pseudotrinodal triangle differ only in the relative amount of pseudophases.

2. Pseudophases and Variance[1]

Studied systems are quaternary or pseudoquaternary systems (brine is commonly considered a pseudocomponent). We are concerned with the gas phase and consequently about pressure (sum of partial pressures of different components) as a variable.

Hence, for a given temperature, variance will be given by:

$$v_t = c + 1 - \Phi \qquad (16)$$

where c is the number of components; ϕ is the number of phases.

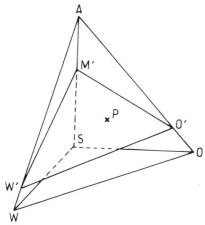

FIGURE 11.   A pseudo three-phase triangle (O′, M′, W′) in the states space for a system WAOS.

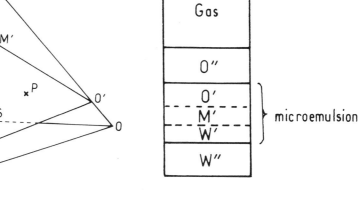

FIGURE 12.   Modeling a Winsor III system in equilibrium with its vapor. The microemulsion is represented as three pseudophases O′, M′, and W′ in equilibrium with the organic and the aqueous phases O″ and W″.

If we consider Winsor III systems (three liquid phases — a microemulsion, an organic O″, and an aqueous system W″), the experiments show that the variance is equal to 1.

$$v_t = 1 \tag{17}$$

The maximum number of phases that should comprise pseudophases O′, W′, and M′, with differing compositions, is therefore equal to 4. This clearly implies (Figure 12):

$$O″ \equiv O′ \quad \text{and} \quad W″ \equiv W′$$

Recall — (1) We are not concerned with interface bending in the extent of our nearly zero interfacial tension assumption; (2) the model is concerned with pseudophase (and now real phases in case of Winsor I, II, and III systems) compositions and not about stability.

3. A Priori Phase Diagrams in Pseudotrinodal Triangles
Preceding statements show that in pseudotrinodal triangles O′, M′, and W′ phase diagrams should look like Winsor diagrams. Peculiarly, relevant tie-lines and trinodal triangles should lie in such triangles (Figure 2).

III. PHASE DIAGRAMS IN PSEUDOPHASE SPACE

Here, we will show that according to the conclusion of Section II, tie-lines and trinodal triangles lie in pseudotrinodal triangles.

A. Experimental Winsor I Diagram
1. Studied System
We have studied systems composed of water, *n*-butanol, toluene, and sodium dodecyl sulfate (SDS) (systems noted WBTS).
Chemicals: Systems were prepared with *n*-butanol *pro analysi* (min. 99.5%, May and Baker), toluene *pro analysi* (min. 99.5%, Merck), SDS (99.9%), and doubly distilled water.
Analyses: The analyses have been accomplished by gas chromatography (Intersmat IGC 120 FB) and proton NMR (Bruker WH 270).
Experimental: Phase diagrams and analyses have been performed at 21°C.

Table 4
NUMERICAL VALUES OF MODEL
CONSTANTS FOR WBTS SYSTEM

	K	K_w	K_m	ϵ
WBTS	55	4	40	0.06

Table 5
COMPOSITION (IN VOLUME) OF
THE VERY WBTS
MICROEMULSION STUDIED

	V_A	V_O	V_W	V_S
WBTS	0.197	0.525	0.209	0.069

Table 6
COMPOSITIONS (IN VOLUME) OF
THE PSEUDOPHASES OF WBTS
MICROEMULSION OF TABLE 5

	W'	O'	M'
V_A	0.068	0.201	0.389
V_O		0.798	
V_W	0.932	0.001	
V_S			0.611

2. Equilibrium Constants of the System under Study

To obtain the values of thermodynamical constants, systems have been studied following the very procedure depicted in Section I.D. This study leads to the values reported in Table 4.

3. Experimental Study of a W', O', M' Diagram for the WBTS System

We have chosen a microemulsion system, the composition of which is given in Table 5.

With the values of constants preliminarily obtained and reported in Table 4, the model allowed us to calculate the pseudophase compositions for **any** point. In particular, we calculated the compositions of the pseudophases associated to the point defined in Table 5. They are reported in Table 6.

We then studied systematically the phase diagram in the plane defined by the three vertices W', O', and M' so defined. The results are given in Figure 13.

The crosses in the liquid-liquid two-phase domain are associated with systems in which the upper phases have been analyzed by gas chromatography as well as by NMR. These analyses do not show any surfactant in this phase, and their results are reported in Figure 14 (oil percentage in the upper phase vs. the aqueous pseudophase percentage in the macroscopic system).

Within the precision of the experimental procedure, it can be deduced that the upper phase of the Winsor I systems, analyzed in this study, is identical to the o' pseudophase: the binodal lines lie in the pseudotrinodal triangle W', O', M', and the systems associated to the vertice W', O', M' act as pseudocomponents.

4. Comparison with Results Obtained for Diagrams with Constant Alcohol/Surfactant Ratio

The phase diagram of Figure 15 has been obtained with the WBTS system for a ratio B/S:2 (weight). The systems are Winsor I systems, with the global compositions noted with crosses in the two-phase zone; their upper phases have been analyzed and the results of these analyses are reported in Figure 16 as a percentage of oil in the upper phase as a function of the percentage of water in the global system. Contrary to the results of Figure 14, one can see a large variation of compositions for these upper phases.

It should also be noted here that the composition variation in the upper phase is fitted

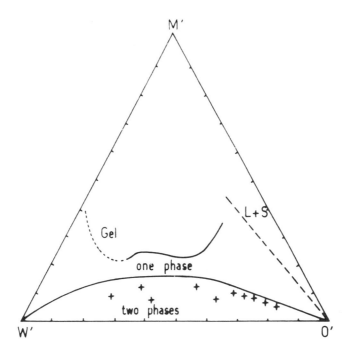

FIGURE 13. Experimental phase diagram for the WBTS system in the W′, O′, M′ pseudotrinodal triangle defined in Table 6. Crosses are Winsor I systems, the upper phases of which have been analyzed.

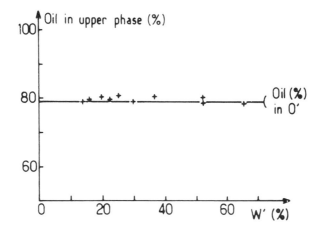

FIGURE 14. Compositions of the upper phases of Winsor I systems studied (see crosses on Figure 13) vs. the amount of water in these systems. The horizontal straight line represents the composition of the O′ pseudophase (Table 6).

very well by the model. Indeed, for several macroscopic systems, Figure 17 shows the agreement observed between the values of γ calculated with the thermodynamical constants of Table 4 (crosses in Figure 15) and the experimental values of γ measured for the upper phases of the same systems.

B. Experimental Winsor III Diagrams

In Section II.A we have shown that tie-lines of Winsor I-like diagrams lie in pseudotrinodal

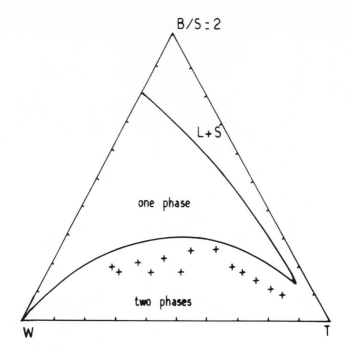

FIGURE 15.   Experimental phase diagram for the WBTS system with a given ratio B/S = 2 (weight). Crosses represent the Winsor I systems, the upper phases of which have been analyzed.

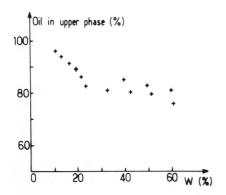

FIGURE 16.   Variation of the composition of the upper phase for the WI systems of Figure 14 (crosses) vs. the amount of water.

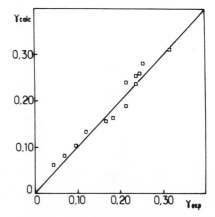

FIGURE 17.   Correlation between calculated γ (values of constants used from Table 4) and experimental for the upper phases of studied systems (crosses in Figure 14).

triangles. In this last section we will show that in Winsor III-like diagrams the pseudotrinodal triangle contains not only the microemulsions associated to the same pseudophases but also the relevant oil-microemulsion and water-microemulsion tie-lines and three-phase triangle.

1. Studied System and Chemical

We have investigated the system composed of: (1) twice distilled water, (2) *n*-pentanol puriss, *pro analysi* (Fluka), (3) decane purum (Fluka), and (4) dodecylbetaine. This surfactant has been synthesized and purified in our laboratory and corresponds to the chemical formula:

Table 7
NUMERICAL VALUES OF
MODEL CONSTANTS FOR
WATER-PENTANOL-
DECANE-
DODECYLBETAINE
SYSTEM

K	K_w	K_m	ϵ
150	2.2	120	0.06

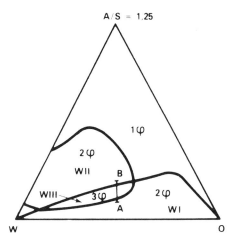

FIGURE 18. Experimental phase diagram for the studied system in a plane defined by the pentanol/betaine ratio equal to 1.25.

$$CH_3 (CH_2)_{11} - N^+ (CH_3)_2 - CH_2 - COO^-$$

The product used contains less than 0.2% of sodium chloride (wt%).

Phase diagrams have been performed at 21°C by means of a robot imagined and constructed in our laboratory; its description will be given elsewhere. In all cases, several samples have been prepared and maintained at 21°C for 1 week in order to confirm the results automatically obtained.

2. A Priori Calculation of the Composition of the Pseudophases

For the studied system, the values reported in Table 7 have been obtained in collaboration with Haouche[26] in a previous study. Before calculating the composition of the pseudophases, we studied the phase behavior in a plane defined by a constant pentanol to betaine ratio, equal to 1.25. The phase diagram is given in Figure 18. The calculation of the composition of the pseudophases has been performed for systems the compositions of which are represented by the segment AB in Figure 18.

In Figure 19, we give the different graphs in terms of the active mixture (AM) percentage (segment AB in Figure 18). The composition of the pseudophases is characterized by:

1. a/s = pentanol over betaine ratio for M'
2. % AO' = pentanol percentage for O'
3. % AW' = pentanol percentage for W'

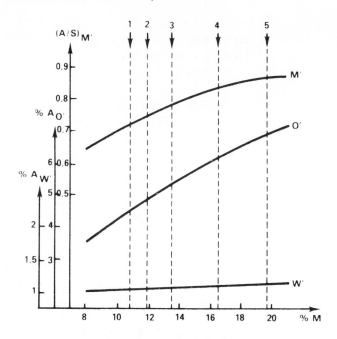

FIGURE 19.   Calculated variations of pseudophases compositions along the AB segment (see Figure 18).

3. W′, O′, M′ Phase Diagrams

Five triangles have been studied. The corresponding compositions of the associated W′, O′, M′ pseudophases are indicated in Figure 19 by five arrows numbered 1 to 5. The different phase diagrams are reported in Figure 20 (a to e).

We must point out that special attention has been devoted to the study of the transition zone marked by a dotted circle and the point T in Figure 20a. In each case (Figure 20 a to e), T appears as a well-defined point and the three-phase domain as a true W′ O′ T triangle. Winsor III type systems have been observed at W′ O′ R = 1 (W′ pseudophase to O′ pseudophase ratio) with only 1.5% of the M′ pseudophase.

4. Discussion and Conclusions

The diagram of Figure 20a to e exhibit the characteristics of typical Winsor phase diagrams; the pseudophases W′, O′, and M′ behave as real pseudocomponents. An increase of the cosurfactant to surfactant ratio a/s leads to a decrease of the interaction energy of the interface with water and consequently, a Winsor I → III → II transition is observed. This appears very clearly if we consider a system the composition of which (W′O′R = 1; 13% M′) is marked by a star on each diagram.

IV. EXTENSION OF THE PSEUDOPHASE MODEL TO THE STUDY OF SALT PARTITIONING IN WINSOR II SYSTEMS

A. Introduction

The pseudophase model allows us to calculate the composition of the water, oil, and interfacial phases or pseudophases coexisting in equilibrium in micellar systems.[8] In several cases, this model has been proved successful when choosing appropriately the pseudocomponents able to draw simplified and tractable two-dimensional phase diagrams (see Section II).

Such simplified phase diagrams, which turned out to exhibit the overall features of ideal

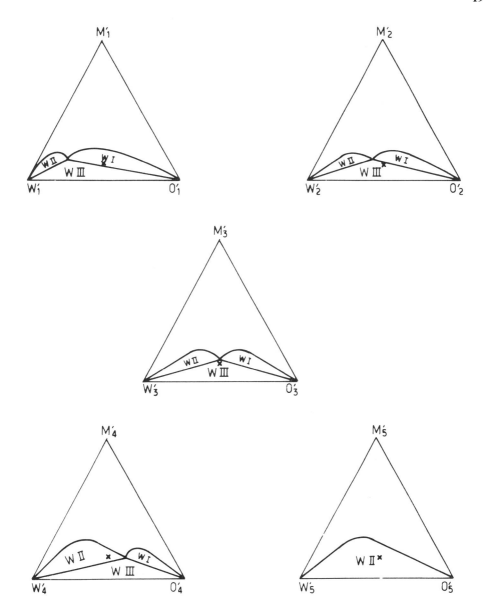

FIGURE 20. Experimental phase diagrams in W', O', M' space. The compositions of W'_i, O'_i, M'_i correspond to the ith arrow on Figure 19. (*) represents the system the composition of which is $w'_i/o'_i = 1$ and $m'_i = 13\%$.

Winsor diagrams, have recently proven convenient for incorporating the phase behavior informations into numerical simulators for enhanced oil recovery applications.[27]

The pseudophase model, however, has so far considered the brine as a pseudocomponent, an approximation not always justified, which has led to some difficulties. It is the purpose of this section to investigate the partitioning of salt between the phases of Winsor type II systems, in the presence of anionic or nonionic surfactants, and to propose a simple model allowing one to account quantitatively for the results.

B. Materials and Experimental Procedures
1. Studied Systems
Two systems have been investigated; one contains a nonionic surfactant (ethoxylated nonyl

Table 8
WEIGHT COMPOSITIONS
OF INVESTIGATED NOP
SYSTEMS

%A	%O	%W	%S
8	40	40	12
12	35	35	18

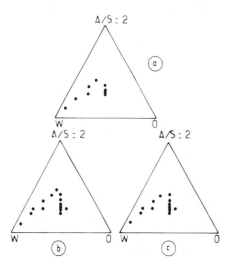

FIGURE 21.   Studied Winsor II systems compositions water (NaCl)-
toluene-butanol-SDS for different salinities: (a) 20 g/ℓ; (b) 40 g/ℓ; (c)
60 g/ℓ.

phenol with 8 ethylene oxide units, noted ENP8), *n*-pentanol, octane, water, and sodium
chloride; the other contains as anionic surfactant (SDS), *n*-butanol, toluene, water, and
sodium chloride. Throughout this chapter, these systems will be referred to as NOP and
SBT, respectively.

ENP8 exhibits the usual distribution of ethylene oxide units. SDS (Touzart et Matignon)
is 99.9% pure. *N*-Pentanol and *N*-butanol were supplied by May and Baker, octane and
toluene by Merck, and sodium chloride by Prolabo. All these chemicals are of >99.5%
purity.

A number of mixtures hve been investigated; all of them are type II in Winsor's nomenclature, i.e., a microemulsion is in equilibrium with an excess aqueous phase.

For the NOP case two systems, the weight compositions of which are given in Table 8,
have been investigated at three salinites S_w: 10, 20, and 40 g/ℓ. All of them have the same
water to oil ratio (1 in weight) and the same surfactant to alcohol ratio (1.5 in weight).

For the SBT system three salinities have been considered — 20, 40, and 60 g/ℓ NaCl.
The studied systems are represented in Figure 21.

2. Experimental

Mixtures were sealed in graduated glass pipettes, shaken several times, and then stored
for 2 days at 21 $\pm$ 0.2°C. Phase volumes were reported: the volume of the excess aqueous
phase $V_{w''}$ and the volume of water ($V_{w'}$) solubilized in the microemulsion phase (see Figure
22). The excess phase was then sampled and analyzed for the Cl$^-$ concentration by poten-
tiometry, using a Mettler DL 40 Memotitrator.

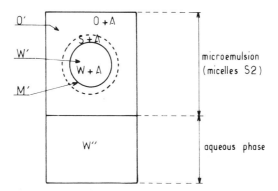

FIGURE 22. Modeling a Winsor II system: the microemulsion is represented as three pseudophases O′, M′, W′, and is in equilibrium with an aqueous phase W″.

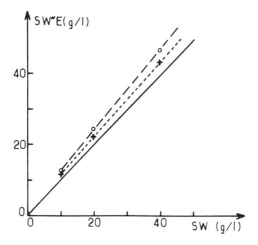

FIGURE 23. Salinity of the excess aqueous phase as a function of the initial brine salinity for the NOP systems defined in Table 8: (○) = 18% NOP; (+) = 12% NOP.

3. Experimental Results

The results of the salinity $S_{w''}$ (g/ℓ NaCl) of the excess aqueous phase measured experimentally are plotted vs. initial salinity S_w for the NOP system (Figure 23), and vs. overall alcohol + surfactant (a + S in wt%) for the SBT system (Figure 24).

Interestingly, it can be seen that, although the surfactant is nonionic type (Figure 23), S_w'' is higher than S_w, and that the difference increases with the + concentration. The scattering of the data in Figure 24 cannot be accounted for by the inaccuracy of the measurements. The magnitude of the effect is much higher than in the case of the NOP system. This obviously indicates that the surfactant ionization has to be taken into account.

C. Partitioning Model

The structure of the micellar phase in a Winsor II system is generally viewed as droplets of water W′ surrounded by a surfactant-alcohol membrane M′ in an oil phase, or pseudophase O′ containing some alcohol (Figure 22).

The hydration of the surfactant polar heads involves some water molecules which are likely salt free. This may be one mechanism causing the nonuniform partitioning of salt

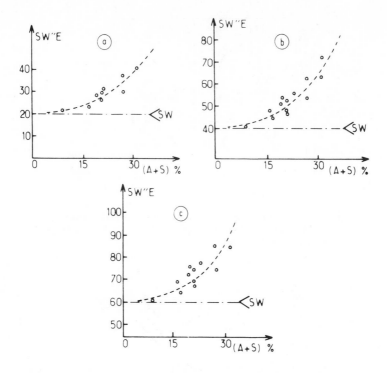

FIGURE 24. Salinity of the excess aqueous phase as a function of overall + concentration for the SBT system. Initial salinities: (a) 20 g/ℓ; (b) 40 g/ℓ; (c) 60 g/ℓ.

between W$'$ and excess aqueous phase W$''$. Another obvious mechanism which has to be taken into account is the association of sodium to micelles when dealing with anionic surfactants.

In line with the pseudophase model, we will represent the system as an equilibrium between the four phases or pseudophases O$'$, W$'$, M$'$, and W$''$ (Figure 12).

It is well known[28-30] that the mean activity coefficient of sodium chloride in aqueous solutions decreases rapidly with the salt concentration and then remains approximately constant (0.69) up to higher than 150 g/ℓ conc. Since we are dealing with salinities within the range 15 to 60 g/ℓ, it will be a good approximation to keep constant the activity coefficient.

1. Hydration of Polar Heads

It is well known that some amount of water is bound to the polar head of nonionic[31] or ionic[32] surfactant. Therefore, the pseudophase W$'$ (Figure 22) now must be divided into two such volumes V_H and V_w^* with

$$V_{w'} = V_H + V_w^* \tag{18}$$

V_H and V_w^* correspond, respectively, to the hydration water and to the free water of the microemulsion.

The pseudophase m$'$ will be noted m* and can be viewed as incorporating the water volume V_H. The water pseudophase, noted w*, is reduced to the volume V_w^*.

If n_s is the number of moles of surfactant in the micellar phase, h the number of moles of water hydrating one surfactant head, and v the water molar volume, then V_H can be written:

$$V_H = n_s \, h \, v \tag{19}$$

2. Partitioning Model

The model will be presented in the case of SBT system. The surfactant is seen here as only partly dissociated, which differs from Adamson's theory.[33] The surfactant ionization is thus described by

$$RNa \leftrightarrows R^- + Na^+ \qquad (20)$$

where RNa and R^- are the associated and dissociated species, respectively, the solubility of which are neglected in both the pseudophase w* and the excess aqueous phase W".

The ionization constant K_i is therefore given by

$$K_i = \frac{(R^-)^* \, (Na^+)^*}{(RNa)^*} \qquad (21)$$

For sake of simplicity, the ratio of the activity coefficient of R^- and RNa is assumed to be constant. It must be pointed out here that (R^-) and RNa) would be surface concentrations; however, Equation 21 deals with their ratio. Thus we are able to express these quantities in moles per liter in the pseudophase W* as in case of Na^+.

By equating the chemical potential of salt in the W* pseudophase and the W" excess phase, respectively, one can write for the partition constant (which will be taken as equal to unity since the standard chemical potentials are identical):

$$K = 1 = \frac{(Na^+)'' \, (Cl^-)''}{(Na^+)^* \, (Cl^-)^*} \qquad (22)$$

where $(X)''$ and $(X)^*$ represent the concentration of the ion X in W" and W*, respectively.

At equilibrium the concentrations must satisfy Equations 21 and 22, as well as the material balance and electroneutrality conditions in Equations 23 and 24 below:

$$(RNa)^\circ = (RNa)^* + (R^-)^* \qquad (23)$$

$$(Na^+)^* = (R^-)^* + (Cl^-)^* \qquad (24)$$

where $(RNa)^\circ$ is the initial concentration of the surfactant in a nonionizated state.

For a given system, the knowledge of the initial composition, of the salinity S_w of the brine used, and of the volume $V_{w''}$ of the excess aqueous phase allows us, with the set of Equations 19 and 21 through 24, to calculate the salinity of the excess aqueous phase in terms of the two parameters K_i and h. In the case of the NOP system, the set of equations is reduced to Equation 22.

D. Comparison Between Experimental and Theoretical Results

1. NOP System

Figure 25 shows the comparison of experimental values of $S_{w''}$ with those calculated from Equation 9 using h = 8. This corresponds to a number of water molecules (h) per ethylene oxide unit of 1, a realistic number, although it is recognized that some compensation may occur: a fraction of the nonionic surfactant is dissolved into the O' pseudophase,[34] and a fraction of the alcohol is present at the interface. Both effects have been neglected.

2. SBT System

Figure 26 shows the comparison of experimentally measured salinities of the excess aqueous phase W" with salinities calculated from the model taking K = 0.9 and h = 6 (6 water molecules per surfactant head).

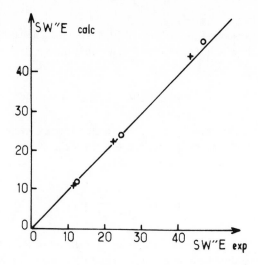

FIGURE 25.   Calculated vs. experimental excess aqueous phase salinity for the same systems as in Figure 23 (hydration model, assuming
1 molecule of salt-free water bound to each EO group).

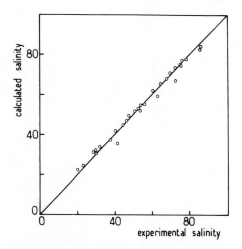

FIGURE 26.   Calculated vs. experimental excess
aqueous phase salinity for the systems as in Figure
24 (hydration model, assuming 6 molecules of salt-
free water bound to each surfactant head, and $K_i =$
0.9).

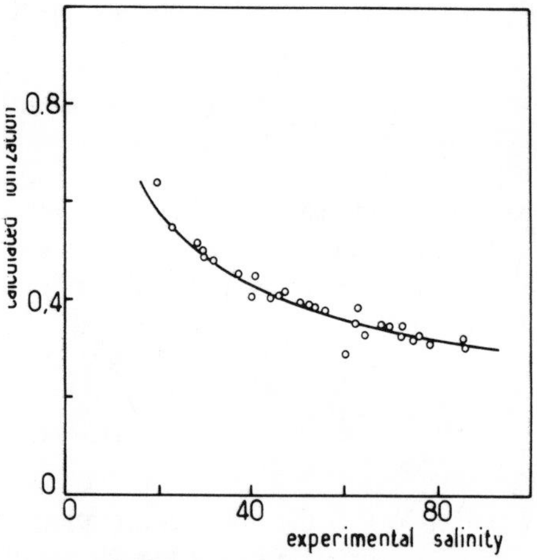

FIGURE 27.   Surfactant ionization degree as a
function of excess aqueous phase salinity (hydration
model as in Figure 26).

The value h = 6 must be viewed as an average since alcohol is also present at the interface.
Its concentration has been calculated with the aid of the pseudophase model and the alcohol
to surfactant molecular ratio has been found close to 2 for all the systems investigated.

With the same values of the parameters K_i and h, we have calculated the ionization degree
τ_i defined by:

$$\tau_i = \frac{(R^-)^*}{(RNa)^\circ} \qquad (25)$$

In Figure 27 the variations of τ_i are plotted against the salinity SW″. The results are in
good agreement with literature data for aqueous solutions of surfactants above the critical

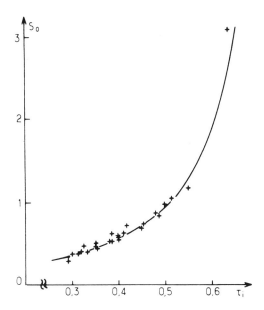

FIGURE 28.   Correlation between the amount of water solubilized in
the micellar phase of SBT systems and the surfactant ionization degree
as calculated in Figure 27.

micellar concentration (CMC). The ionization degree decreases when the salinity increases,
and increases with the alcohol content.[35-39]

If we call S_o the water solubilization in the microemulsion phase, S_o may be defined by

$$S_o = \frac{1}{(RNa)^o} \tag{26}$$

Figure 28 shows the variation of S_o in terms of the ionization degree τ_i. The results are
in line with the Winsor theory.[4] Indeed the solubilization increases with the ionization degree,
that is to say, with the charge density of the pseudophase m* that leads to an increase of
the interaction energy of the surfactant with water.

V. THEORETICAL AND EXPERIMENTAL STUDY OF VAPOR PRESSURES

In a recent study,[15] we have been able to corroborate the pseudophase assumption without
referring to any model of solution. This has been done by comparing vapor pressure of O
+ A measured for microemulsion systems to those observed with W, O, A ternary systems.
This qualitative aspect has been extensively discussed and we shall not reiterate. The aim
of this section is to discuss the kind of model of solution we have to use to quantitatively
account for experimental results.

A. Models of Solutions

We shall deal with some models of solutions within the frame of pseudophase assumption:
we shall deal only with the oil pseudophase and, for sake of simplicity, we shall neglect
the water pseudophase.

1. Prigogine's Ideal Associated Solutions Model

In this well-known model, the chemical potential of ith order self-associated alcohol
complex (of Part I.B.1) is given by:

$$\mu_i = \mu_i^* + RT \, Ln \, x_i \tag{27}$$

and the one of oil by:

$$\mu_o = \mu_o^* + RT \, Ln \, x_o \tag{27a}$$

where x_i and x_o are real molar fractions defined as

$$x_i = \frac{n_i}{n_o + \sum\limits_i n_i} \qquad x_o = \frac{n_o}{n_o + \sum\limits_i n_i}$$

where n_o and n_i are the number of moles of oil and of i species, respectively.

2. Model with Simple Flory's Entropic Correction

This is the model we presented in Section I.B.1. We shall recall here only the expressions accounting for chemical potentials.

$$\mu_i = \mu_i^* + RT \, Ln \, \phi_i$$

$$\mu_o = \mu_o^* + RT \, Ln \, \phi_o \tag{28}$$

3. Model with Complete Flory's Entropic Correction

Actually applying Flory's model[19] to self-associated alcohol solutions[17] leads to a mixing free enthalpy ΔG_m.

$$\Delta G_m = kT[n_o \, ln \, \phi_o + \sum\limits_i n_i \, Ln \, \phi_i] \tag{29}$$

This formula allows us to define chemical potentials as

$$\mu_i = \mu_i^* + RT\left[Ln \, \phi_i + 1 - \frac{i(n_o + \sum\limits_j n_j) \, v_A}{n_o v_o + \sum\limits_j j \, n_j \, v_A} \right]$$

$$\mu_o = \mu_o^* + RT\left[Ln \, \phi_o + 1 - \frac{(n_o + \sum\limits_j n_j) \, v_o}{n_o v_o + \sum\limits_j j \, n_j \, v_A} \right] \tag{30}$$

where v_A and v_o, are the molar volumes of a + o, respectively. Excess volumes are neglected and molar volumes of the ith order complex are supposed to be equal to i times the molar volume of alcohol. To avoid confusion, summation index used is j.

B. Microemulsion Study

1. Liquid Phase Study — Equilibria Between O', W', and M' Pseudophases

Alcohol self-association — In the case of Mecke's hypothesis (Section I.B.1), the association statistics are not modified by using Equation 30 for chemical potentials in place of Equation 28. Actually, the correction term in Equation 30 disappears when one writes the equilibrium condition:

$$\mu_{i+1} = \mu_i + \mu_1$$

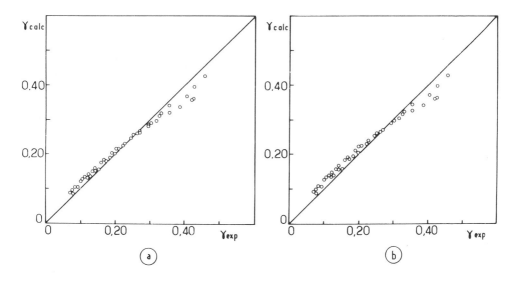

FIGURE 29. Correlations observed between values calculated ($\gamma = V_a/V_o$) and experimentally observed when γ is calculated via (a) formula 28 and (b) formula 30.

Equilibrium between O′, W′, and M′ pseudophases — We recall here the relation (Equation 11) of Section I.C.1; the alcohol partition obeys:

$$\mu_{A_1}^\circ = \mu_A^M = \mu_A^W$$

In the case where, in M′ and W′ pseudophases, the alcohol is supposed to be a monomeric species, the values of μ_A^M and μ_A^W obtained by Equation 30 are very close to those obtained by Equation 28. On the other hand, the value of $\mu_{A_1}^\circ$ is very different, depending upon use of Equation 30 or of Equation 28. Nevertheless, we present in Figures 29a and b the correlations observed between γ_{calc} obtained via Equation 28 or Equation 30. Studied systems are saturated oil-rich microemulsions (*n*-butanol-toluene-SDS-w). The values γ_{exp} have been obtained through dilution experiments.[23]

The results are satisfactory in both cases; we cannot differentiate between the two models when studying properties of condensed systems in the low alcohol concentration range. The only difference lies in numerical values of constants K_W and K_M for each case.

2. Vapor Pressure Study

We shall deal here with results obtained for the w-dodecane-*n*-hexanol-SDS system.

a. Vapor Pressures of Dodecane-Hexanol Binary Systems

Experimental and calculated vapor pressures presented hereafter are, in fact, relative pressures defined as:

$$\frac{P_i}{P_i^\circ} \tag{31}$$

where P_i holds for vapor pressure of i in the given system and P_i° for the pressure of pure i.

In Figures 30 through 32 we present a comparison between experimental and theoretical results obtained by applying Formulas 28, 30, and 27, respectively. It should be noted that the ''simple'' Flory's model should be rejected. Calculated vapor pressures for oil do not

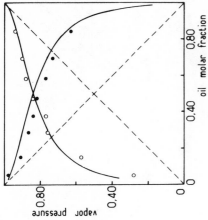

FIGURE 32.   Comparison between observed ([●] = alcohol; [○] = oil) and theoretical values of relative pressure when the calculus is run using formula 27.

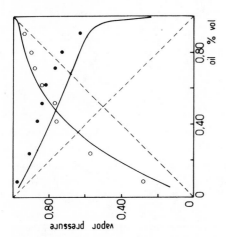

FIGURE 31.   Comparison between observed ([●] = alcohol; [○] = oil) and theoretical values of relative pressure when the calculus is run using formula 30.

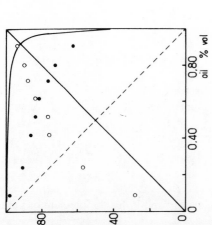

FIGURE 30.   Comparison between observed ([●] = alcohol; [○] = oil) and theoretical values of relative pressure when the calculus is run using formula 28.

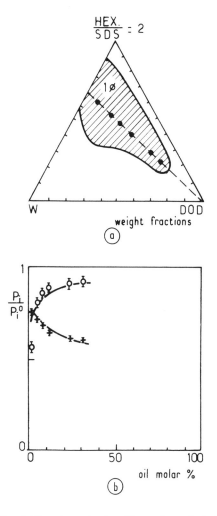

FIGURE 33. (a) Experimental points; (b) experimental relative vapor pressures measured for points represented on Figure 33a.

depend upon self-association; they are identical to those of a perfect system. In spite of more satisfactory results with the complete Flory's model (Equation 30, Figure 31), the best fit is obtained with the molar fraction formalism (Equation 27).

b. Vapor Pressure of Microemulsion

We shall give a brief insight on the last results obtained in our laboratory.

It should be noted that beyond measurements on A, O binary systems, studying microemulsions requires measurements on W, A, O ternary systems. To perform the theoretical analysis, we had to take into account W, a interactions within organic phases (i.e., in O' pseudophases). We treated them as complexing equilibria within the molar fraction formalism. We shall not reiterate the whole model used and shall present only in Figure 33b the correlation between calculated and experimental vapor pressures for systems defined in Figure 33a.

The excellent fit observed in Figure 33b clearly shows that the pseudophase model, after having been somewhat refined, allows one to interpret results obtained via analyzing properties as subtle as vapor pressure.

VI. CONCLUSIONS

The results reported in Sections II to IV clearly show that the pseudophase model described in Section I is an essential tool for studying microemulsions.

Indeed, we have been able to establish that:

1. W'M'O' pseudophases can be looked at as pseudocomponents (see 2). In W'O'M' triangles, phase diagrams appear as ideal Winsor triangles. It must be pointed out here that in the case of multiphase Winsor I, II, or III systems, the analyses of O or W excess phases constitute a determination of the pseudophase compositions of the relevant single-phase microemulsion.

2. The pseudophase model can be extended to salted systems; a quantitative interpretation of salt partition, favoring the excess water phase of Winsor II (or III) systems, has been proposed by considering
    (a) the hydration of surfactant polar heads in m' pseudophase
    (b) the surfactant ionization in the case of ionic pseudophase
    Furthermore, the model allowed us to establish a correlation between the interface ionization degree (that depends upon salinity) and the water solubilization in the microemulsion.

3. From the results of Section IV, an experimental vindication of the pseudophase model can be obtained via studying quaternary microemulsion vapor pressure for ionic surfactant systems. Thus, we have been able to propose a quantitative interpretation of A + O partial vapor pressures variation with the global composition of the microemulsion.

REFERENCES

1. **Bellocq, A. M., Biais, J., Bothorel, P., Clin, B., Fourche, G., Lalanne, P., Lemanceau, B., and Roux, D.,** Microemulsions, *Adv. Colloid Interface Sci.,* 20, 167, 1984.
2. **Winsor, P. A.,** Binary and multicomponent solutions of amphiphilic compounds. Solubilization and the formation and theoretical significance of liquid crystalline solutions, *Chem. Rev.,* 68, 1, 1968.
3. **Bellocq, A. M. and Roux, D.,** Phase diagram and critical behavior of a quaternary microemulsion system, *Microemulsions: Structure and Dynamics,* Friberg, S. and Bothorel, P., Eds., CRC Press, Boca Raton, Fla., 1986, chap. 2.
4. **Winsor, P. A., Ed.,** *Solvent Properties of Amphiphilic Compounds,* Butterworths, London, 1954.
5. **Kunieda, H. and Friberg, S. E.,** Critical phenomena in a surfactant/water/oil system. Basic study on the correlation between solubilization, microemulsion and ultra low interfacial tensions, *Bull. Chem. Soc. Jpn.,* 54, 1010, 1981.
6. **Bourrel, M., Lipow, A. M., Wade, W. H., Schechter, R. S., and Salager, J. L.,** Properties of Amphiphile/Oil/Water Systems at an Optimum Formulation for Phase Behavior, SPE Paper, no. 7450, Fall Meeting of the Society Petroleum Engineers, Houston, 1978.
7. **Reed, R. L. and Healy, R. N.,** Some physicochemical aspects of microemulsion flooding: a review, in *Improved Oil Recovery by Surfactant and Polymer Flooding,* Shah, D. O. and Schechter, R. S., Eds., Academic Press, New York, 1977, 383.
8. **Biais, J., Bothorel, P., Clin, B., and Lalanne, P.,** Theoretical behaviour of microemulsions: geometrical aspects, dilution properties and role of alcohol, *J. Dispersion Sci. Technol.,* 2, 67, 1981.
9. **Fourche, G. and Bellocq, A. M.,** Diffusion de la lumière par 2 microémulsions: eau-dodécane-oléate de potassium-hexanol et eau-toluène dodécylsulfate de sodium-butanol, *C.R. Acad. Sci. Paris, Ser. B,* 261, 289, 1979.
10. **Dvolaitzky, M., Guyot, M., Lagües, M., Le Pezan, J. P., Ober, R., Sauterey, C., and Taupin, C.,** A structural description of liquid particle dispersions: ultra centrifugation and small angle neutron scattering studies of microemulsions, *J. Chem. Phys.,* 69(7), 3279, 1978.

11. **Cazabat, A. M., Langevin, D., and Pouchelon, A.,** Light scattering study of water in oil microemulsions, *J. Colloid Interface Sci.,* 73(1), 1, 1980.

12. **Valeur, B. and Keh, H.,** Determination of the hydrodynamic volume of inverted micelles containing water by the fluorescence polarization technique, *J. Phys. Chem.,* 83, 3305, 1979.

13. **Robbins, M. L.,** The theory of microemulsion, paper presented at Nalt. AIche Meet. on Interfacial Phenomena in Oil Recovery, Tulsa, 1974; Microemulsion, at 48th Nalt. Colloid Symp., ACS Preprints, Austin, 1974, 174.

14. **Ruckenstein, E. and Chi, J. C.,** Stability of microemulsions, *J. Chem. Soc.,* 71, 1690, 1975.

15. **Biais, J., Ödberg, L., and Stenius, P.,** Thermodynamic properties of microemulsions: pseudophase equilibrium vapor pressure measurements, *J. Colloid Interface Sci.,* 86(2), 350, 1982.

16. **Prigogine, I. and Defay, R.,** Solutions associées, in *Thermodynamique Chimique,* Desoer, Liège, 1950.

17. **Prigogine, I. and Defay, R.,** Solutions associées, in *Thermodynamique Chimique,* Dosoer, Liège, 1950.

18. **Biais, J., Lemanceau, B., and Lussan, C.,** Essais de détermination de la structure associée des solutions alcooliques en solvant inerte par R.M.N., *J. Chim. Phys.,* 64(6), 1018, 1957.

19. **Flory, P. J., Ed.,** *Principle of Polymers Chemistry,* Cornell University Press, Ithaca, N.Y., 1953.

20. **Biais, J., Dos Santos, J., and Lemanceau, B.,** Etude par résonance magnétique nucléaire de l'auto-association de quelques alcools à l'aide d'un modèle de solution tenant compte de la différence de taille des particules, *J. Chim. Phys.,* 67(4), 806, 1970.

21. **Mavridis, P. G., Servanton, M., and Biais, J.,** Etude par résonance magnétique nucléaire de l'effet de solvant sur la structure associée du *n*-heptanol, *J. Chim. Phys.,* 69(3), 436, 1972.

22. **Mecke, R. and Kempter, H.,** Spectroscopic determination of association equilibria, *Naturwissenschaften,* 27, 583, 1939.

23. **Graciaa, A.,** Thèse Docteur ès Sciences, Université de Pau, 1978.

24. **Gerbacia, W.,** Microemulsion formation, Thesis, CUNY, New York, 1974.

25. **Bavière, M., Wade, W. H., and Schechter, R. S.,** Effets des alcools sur le type de certains systèmes micellaires, *C.R. Acad. Sci. Paris, Ser. C,* 289, 353, 1979.

26. **Haouche, G.,** Micelle et microémulsion dans les systèmes amphotères: la *n*-dodécylbétaine, Thesis, Université de Montpellier, 1984.

27. **Prouvost, L. and Pope, G. A.,** Modelling of phase behavior of micellar systems used for EOR, 5th Int. Symp. on Surfactant in Solution, Bordeaux, July 1984.

28. **Kielland, J.,** Individual activity coefficients of cations in aqueous solutions, *J. Am. Chem. Soc.,* 59, 1675, 1937.

29. **Weast, R. C., Ed.,** *Handbook of Chemistry and Physics,* 48th ed., CRC Press, Boca Raton, Fla., 1967.

30. **Harned, H. S. and Owen, B. B., Eds.,** *The Physical Chemistry of Electrolytic Solutions,* Reinhold, New York, 1950, 557.

31. **Rendall, K. and Tiddy, G. J. T.,** Interaction of water and oxyethylene groups in lyotropic liquid-crystalline phase of poly (oxyethylene) *n*-dodecyl ether surfactants studied by ^{2}H nuclear magnetic resonance spectroscopy, *J. Chem. Soc. Faraday Trans. 1,* 80, 3339, 1984.

32. **Boicelli, C. A., Giomini, M., Giuliani, A. M., and Trotta, E.,** Communication at Int. Symp. on Colloid and Surface Science, Interlaken, October 1984.

33. **Adamson, A. W.,** Model of micellar emulsions, *J. Colloid Interface Sci.,* 29, 261, 1969.

34. **Graciaa, A., Lachaise, J., Sayous, J. G., Grenier, P., Schechter, R. S., and Wade, W. H.,** The partitioning of complex surfactant mixtures between oil/water/microemulsion phases at high surfactant concentrations, *J. Colloid Interface Sci.,* 93, 474, 1983.

35. **Lindman, B. and Wennerstrom, H.,** *Solution Behavior of Surfactants,* Vol. 1, Mittal, K. L., Ed., Plenum Press, New York, 1982, 10.

36. **Philipps, J. N.,** The energetics of micelle formation, *Trans. Faraday Soc.,* 51, 561, 1955.

37. **Zana, R., Yiv, S., Trazielle, C., and Lianos, P.,** Effect of alcohol on the properties of micellar systems, *J. Colloid Interface Sci.,* 80(1), 208, 1981.

38. **Zana, R.,** Ionization of cationic micelles: effects of the detergent structure, *J. Colloid Interface Sci.,* 78(2), 330, 1980.

39. **Jain, A. K. and Singh, R. P. B.,** Effects of alcohols on counterions association in aqueous solutions of SDS, *J. Colloid Interface Sci.,* 81(2), 536, 1981.

Chapter 2

PHASE DIAGRAM AND CRITICAL BEHAVIOR OF A QUATERNARY MICROEMULSION SYSTEM

Anne-Marie Bellocq and Didier Roux

TABLE OF CONTENTS

I. INTRODUCTION

It has been clearly and intensively shown by Ekwall and co-workers[1] that phase equilibria in multicomponent aqueous mixtures of amphiphilic molecules can be richly diverse and intricate. Due to their considerable potential for aggregation, the surfactant solutions show a multiplicity of structures (bilayers, cylinders, spherical micelles) which can organize and produce a great variety of phases.[1-6] In addition to liquid isotropic micellar phases, either optically anisotropic or optically isotropic mesophases occur. The most commonly observed mesophases are lamellar smectic phases and hexagonal phases of infinite rodlike aggregates. Cubic phases are detected in numerous binary and ternary mixtures. There also exist anisotropic nematic phases of limited size aggregates. Two types of uniaxial nematic phases[7,8] and a biaxial nematic phase[9] have been recently discovered in some ternary mixtures.[10,11] The phase diagrams of several water-surfactant-alcohol (W, S, A) and water-surfactant-oil (W, S, O) mixtures have been investigated in detail.[1-12] They display a rich variety of phases and also complex multiphase regions where two or three liquid isotropic and mesomorphic phases are in equilibrium. Quaternary mixtures of water-oil-surfactant-alcohol (W, O, S, A) are also expected to exhibit a complex range of phase behavior and to offer an interesting area for observing a large variety of phase transitions including critical phenomena.

In some favorable cases, quaternary mixtures of oil, water, surfactant, and alcohol exist as a fluid, transparent isotropic liquid phase called a microemulsion. As we will see later in this paper, the phase diagrams of these quaternary mixtures are not qualitatively different from those of ternary mixtures described by Ekwall.[1-12] One still encounters ordered phases but their extent is considerably reduced. Generally mesomorphic regions are replaced over a broad domain of water and oil concentrations by an isotropic liquid microemulsion phase where no long range order occurs.

Up to now only very restricted portions of the phase diagrams of quaternary mixtures have been explored. Very early on the Swedish school attempted to determine the extent and shape of the region of existence of microemulsion in quaternary systems.[13-15] By examination of sections of the phase diagram at several levels of oil, the Swedish authors have established a direct connection between the microemulsion areas and the inverse micellar solutions described by Ekwall et al.[1-12] The multiphase regions occurring in these mixtures are little known. In contrast the multiphase regions of quinary mixtures containing salt have received much interest in connection with their potential use in oil recovery.[16,17] A great number of studies have focused attention on the so-called Winsor III three-phase region where a microemulsion is in equilibrium with both an organic and an aqueous phase.[18] Previous studies of mixtures of aqueous solutions of surfactant in presence of alcohol, salt, and hydrocarbon have also shown that a multiplicity of equilibria between isotropic liquid phases can be produced by these systems.[19] Then, for example, in addition to the Winsor III equilibria two three isotropic phase equilibria have been observed in the system water-dodecane-pentanol-NaCl-sodium octylbenzenesulfonate.[19] These studies have also established that microemulsion systems can give rise to critical points and critical end points (CEP).[18,19] Besides, recent results suggest that surfactant multicomponent systems can exhibit tricritical points.[20,21]

Because phase equilibria and the location of phase boundaries can dominate the physical properties of multicomponent systems, knowledge of the phase diagram is a necessary fundamental basis toward understanding the physics of surfactant systems. Therefore the first purpose of this chapter is to present a detailed description of the phase diagram of a quaternary mixture. The system investigated consists of water-dodecane-pentanol-sodium dodecyl sulfate (SDS). The study of the diagram provides characterization of several isotropic and mesomorphic phases. The emphasis will be on the evolution of the corresponding single phase domains and on the change of tie-lines and tie-triangles as the relative concentrations of the components are varied. Examination of various multiphase regions allows one to locate two critical end points and a line of critical points P_C^1. This study also provides evidence that the water over surfactant ratio (X) behaves as a field variable. This last finding serves to propose an entire description of the diagram.

Our interest in this chapter is also devoted to the study of several physical properties in the vicinity of critical points. Recent experimental interest has been focused on the critical behavior of micellar solutions and microemulsions.[22-32] Experimental works on these systems have led to puzzling results. For example, it has been found that the critical exponents measured in binary mixtures of water and nonionic surfactant are dependent on the surfactant.[30,31] These data address the question of whether surfactant solutions are relevant to the Ising model or not. This important problem prompts us to investigate the critical behavior of the water-dodecane-pentanol-SDS system. We have determined the exponents ν, γ, and β which characterize a critical point.[33] The critical behavior was studied on several distinct positions of the critical line P_C^1 and near a critical end point. The critical points were approached along different paths either by raising the temperature at fixed composition or by varying the water over surfactant ratio at constant temperature. The values of the critical exponents ν and γ measured along the critical line P_C^1 deviate largely from the universal Ising values.[34] The origin of the special dependence observed along this line will be discussed as well as an anomalous behavior of viscosity in relation with the phase diagram findings.

An outline of the remainder of this paper is as follows. In Section I, we describe the diagrams that we experimentally obtained. In Section II, we give a representation of these diagrams in a field space. The most important features of the diagram are summarized in Section III. Critical behavior results are presented in Section IV and discussed along with viscosity data in Section V.

II. PHASE DIAGRAMS

A four-component mixture has five independent thermodynamic variables. Our experiments have been carried out under atmospheric pressure and at fixed temperature; therefore, the representation of the phase diagram requires three independent variables, which can be either densities or fields. In 1970, Griffiths and Wheeler[35] showed that it is useful in discussions of phase transitions in multicomponent mixtures to distinguish two classes of intensive thermodynamic variables. According to these authors, the variables such as pressure, temperature, and chemical potential which are always identical in the coexisting phases are called ''fields'' and those which are generally different in the coexisting phases such as concentration and refractive index are called ''densities''. In practice, compositions are easily accessible, and the phase diagrams are constructed in a three-dimensional density space. However, there are at least two advantages in using fields alone as independent variables when one wishes to obtain a qualitative understanding of some phase transition phenomenon. The first is that the phase diagrams are simpler to draw (and hence to imagine) in a field space than they are in a density space.[36] Besides the benefit of the simplification of the appearance of the phase diagram, field variables appear very useful when considering the relative magnitudes of the divergences at critical points. Indeed, Griffiths and Wheeler[35] have shown that as soon as one follows a path in the field space tangent to the coexistence curve, the critical behavior is described by the same exponents whatever the field considered to approach the critical point.

Experimentally we have constructed the phase diagram in the compositional space and we have depicted it in a tetrahedron (Figure 1). Previously, using the concept of variance and the Gibbs phase rule as guides, we have described the topology of the single and multiphase regions of the diagram of a quaternary mixture at fixed pressure and temperature.[37] In practice, we have investigated sections of the tetrahedron defined by a fixed value of the

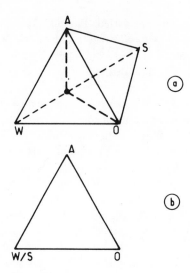

FIGURE 1.   (a) Three-dimensional representation of the phase diagram of the quaternary water (W) - oil (O) - alcohol (A) - surfactant
(S) system in a tetrahedron; (b) pseudoternary diagram at constant water
over surfactant ratio (W/S).

water over surfactant ratio (Figure 1). In the remainder of this paper this variable is designated
by X. The choice of this variable allows a good description of the oil-rich part of the phase
diagram. Compositions in the diagrams are expressed in weight fractions. Because the
diagrams obtained in the density space are very complex, we have systematically depicted
them in a field space. These latter diagrams are purely schematic since we do not know the
chemical potentials of the components.

A. Experimental Procedure and Materials

Phase boundaries were determined according to the following procedure. For each X
section, sealed tubes containing 2 or 3 mℓ of mixture of designated composition were
prepared by addition of oil to a microemulsion. Samples in the tubes were mixed by vortex
mixer and allowed to equilibrate in a chamber at a constant temperature (T was controlled
within 0.1°C). After 24 hr of storage, the number of phases and the level of the interface(s)
were measured, and examination between crossed polars was made to appreciate the birefringence of the phases. After these first observations were carried out, the tubes were again
mixed and reexamined a few days later in order to check whether the new equilibrium
obtained was the same as the first observed. Phase equilibrium was said to have been achieved
when no further change with time in solution appearance or phase volumes was apparent.
The time for equilibrium to be attained varied and was dependent on composition, viscosity,
structure, and vicinity of critical points. In the oil-rich part of the phase diagram, phase
equilibrium was reached after a few hours except near critical points. Tubes were maintained
at constant temperature for several months and no change was detected. For the parts of the
phase diagram where a great complexity is found, several hundreds of tubes have been
prepared. Consistency in the data has been achieved by showing continuity of the multiphase
regions which surround a given region.

The rich water plus surfactant parts of the sections defined by X below three were not
investigated precisely. Due to the high viscosity of the phases occurring in these regions,
phase separation was very difficult and the boundaries marking the phase transitions were
not determined accurately.

Identifications of mesomorphic phases were made by observation of optical textures in a
polarized microscope.

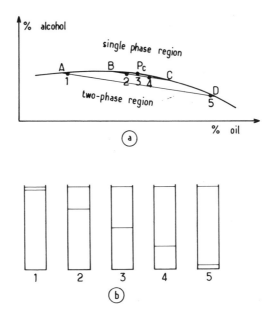

FIGURE 2. (a) Schematic representation of the coexistence curve around a critical point P_C; (b) variation of the phase volumes for the equilibria (labeled 1 to 5 in Figure 2a).

In the literature a variety of symbols have been used to represent different phase structures; a summary of these has been given in Reference 4. In the following, we have used the Ekwall's system of nomenclature. The isotropic phases are designated by the letter L. The subscripts 1 and 2 indicate that the phase protrudes from the water and the oil corners of the diagram, respectively. In some cases the L_1 and L_2 regions are connected and there is a single isotropic solution phase extending from the water corner to the oil or alcohol corner in the diagrams; this phase is denoted L. The lamellar phase is designated D, the hexagonal phase E, and the rectangular phase R.

The determination of the compositions of the coexisting phases in multiphase equilibria was performed by gas chromatography and gravimetric analysis. Prior to analysis, the samples were allowed to equilibrate in cells for several days or weeks in a chamber at 21°C. T was controlled to 0.1°C. Water, dodecane, and pentanol concentrations were determined by gas chromatography. Heptanol-1 has been used as an internal reference. Their separation was effected on a 1-m long Porapack® P column at 190°C. The surfactant was weighed after evaporation under vacuum of all liquids. The relative precision obtained on the concentrations depends on the component. It is about 1.5% for water, 1% for pentanol and dodecane, and 2% for SDS.

In the mixtures investigated, critical points were observed. Their precise determination is difficult since it requires the simultaneous control of temperature and two or three density variables. In the vicinity of a critical point, a phase separates into two others in which volumes and compositions are very close. The evolution of the volumes for the two-phase equilibria lying in the near vicinity of the coexistence curve is shown schematically in Figure 2.

Therefore, accurate determination of the composition C_C of a critical point was carried out by careful examination of a series of samples obtained by mixing two stock microemulsions B and C whose compositions are very close to C_C (Figure 2). In some cases, composition analyses of each of the phases in equilibrium were made to confirm the location of the critical point. Such methods permit the location of the critical point with an absolute precision of 0.2% on the compositions.

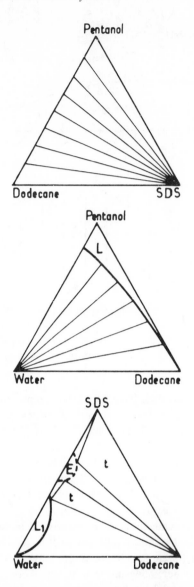

FIGURE 3.   Phase diagram at 21°C of the ternary systems: pentanol-dodecane-SDS, water-dodecane-pentanol, and water-SDS-dodecane. L and L_1 are isotropic phases, E is an hexagonal phase; regions t are three-phase regions.

Alcohol and dodecane were Aldrich products (''purum'' reagents) and SDS was purchased from Touzart et Matignon (''pur'' reagent).

B. Diagrams of Ternary Mixtures

Prior to describing the phase diagram of the quaternary system we investigated those of the four ternary systems which bound the tetrahedron of representation for the entire system.

The phase diagrams of the three ternary systems pentanol-dodecane-SDS, pentanol-dodecane-water, and water-SDS-dodecane are shown in Figure 3. The first one is very simple; indeed, SDS being immiscible in dodecane and pentanol, a two-phase region, takes up almost the entire diagram. The second diagram is not very different; pentanol is slightly miscible with water; therefore, the ternary system is the extension of that of the binary

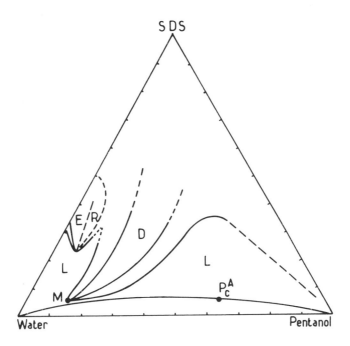

FIGURE 4. Phase diagram at 25°C of the ternary water-pentanol-SDS system. L is an isotropic phase, and E, R, and D are mesomorphic phases. The dashed boundaries have not been determined accurately. The point M is an azeotrope-like point; P_C^A is a critical point.

system water-SDS.[5,38] The micellar phase L_1 and the hexagonal phase E dissolve a small amount of dodecane.

The diagram of the fourth ternary system at 25°C is represented in Figure 4. Four one-phase regions are observed; in three of them (D, E, R) the mixtures are birefringent, and in the last one they are isotropic. At 25°C, the ternary mixtures water-pentanol-SDS display a continuous band of isotropic solution L between pure water and pure pentanol. Provided that the surfactant concentration is adjusted, it is possible to go continuously from the water corner to the pentanol corner without any phase separation occurring. The main result of this continuity is the occurrence of a critical point P_C^A in this diagram. The accurate location of this point has been determined by phase composition analysis of several two-phase equilibria. The phase E corresponds to the hexagonal phase observed in the binary system H_2O-SDS. The new evidenced phases D and R are lamellar and rectangular phases, respectively. Phases E and R are only obtained at high surfactant content ($\sim$30%). The boundaries of the regions of existence of these phases are not easy to define precisely because of the difficulty in separating them. The lamellar mesophase D occurs over a wide range of surfactant concentration. Its region of existence extends down toward the water corner, up to 79% of water. The isotropic lamellar phase transition is a first order transition. As a result, the phases L and D are separated by a two-phase region whose extent is found to decrease as the surfactant content decreases. At the point M, the boundaries of the isotropic and lamellar phases are tangential. The phase situation in this point resembles that at the azeotropic point in the liquid gas equilibria. Analysis of several two-phase equilibria located on both sides of the lamellar phase shows that the length of the tie-lines vanishes at the point M. In this point the two phases D and L coexist with the same composition for each of the three components. While at a critical point, certain thermodynamic properties have a universal behavior and follow power laws; at an azeotropic point, no particular behavior is expected.

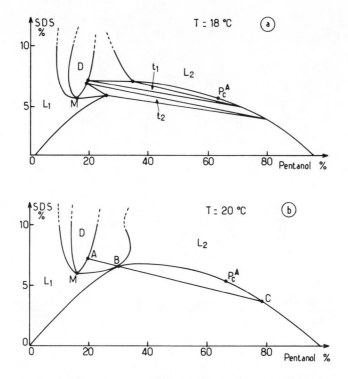

FIGURE 5.   Phase diagrams of the ternary water-pentanol-SDS system at
18°C and 20°C. L_1 and L_2 are isotropic phases, t_1 and t_2 are two three-phase
regions. The line ABC is an indifferent three-phase state.

Significant changes in the phase equilibria occur on lowering the temperature. Below
20°C, the isotropic domain L is no longer continuous; it has split into two one-phase regions
L_1 and L_2 which expand from the water and pentanol corners, respectively (Figure 5a).
These two isotropic regions are separated by a complex multiphase region which involves
two three-phase domains, t_1 and t_2. In these equilibria, two of the phases are isotropic, and
the third one is the lamellar phase D. Composition analysis of the coexisting phases in the
equilibria t_1 and t_2 at different temperatures shows that these equilibria originate in an
indifferent state at 20°C[39] (Figure 5b). Experimental data are reported in Figure 6. Two
major consequences of increasing temperature are found:

1.   The concentration of each component (water, pentanol, SDS) in the corresponding
       phases of the equilibria t_1 and t_2 tends toward the same value
2.   The tie-triangles become thinner

At 20°C, the points A, B, and C representative of the three coexisting phases of both
equilibria are located on the same straight line. The system is then in an indifferent state.
Let us note that here too, as for the azeotropic state, the system does not show a particular
behavior. A schematic picture of the temperature dependence of the t_1 and t_2 regions is given
in Figure 7. At 20°C, the regions L_1 and L_2 have one common point; they merge above this
temperature.

C. Diagram of the Quaternary System

In the description of the quaternary mixture, we will emphasize the details of the evolution
of the phase equilibria as the X ratio is varied. We have focused our attention not only on
the characterization and the location of the boundaries of the various phases but also on the

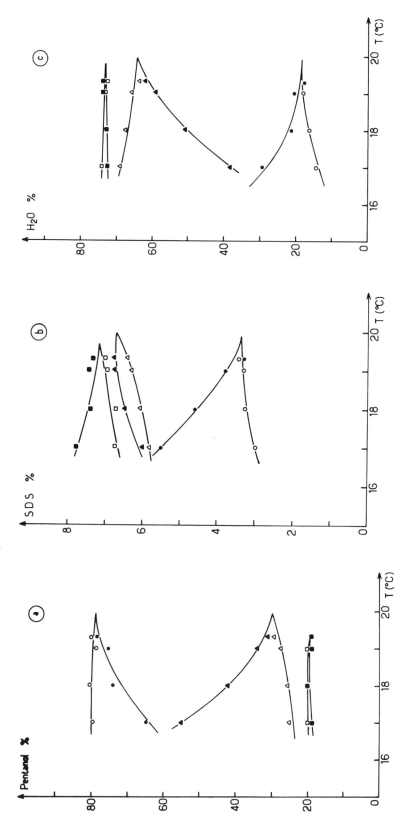

FIGURE 6. Temperature dependence of the pentanol (a), SDS (b), and water (c); concentrations in each phase of the equilibria t_1, and t_2, $\bullet$, $\blacktriangle$, $\blacksquare$, correspond to the upper, middle, and lower phases of the equilibria t_1, respectively; the open symbols ($\circ$, $\triangle$, $\square$) correspond to the same phases for the equilibria t_2.

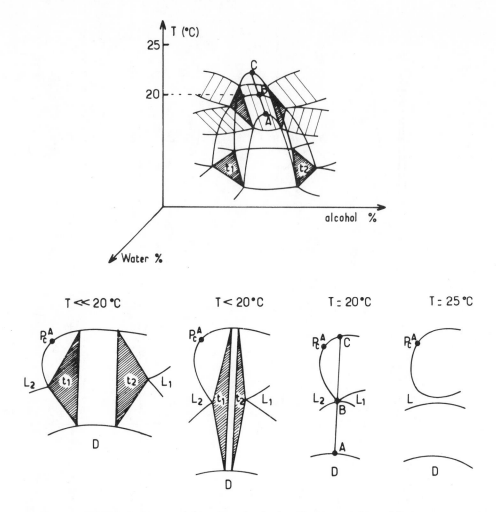

FIGURE 7.   Two- and three-dimensional schematic representations of the
tie-triangles t_1 and t_2 at different temperatures.

equilibria between the phases, i.e., we have determined the changes in direction of the tie-
lines and tie-triangles as X is varied.

In the planes X below X = 0.76 in the oil rich region, only one wide isotropic region,
termed L_2, is detected. This region corresponds to the microemulsion region. For X = 0.76
a birefringent phase appears (Figure 8). This phase, termed D, has a lamellar structure; it
can contain up to 98% of dodecane and alcohol in volume. In this plane, the polyphasic
region corresponds to two-phase equilibria between the isotropic microemulsion and the
liquid crystalline phase. In the plane X = 0.76, for the high oil content, the two-phase
region becomes very thin. It seems probable that the microemulsion L_2 phase and the lamellar
phase D exhibit an azeotropic-like point (Figure 8). As one can see in Figures 9 and 12 to
16, both regions L_2 and D exist in all the diagrams X >0.76. Their extent obviously depends
on X. In the section corresponding to X = 1.034 (Figure 9) there are, in addition to the
two one-phase domains occurring at low X, two phases L_1 and L_3. While motionless these
phases are isotropic; they exhibit flow transient birefringence as soon as any disturbance is
created. In particular, flow birefringence is easily generated by shaking the sample tube.
Besides, both phases L_1 and L_3 scatter light. One of the most important features occurring
in the plane X = 1.034 is a critical point P_C along the coexistence curve of the microemulsion
L_2 region. The direct consequence of this critical point is the existence of a two-phase region

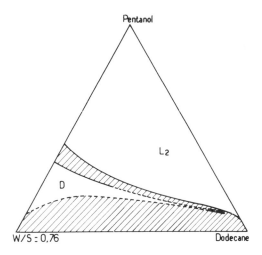

FIGURE 8. Section at constant water over surfactant ratio X = 0.76 of the phase diagram of the water-dodecane-pentanol-SDS system at 21°C. L_2 and D are liquid isotropic and lamellar liquid crystalline regions, respectively. The hatched domain is the multiphase region.

(L_2L_2') where two isotropic microemulsions are in equilibrium. The richest microemulsion in oil is termed L_2'. Appearance of new phases, on the one hand, and that of the critical point on the other generate several multiphase regions. In particular, a four-phase region (T) is seen where the four phases L_2, L_2', D, and L_1 coexist simultaneously. Because of their density, the sequence of phases in a tube is L_2'-L_1-D-L_2; L_2' is the upper phase, L_2 is the lower phase. In a four-component system, at constant pressure and temperature, such a region is invariant. In the density space, this region is a tetrahedron (Figure 10); it is surrounded by the four following three-phase regions: L_2' L_1 D (t_1), L_1D L_2 (t_2), L_2' L_1 L_2 (t_3), and L_2' D L_2 (t_4). These domains are themselves separated by the six two-phase regions: L_2' L_2(d_1), L_2 L_1(d_2), L_2 D(d_3), L_2'D (d_4), L_1D (d_5), and L_2' L_1 (d_6). Each of these six regions starts on the sides of the tetrahedron. In the plane X = 1.034, the three-phase regions t_1, t_2, and t_3 are observed. In t_3 the three coexisting phases are isotropic; in t_1 and t_2 one of them is birefringent. At low alcohol content, two three-phase regions, t_5 and t_6, are seen; they connect the phases L_2', L_1, D (t_5), and L_2', L_3, D (t_6), respectively (Figure 9). Due to the very small extension of most of the multiphase regions observed in the plane X = 1.034, it has been difficult to determine accurately their boundaries. Figure 11 shows the sequence of equilibria found along the AA' line of Figure 9.

The diagram found for X = 1.207 is quite similar to that observed for X = 1.034 (Figure 12). Whereas in the two planes, X = 1.034 and X = 1.207, the regions L_1 and L_3 are separated; as X increases above X = 1.38 they form a single region which is referred to in the following as the phase L_1 (Figure 13). This latter displays shear birefringence which rapidly decreases as the oil content decreases and as X increases. For X = 1.89, the mixture water-SDS becomes homogeneous; it can dissolve a small amount of dodecane and alcohol. For X = 2.586, the region L_1 extends up to the water-SDS corner (Figure 14). Finally, for X = 3.017, the regions L_1 and L_2 have merged and the diagram presents a very large isotropic one-phase region named L (Figure 15). This domain remains continuous up to X = 5.3 (Figure 16). For this last value, the amount of surfactant with respect to water is no longer sufficient to achieve the continuity of the single-phase domain. For X above 5.3, this latter has split into two regions, one rich in oil and alcohol and the other rich in water and surfactant.

As mentioned earlier, the lamellar phase D occurs in all the X planes above X = 0.76.

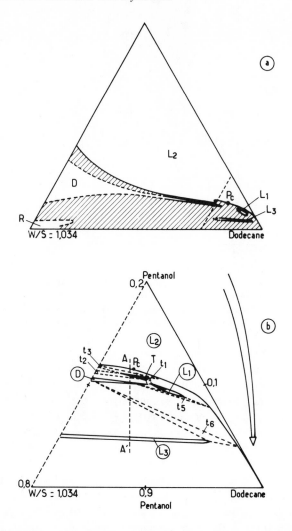

FIGURE 9. (a) Section X = 1.034 of the water-dodecane-pentanol-SDS system at 21°C. R is a rectangular liquid crystalline phase, L_1 and L_3 are two flow birefringent microemulsion phases. P_C is a critical point. Other abbreviations are the same as in Figure 8. (b) Magnification of the lower right corner of Figure 9a. t_1, t_2, t_3, t_5, and t_6 are three-phase regions, T is a four-phase region.

As X increases, its regions of existence becomes smaller and is progressively shifted toward the alcohol-water plus surfactant side of the diagram. This mesophase which appears as an extension of that observed in the ternary system water-pentanol-SDS can contain up to 98% of oil and alcohol. A second liquid crystalline phase occurs in the planes X below 2.8. It is probably the extension of the hexagonal or rectangular phases found for the ternary system water-SDS-pentanol. The limits of this mesophase have not been determined accurately.

In each X plane above X = 0.95 a critical point P_C exists on the coexistence curve of the microemulsion region as well as a two-phase domain where two microemulsions are in equilibrium. The extent of this domain sharply increases with X. The set of the critical points generate a critical line P_C^l which extends from X = 0.95 up to X = 6.6. However, we are not sure that this last point (P_C^A) which is located on the face of the tetrahedron water-pentanol-SDS belongs to the critical line P_C^l. Indeed all the points of the line defined by the values of X ranging from 0.95 to 6.47 are lower critical points whereas the point P_C^A is an

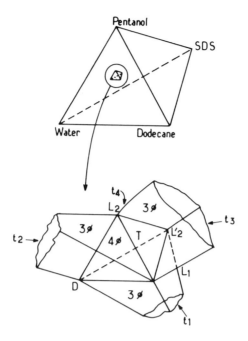

FIGURE 10. Schematic representation of the four-phase region T and
the four three-phase regions t_1, t_2, t_3, and t_4 in the tetrahedron which
represents the overall system. $T = L_2' L_1 D L_2$; $t_1 = L_2' L_1 D$; $t_2 =$
$L_1 D L_2$; $t_3 = L_2' L_1 L_2$; $t_4 = L_2' D L_2$.

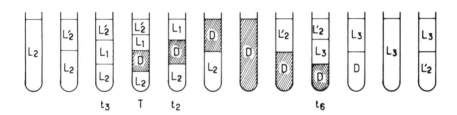

FIGURE 11. Pattern of phase behavior observed in the section $X = 1.034$ along the line AA′
drawn in Figure 9b.

upper critical point. A detailed study of the diagrams for X ranging between 6.47 and 6.6
is currently under progress in order to clarify this point.

The multiphase region found in the plane $X = 1.034$ undergoes significant changes as
X is varied. In the plane $X = 1.55$, the four-phase region T and the three-phase regions t_1,
t_5, and t_6 do not exist. The equilibria t_2 has disappeared in the plane $X = 3.017$ where the
phases L_1 and L_2 have merged. In this plane, the equilibria t_3 is still seen as well as another
three-phase region (t_7) where the coexisting phases are all isotropic. The regions t_3 and t_7
have both disappeared for $X = 3.5$. So far we do not know for which value of X the t_7
equilibrium appears.

D. Compositions of the Coexisting Phases in Multiphase Equilibria

Determination of the composition of the coexisting phases in multiphase equilibria gives
access to the shape of the coexistence phase surfaces and to the directions of tie-lines and
tie-triangles. We have measured the concentrations of the four components in each phase
of one four-phase sample T and of several two- and three-phase samples.

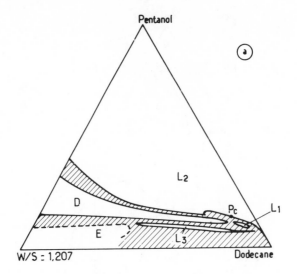

FIGURE 12a.   Section X = 1.207 of the phase diagram of the water-dodecane-pentanol-SDS system at 21°C. E is an hexagonal phase.

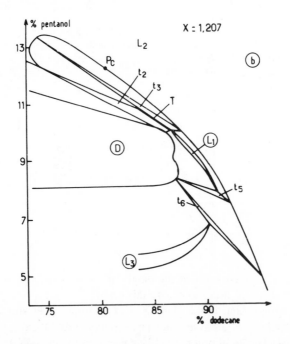

FIGURE 12b.   Magnification of the lower right corner of Figure 12a. Abbreviations are as in Figures 8 and 9.

Table 1 reports the data for the equilibrium T. The compositions of the four phases are found very close which explains the very small extent of the four-phase tetrahedron in the phase diagram at 21°C. It should be very interesting to investigate the temperature dependence of this volume and examine how it disappears.

We have seen earlier that in the plane X = 1.55 the phases L_2, L_1, and D are separated by a complex multiphase region which includes, in particular, the three-phase regions t_2 and t_3 and the two-phase region d_1 where two microemulsions are in equilibrium (Figure 13). Measurements were made on ten samples denoted D_1 to D_{10} lying in region d_1, one sample

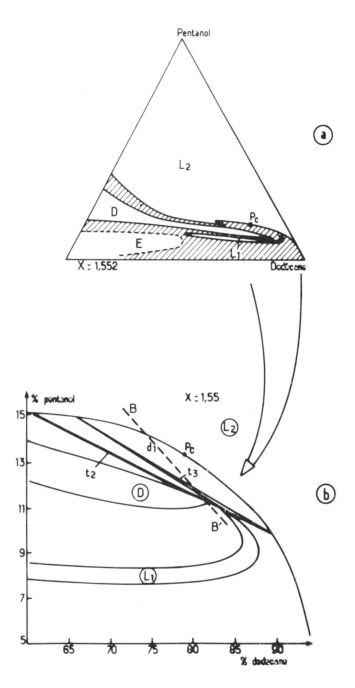

FIGURE 13. (a) Section X = 1.55 of the phase diagram of the water-dodecane-pentanol-SDS at 21°C; (b) magnification of the lower right corner of Figure 13a. Abbreviations are as in Figures 8 to 10 and 12.

in region t_2, and one in region t_3. The samples D_1 to D_{10} were located along the line BB' drawn in Figure 13b. In all these samples the global water to surfactant ratio is 1.55.[40] Due to the very small extent of both three-phase regions t_2 and t_3, very careful preparations are required. Three-phase samples were obtained by adjusting the amounts of alcohol and oil. The water over surfactant ratio in each phase was determined from the data of the phase analysis. Figure 17 gives the value of X measured (denoted X_i) in each phase (i) in equilibrium

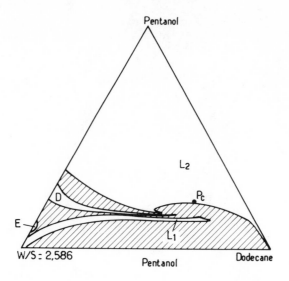

FIGURE 14.   Section X = 2.586 of the phase diagram of the water-dodecane-pentanol-SDS at 21°C. Abbreviations are as in Figures 8 to 10 and 12.

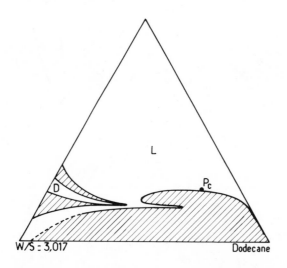

FIGURE 15.   Section X = 3.017 of the phase diagram of the water-dodecane-pentanol-SDS at 21°C. Abbreviations are as in Figures 8 to 10 and 12.

for the three-phase equilibria t_2 and t_3 and for the ten two-phase equilibria (denoted D_1 to D_{10}). In every case, the value of X_i is close to 1.55 which is the value of the global preparation. This result is an indication that the two triangles, which represent the two three-phase equilibria, and the tie-lines associated with the ten two-phase equilibria all lie on the plane X = 1.55.

To examine whether this property holds for other values of X we have prepared and analyzed further three-phase samples t_2[25,40] and two-phase mixtures D_1. The data of the analysis of all the components in each phase for the whole X range where t_2 exists are plotted in Figure 18. The value of X measured in each phase (X_i) is compared in Figure 19 with

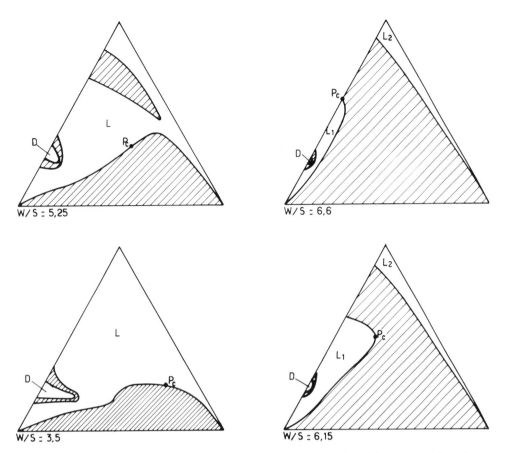

FIGURE 16.   Boundaries of the isotropic L and lamellar D regions in the sections of the quaternary phase diagram defined by X above 3. The hatched domains are multiphase regions.

Table 1
COMPOSITIONS OF THE FOUR PHASES L'_2, L_1,
D, AND L_2 OF THE EQUILIBRIUM T AT 21°C

	L'_2	L_1	D	L_2	Absolute uncertainty
Water	2.0	3.4	4.6	6.9	±0.4
Pentanol	10.7	10.9	10.9	12.5	±0.5
Dodecane	85.1	82.8	80.4	74.5	±1
SDS	2.2	3.0	4.1	6.1	±0.4

Note: Results are in wt%.

the value of the overall mixture (X). The conservation of X in each phase is better than 1% which is within the experimental accuracy. This remarkable result indicates that the X ratio has the characteristics of a field variable. The major consequence of this property is that the phase diagrams experimentally determined in which X is maintained constant are true pseudoternary diagrams. The three-phase volume t_2 consists, then, of a stack of tie-triangles, each lying in an X plane. As already mentioned, analysis of several three-phase samples allows us to determine the boundaries of the volume in the three-dimensional phase diagram.

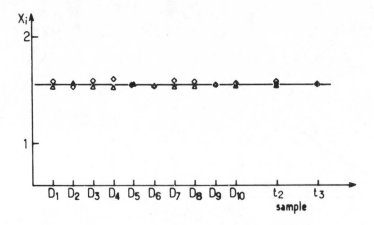

FIGURE 17.   Value of the water over surfactant ratio (X_i) measured in each phase (i) for ten two-phase equilibria (termed D_1 - D_{10}) lying along the line BB′ in region d_1 and for two three-phase equilibria of the regions t_2 and t_3 (Figure 13b). For all these samples, the global value of X is equal to 1.55 (T = 21°C).

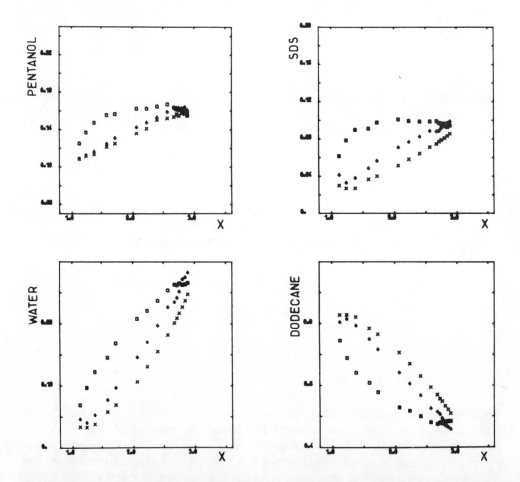

FIGURE 18.   Variation as a function of X of the water, pentanol, SDS, and dodecane concentrations in the three coexisting phases of several t_2 equilibria (T = 21°C). x, ◇, and □ correspond, respectively, to the upper phase (L_1), the middle phase (D), and the lower phase (L_2).

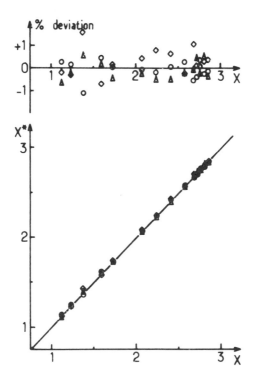

FIGURE 19.   Value of the water over surfactant ratio (X*) measured in each phase of the three coexisting phases of the equilibria t_2 as a function of the value X of the global mixture. The standard deviation is also reported (T = 21°C). $\Diamond$, $\circ$, and $\blacktriangle$, respectively, represent the value of X in the lower, middle, and upper phases.

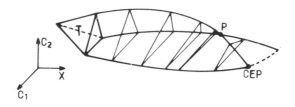

FIGURE 20.   Schematic representations of the three-phase volume t_2 in the $C_1 C_2$-X space. C_1 and C_2 correspond, respectively, to the alcohol and oil percentages. The volume t_2 is limited at low X by the four-phase tetrahedron T and at high X by the CEP. At the point P, the lamellar phase and the isotropic L_2 phase have the same composition (azeotropic point).

Data reported in Figure 18 show that the volume t_2 extends from the four-phase region T at X = 1.05 to a CEP at X = 2.98 (Figure 20). At low X, each of the three phases has the same composition as the corresponding phase of the four-phase equilibrium T (see Table 1). The volume t_2 is situated in the oil-rich part of the phase diagram (the water content never exceeds 25% in weight). Between X = 1.05 and X = 2.80 the two isotropic phases are the upper and the lower phases, and the middle phase is the liquid crystalline phase D. For low values of X, the isotropic upper phase (L_1) exhibits flow birefringence. Above X = 2.80, due to the increasing proximity of the two isotropic phase densities, the liquid

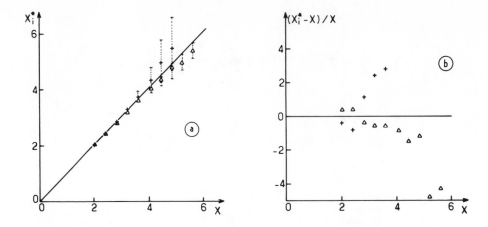

FIGURE 21.   (a) Value of the water over surfactant ratio (X*) measured in both coexisting phases of the equilibria d_1 observed in various X sections as a function of the value X of the global mixture (T = 21°C). (b) Deviation between X_i^* and X.

crystalline phase becomes the lower phase and the two isotropic phases are adjacent. At X = 2.98 these two phases merge at a CEP. A schematic representation of the volume t_2 in the space alcohol concentration-oil concentration-X ratio is given in Figure 20. At the point P, the lamellar phase and the isotropic L_2 phase have the same composition for the four components. This point corresponds to an azeotropic-like point.

Measurements of compositions of conjugate phases of ten samples occurring in region d_1 have been carried out in order to determine in which region of the phase diagram the field character of X is altered or lost. In the mixtures investigated, X varies between 2 and 5.6, the global percentage of oil between 71 and 40.35, the global percentage of water between 8.17 and 21.65, and the global percentage of pentanol between 16.89 and 34.14. The ratio X_i measured in each phase is reported vs. the value X of the overall mixture (Figure 21). For X < 4, the conservation of X in each phase is within the experimental accuracy. For X > 4, the volume of the upper phase of the samples investigated is very small in comparison to that of the lower phase; besides, the upper phase contains very small amounts of water and surfactant. Taking into consideration these two effects, our data do not have the precision required to conclude whether X behaves as a field variable when its value is above 4.

In summary, phase analysis of a large number of two- and three-phase equilibria which involve both isotropic and mesomorphic phases has provided evidence that the X ratio is field-like in the oil-rich mixtures.

III. REPRESENTATION OF THE PHASE DIAGRAMS IN A FIELD SPACE

As pointed out by Griffiths and Wheeler,[35] one way to simplify the representations of the phase diagrams of multicomponent mixtures is to use field variables. Indeed the replacement of a density by a field variable reduces the dimensionality of the coexistence phase regions. Then let us consider, for example, a ternary mixture, at fixed temperature and pressure, its diagram is two-dimensional. In a density space, all the regions of existence and coexistence of phases appear as surfaces whatever their variance; in a field space, phases appear as surfaces, two-phase regions as lines, and three-phase regions as points (Figure 22). Because fields take the same value in the coexisting phases, multiphase regions are reduced to the common boundaries of the regions of existence of the concerned phases. In this representation, the dimensionality of each region is equal to its number of degrees of freedom. This property is of great interest for investigations of multicomponent mixtures. Then, for a

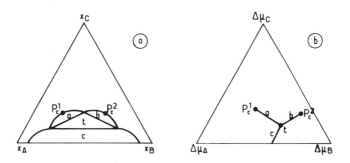

FIGURE 22. Phase diagrams of a ternary mixture ABC in the density space, x_A, x_B, x_C, and the field $\Delta\mu_A$, $\Delta\mu_B$, $\Delta\mu_C$. The field variables are related to the chemical potentials of A, B, and C; a, b, and c are two-phase regions, t is a three-phase region.

system defined by several independent variables, it is convenient to describe the global diagram in terms of sections obtained by holding n field variables constant. The diagram in such sections displays the same qualitative features as a system with n less degrees of freedom. Consequently, the phase diagram of a quaternary mixture obtained at fixed temperature, pressure, and constant value of the chemical potential of one of the components has features similar to those of a ternary mixture at fixed P and T.

One of the essential consequences of the simplification of the appearance of the diagrams is to make easier the understanding of the complicated evolution of phase equilibria occurring in multicomponent systems. Unfortunately, in practice, representation of the diagram in a field space encounters several difficulties related to the nature itself of these variables and to their experimental measurements. A convenient choice of fields might be temperature, pressure, and the chemical potentials of the components. The determination of these requires additional experimental investigations. However, in most cases, it is not necessary to draw quantitatively the phase diagram. A schematic representation is often sufficient and can serve as a guide to follow the evolution of phase equilibria. The experimental observation that in the quaternary mixtures water-dodecane-SDS-pentanol, the water to surfactant ratio X behaves as a field is of fundamental importance for the transcription of experimental density diagrams in a field space. As mentioned earlier, in this last representation, the X sections should present the same features as a ternary system at constant pressure and temperature. Therefore, as an example we will first focus attention on the phase diagrams of the ternary water-pentanol-SDS system experimentally found at 25, 20, and 18°C. These diagrams shown in Figure 23 contain some illustrative features occurring in the quaternary mixture. Each diagram obtained at constant P and T is drawn schematically in a triangle. In this representation, phases are represented by surfaces, two-phase coexistence by lines, and three-phase coexistence by points. For the three temperatures, the coexistence line (a) between region L and D shows a minimum which corresponds to an azeotropic-like point (point M) and the coexistence line (b) terminates at a critical point (P_C^A). As temperature decreases the coexistence lines (a) and (b) move closer together until they become tangential at the point N (T = 20°C). As temperature is further decreased, the lamellar phase intersects the coexistence line (a). This intersection gives rise to the two three-phase regions t_1 and t_2 as well as the two single-phase regions L_1 and L_2. The quadruple point N represents the indifferent state. At this point the regions L_1 and L_2 coalesce and the three-phase regions t_1 and t_2 are degenerated.

Figure 24 depicts in a field space the diagrams of the X sections of the quaternary mixture experimentally investigated. In the following we present an overall view of the phase diagram showing its qualitative behavior in the entire space considered. For that, we have attempted

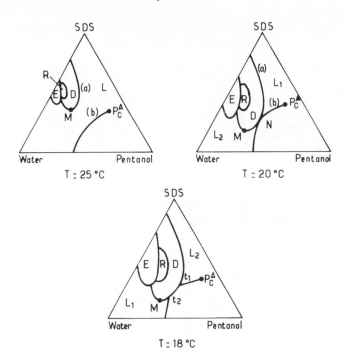

FIGURE 23.   Schematic phase diagrams of the water-pentanol-SDS system
at 18°C, 20°C, and 25°C in the field space consisting of the chemical potentials
of the components. This figure is to be compared with Figures 4 and 5.

to imagine a simple gradual conversion of the equilibria as X varies between zero and infinite
consistent with the experimental diagrams. In Figures 24 to 27, triple points are denoted by
t, critical points by P_C^i, and critical end points by P_{CE}^i.

The increase of X from 0.76 to 1.034 produces strong changes in the diagram. The two
different sequences of diagrams shown in Figure 25 yield the correct qualitative features of
the diagram X = 1.034. In the first one, the creation of the island of the phase L_1 results
from the existence of a line of critical points P_C^2 ended by two critical end points P_{CE}^2 and
$P_{CE}^{2'}$. In the second scheme of evolution, the creation of the phase L_1 in the interior of the
triangle is explained by the occurrence of an indifferent state involving the equilibria t_1 and
t_5. Both solutions have in common some salient features. In particular, the line of critical
points P_C^1 and the critical end point P_{CE}^1 occur in both sequences. They also contain a point
of four-phase coexistence labeled T. This point appears at the merging of the four triple
points t_1, t_2, t_3, and t_4. As we will see later in this chapter, correlation length measurements
in the phase L_2 of the equilibrium t_1 support the first sequence of diagrams.

The essential feature found between the planes X = 1.207 and 1.552 is the connectivity
of the phases L_1 and L_3. In the proposed development of equilibria (Figure 26), as X increases,
the boundary between the phases D and L_2' contracts until the two triple points t_5 and t_6
coalesce into a quadruple point where the phases L_1 and L_3 merge.

Between the planes X = 1.55 and X = 3.017, the coalescence of the phases L_1 and L_2
is achieved by the shrinking of the three-phase region t_2 at the critical end point P_{CE}^3 (Figure
26); the critical line P_C^3 appearing in this last point goes to the critical end point $P_{CE}^{3'}$ where
the line of triple points t_3 ends. The last line of triple points t_7 might extend between the
two critical end points P_{CE}^4 and $P_{CE}^{4'}$.

In conclusion, two descriptions of the overall diagram of the quaternary mixture water-
dodecane-pentanol-SDS at 21°C are proposed. Each of them includes a four-phase region
and seven lines of triple points. Figure 27 summarizes the evolution of the lines of triple

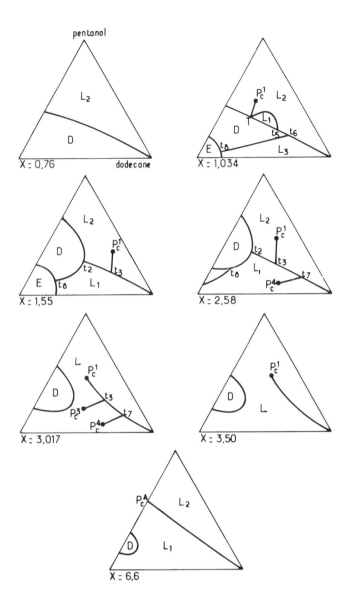

FIGURE 24. Schematic representations in a field space of the diagrams shown in Figures 8, 9, 12-16. The field variables consist of X and the chemical potentials of pentanol and dodecane.

points and of the lines of critical points as a function of X. In the first model, four lines of critical points and seven critical end points occur. In the second model, only three lines of critical points and five critical end points are required to explain the development of equilibria. Let us note that both descriptions imply the occurrence of degenerate triple points. Such points have been found to exist in the ternary mixture water-SDS-pentanol. Their existence has also been suggested by Lang and Morgan[20,22] to interpret the formation of the anomalous phase observed in the diagrams of certain water nonionic surfactant mixtures.

Most of the features predicted in the above scheme have been observed. The three-phase equilibria experimentally observed are given in Table 2. Apart from the equilibria t_4 and t_8, all the three-phase equilibria have been observed. Concerning critical points, only the line P_C^1 and the CEP P_{CE}^1 and P_{CE}^3 have been accurately located. The other lines still have to be searched, but it is likely that their extent in the compositional diagram is very short.

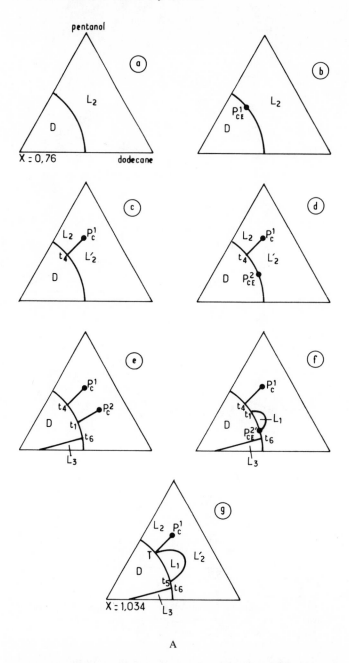

FIGURE 25.   Schematic representations in a field space of two sequences of diagrams which account for the development of phase equilibria between X = 0.76 and X = 1.034. In the sequence (A), three critical end points are involved. In the sequence (B), the creation of the phase L_1 results from a bitriple point (diagram d). In each triangle, the field variables are X (lower left corner), the chemical potentials of dodecane (lower right corner) and pentanol (upper corner).

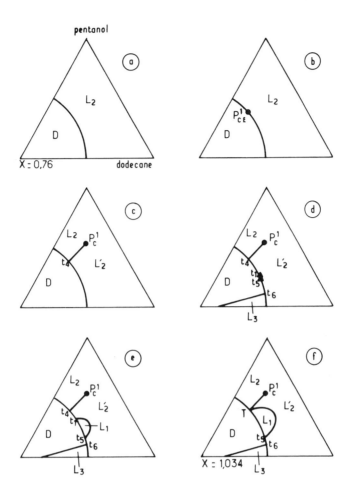

FIGURE 25B.

It is worth emphasizing that the complete presentation of the diagram given has been possible only because of the field character of the variable X. The choices of phase equilibria that we propose, in particular between the sections X = 0.76 and X = 1.034, are not the only possibilities which exist, but they are likely some of the simplest. One might expect that further accurate investigations of certain sections ranging between X = 0.9 and 1.034 should allow identification of features anticipated in the various possible progressions of equilibria and the ability to distinguish the correct one. However, such a study in the density space presents practical difficulties related to the large number of equilibria expected on one hand, and the very small extension of most of the multiphase regions found in this part of the diagram on the other hand. This particular situation might arise from the fact that the temperature of study, 21°C, is close to that at which some multiphase regions, and especially the four-phase region T, disappear. Preliminary study of the temperature dependence of region T indicates that its extent increases with temperature. Therefore, in order to test our expectations, it will be more appropriate to perform further experimental investigations at a temperature above 21°C.

IV. CONCLUDING REMARKS ON THE PHASE DIAGRAM

In the course of the study of the phase diagram of the quaternary mixture water-pentanol-dodecane-SDS we have identified, in addition to the isotropic microemulsion phase, two

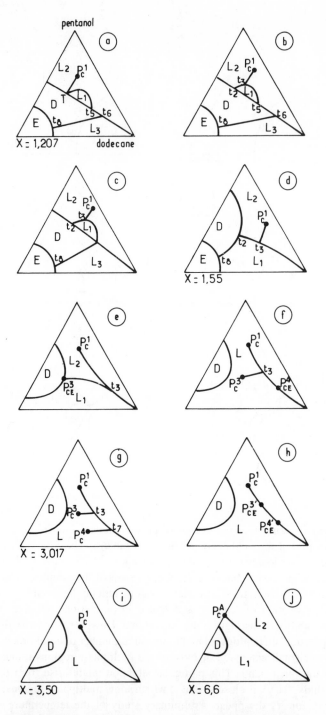

FIGURE 26.   Schematic representation of a sequence of diagrams which accounts for the development of phase-equilibria between X = 1.207 and X = 6.6. The field variables are the same as in Figure 25.

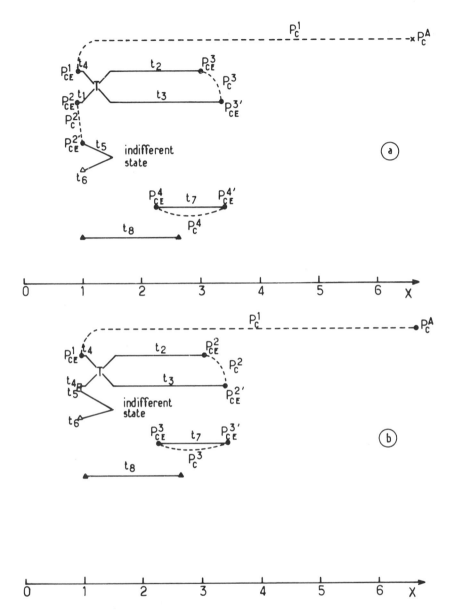

FIGURE 27. Evolution of the lines of triple points (t_1 - t_8) and of the lines of critical points P_C as a function of X. Model (a) and (b) correspond respectively to the sequences (a) and (b) given in Figures 25A and 25B.

mesomorphic phases. These phases appear as an extension of those found for the ternary mixture water-pentanol-SDS. The lamellar phase D possesses the exceptional property to dissolve considerable amounts of alcohol and oil up to 98%. A structural study of similar phases has revealed the existence of a large number of structural defects.[41] In certain regions of the phase diagram the microemulsion presents particular characteristics; so, for example, some mixtures rich in oil are flow birefringent. The phases are connected by a complex multiphase region where two, three, or four phases coexist. These equilibria are different from the so-called Winsor equilibria.

The field-like character of the water to surfactant ratio is shown to be a prominent feature of the quaternary system in the oil-rich part of the diagram. The major consequence of this

Table 2
SUMMARY OF THE THREE- AND FOUR-PHASE EQUILIBRIA EXPERIMENTALLY OBSERVED IN THE VARIOUS X PLANES STUDIED

Equilibria	X 1.034	1.207	1.37	1.55	2.558	3.017	3.5
T $L_2'\,L_1\,D\,L_2$	x	x	—	—	—	—	—
t_1 $L_2'\,L_1\,D$	x	x	x	—	—	—	—
t_2 $L_1\,D\,L_2$	x	x	x	x	x	—	—
t_3 $L_2'\,D\,L_2$	x	x	x	x	x	x	—
t_4 $L_2'\,D\,L_2$	—	—	—	—	—	—	—
t_5 $L_2'\,L_1\,D$	x	x	x	—	—	—	—
t_6 $L_2'\,L_3\,D$	x	x	x	—	—	—	—
t_7 $L\,L\,L$	—	—	—	—	x	x	—
t_8 $D\,E\,L_1$	—	—	—	—	—	—	—

Note: x: equilibria observed; —: equilibria not observed.

result has been to permit a complete description of the diagram. In this description, the considerable complexity of the multiphase region appears to be generated by the occurrence of several lines of critical points. One of them, the line P_C^1, as well as two CEPs have been located.

In the quaternary mixture under study, the two-phase region which bounds the microemulsion phase depends on the value of X. When X is below 0.95, the microemulsion separates with the lamellar phase D, while when X is above 0.95, two microemulsion phases are in equilibrium and a critical point exists. These two types of phase separation are also observed in ternary mixtures.[42] Indeed the phase diagrams of the mixtures H_2O-AOT-decane or isooctane shown in Figure 28 present strong similarities with those of the quaternary mixture. At 25°C, the diagram of the system containing isooctane does not present a critical point and the micellar inverse phase is in equilibrium with the lamellar phase D. As isooctane is replaced by decane, a critical point appears and the region of microemulsion-liquid crystal phase coexistence is conversed into a region of microemulsion-microemulsion coexistence.

In conclusion, the same phase behavior is found as one changes the X ratio in a quaternary mixture, the oil or the temperature in ternary ones. This phase behavior is characterized by two types of phase diagrams which are schematically shown in Figure 29.

V. CRITICAL BEHAVIOR

The study of the phase diagram of the water-dodecane-pentanol-SDS system has permitted us to locate a line of critical points P_C^1 and two CEPs. As mentioned in the introduction critical phenomena in micellar solutions and microemulsions are not fully understood and deserve further study. In the following, we present results obtained along the line P_C^1 and at the CEP P_{CE}^3.

One of the central ideas of modern critical phenomena theory is the concept of universality.[43] The critical exponents are expected to have the same value for all systems in the same universality class. A universality class is specified by only two parameters, namely the dimensionality of the space (d) and the dimensionality of the order parameter (n) of the transition. Over the last 15 years, the main experimental works have dealt with the liquid-gas critical point of pure fluids and with the liquid-liquid critical point of binary liquid mixtures.[34,43-45] Recently, the interest has been focused on the critical behavior of micellar solutions and microemulsions.[22-32]

In most cases, the aim of the experiments consisted of measuring the critical exponents,

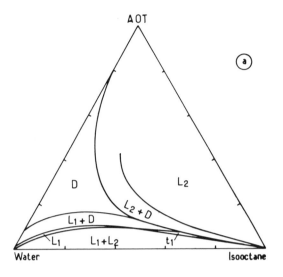

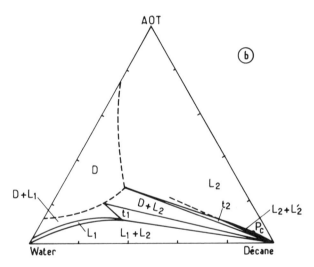

FIGURE 28. Phase diagrams at 25°C of the ternary systems: a = water-AOT-isooctane; b = water-AOT-decane. Abbreviations are the same as in Figures 8 to 10 and 12.

since their values are characteristic of the universality class of the critical point. Experimental data clearly indicate that the critical points of pure fluids and molecular binary mixtures belong to the same universality class as the Ising magnet model.[34,43-45] In these experiments, temperature is used as a field to approach the critical point. In binary mixtures, measurements can be made either in the single-phase region along a path with a constant composition or in the two-phase region along the coexistence curve. The main characteristic of these paths is that a density-like variable is kept constant. As pointed out by Griffiths and Wheeler,[35] these special paths which are asymptotically parallel to the coexistence line in the space of fields must lead to the exponents related to the critical point. When the critical point is approached in the one-phase region along another direction as, for example, at constant temperature, the exponents are about 50% weaker.

Griffiths and Wheeler[35] have generalized these considerations to the case of systems with more than two independent variables. They predict that the critical behavior will be essentially the same as in pure fluids provided that the critical exponents are measured along equivalent

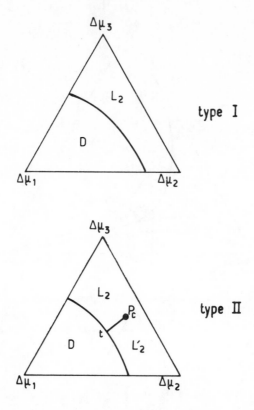

FIGURE 29.   Schematic representations in the field space $\Delta\mu_1$, $\Delta\mu_2$, $\Delta\mu_3$ of the two types of diagrams found in the oil-rich part of the phase diagrams. The field variables are related to the chemical potentials of the water, oil, and surfactant molecules.

directions in the space of independent variables of both systems. Their theoretical predictions also state that along a path parallel to the coexisting surface in the space of fields the critical behavior is described by the same exponents whatever the field considered. Up to now, in multicomponent systems, there is no experimental evidence for the possibility of approaching a critical point by varying a field variable at constant temperature and pressure. The main difficulty is to find an appropriate field variable, different from temperature or pressure, which may be experimentally controlled.

In multicomponent systems, there is, in addition to the directions along and intersecting the coexistence surface, another way of approaching the critical point, namely along the critical surface. Along this last path, the form of divergence depends on the rate to approach to the critical surface. A particular case has been considered by Fisher[46] who has shown that in ternary systems critical exponents measured along the direction, which keeps two density-like variables constant, are about 10% stronger than those taken along other paths in the coexistence surface. The critical behavior in ternary and multicomponent systems has been less studied than in binary mixtures. Experimental studies of the plait point of several ternary mixtures still indicate an Ising behavior and confirm the Fisher renormalization of the exponents by the factor $1/(1 - \alpha)$ with $\alpha \cong 0.1$.[47-51] Another particular case of exponent renormalization caused by the manner in which the experimental path approaches a line of critical point has been experimentally investigated.[52]

Over the past few years a large number of papers have been published on the experimental study of critical phenomena in mixtures involving surfactants. Two-, three-, four-, and even five-component systems were investigated.[22-32] This current experimental interest in the

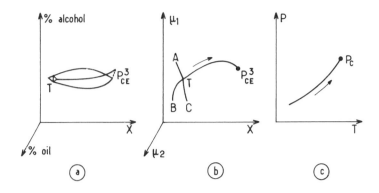

FIGURE 30. Representations of the three-phase volume t_2. (a) In the density space pentanol concentration, dodecane concentration, X. (b) In the field space consisting of X and the chemical potentials μ_1 and μ_2 of alcohol and oil. The arrow shows the path followed to approach the CEP P_{CE}^3. This path is geometrically equivalent to the approach of the liquid-gas critical point of a pure fluid in the two-phase region (C).

critical behavior of micellar and microemulsion solutions is due, to a great extent, to the very intricate results obtained so far. Indeed if some experimental results[22-29] are in agreement with those found in pure or usual binary fluids, other data indicate a more complex behavior.[30-32] Excepting Reference 22, the critical phenomenom, in most cases, is measured by its critical exponents ν and γ which characterize the divergences of the correlation length and of the osmotic compressibility of the solution, respectively. In some surfactant solutions, the values measured correspond to either negative[31] or positive[32] deviations from the Ising indexes ($\nu = 0.63$; $\gamma = 1.24$). Then, for example, the values of the critical exponents ν and γ found for a series of nonionic surfactant aqueous solutions seem to be dependent on the surfactant.[30,31] One common feature to all the critical points investigated so far is that they are lower critical points; the phase separation occurs as temperature is raised.

In most of the surfactant systems investigated, the critical point was approached by raising temperature. Data presented in the preceding section have provided evidence that in the oil-rich part of the phase diagrams, the water to surfactant ratio (X) has the property of a field variable. Therefore, this result offers us the possibility to approach a critical point at constant temperature by using X as a field.[25,27,40,53,54]

In order to improve our understanding of the critical behavior in surfactant solutions, we have undertaken a systematic investigation of several critical points of the water-dodecane-pentanol-SDS system. In this study we have determined, from light scattering intensity and refractive index measurements, the critical exponents ν, γ, and β. Results were obtained in different points of the critical line P_C^1 and also for the CEP P_{CE}^3. Critical points have been approached both in the single and in the multiphase regions along different paths either by raising temperature at fixed composition or by increasing X at constant temperature.

A. Study of the CEP P_{CE}^3

The three-phase region t_2 expands from the four-phase region T to the CEP P_{CE}^3. The representations of this region in both density and field spaces are given in Figure 30. We have approached the CEP P_{CE}^3 along the coexistence curve at constant temperature by varying X. This path is equivalent, from a topological point of view, to the approach along the coexistence curve of the liquid-gas critical point of a pure fluid. Therefore, according to the theoretical predictions one expects to measure the critical exponents related to the CEP P_{CE}^3.

Determinations of ν and γ were obtained from light scattering measurements in 21 three-phase samples corresponding to different values of X. Each three-phase sample was prepared

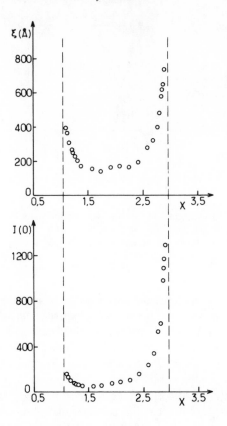

FIGURE 31.   Variation as a function of X of the scattered intensity
I(o) and of the correlation length ξ measured in the lower phase L_2 of
the equilibrium t_2 (T = 21°C).

in a cylindrical sealed cell and was equilibrated for several weeks in a constant temperature
chamber (controlled to 0.1 K). After equilibrium light scattering measurements were performed. Experimental conditions are given in References 25, 27, 40, and 53.

The total intensity of the scattered light was measured in the two isotropic phases L_1 and
L_2 by varying the position of the cell in the laser beam. For all the samples investigated,
the angular dissymmetry of the total scattered intensity is well fitted by the Ornstein-Zernike
(O-Z) law: $I(q) = I(o)/(1 + q^2 \xi^2)$ where ξ is the correlation length of the refractive index
fluctuations. In the lower microemulsion phase L_2, both I(o) and ξ show a large variation
as X is varied. As the CEP is approached, the large increase in the two measured quantities
is characteristic of a critical behavior. ξ varies from 140 Å for X = 1.7 to 700 Å for X =
2.90 (Figure 31). An increase of the intensity and of the correlation length is also observed
in the proximity of the four-phase equilibrium T (toward low values of X). This "critical-
like" behavior observed in the vicinity of the four-phase equilibrium is consistent with the
very close values of the compositions of the phases in equilibrium (Figure 18). This result
suggests that the four-phase equilibrium might disappear by a critical point reached by
changing temperature.

In order to measure the critical exponents ν and γ, we have investigated, in a log-log
plot, the behavior of the intensity I(o) and of the correlation length ξ as a function of the
reduced field variable ϵ. ϵ indicates the distance from the critical point and is defined as ϵ
$= (X_c-X)/X_c$. X_c is the value of X at the critical point where the three-phase region t_2
disappears. The results given in Figure 32 show that I(o) and ξ follow power laws:

$$I(o) = I_o \cdot \epsilon^{-\gamma_x} \qquad \xi = \xi_o \cdot \epsilon^{-\nu_x} \qquad (1)$$

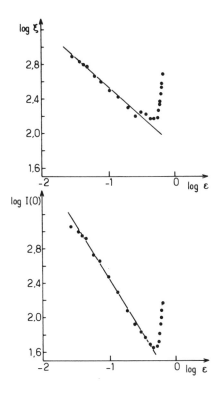

FIGURE 32.   Log-log plots of total intensity I(o) and correlation length
ξ measured in the lower phase of equilibrium t_2 as a function of the
reduced variable ϵ, ϵ indicates the proximity of the CEP. $\epsilon = (X - X_C)/X_C$ where X_C is the value of X at the CEP ($X_C = 2.98$).

The x index indicates that the path we have chosen is not the classical isochore path. A least squares fitting of the data gives: $\nu_x = 0.65 \pm 0.05$, $\gamma_x = 1.20 \pm 0.08$, and $X_c = 2.98 \pm 0.02$. These exponents are very close to the universal exponents found in binary mixtures.[43] The power laws dependence of the thermodynamic behavior clearly confirms that X can be considered as a field variable. The values of γ_x and ν_x are the "liquid-gas" values predicted by an Ising model. This result was not obvious in regards to the existence of a liquid crystalline phase in the equilibrium.

It is worthwhile to point out that Ising values are also obtained in the ternary mixture AOT-water-decane ($\nu_x = 0.61 \pm 0.06$ and $\gamma_x = 1.26 \pm 0.10$) as the critical point is approached in the single-phase region by varying the field X at constant temperature and constant decane concentration.[27] This result is in accordance with the theoretical predictions of Griffiths and Wheeler for ternary mixtures since along the path a field (temperature) and a density (oil concentration) are kept constant.

Measurements in the upper phase, L_1, which is flow birefringent for the low values of X, give different results. In this phase, the critical behavior is only observed very close to the CEP (for X values larger than 2.8). The particular behavior observed in this phase is most likely related to the existence of anisotropic particles.

B. Study in Several Points of the Line P_C^1

1. Critical Behavior at Fixed Temperature

Phase diagrams analysis has shown the existence of a critical line on the boundary of the microemulsion region. This line extends from the CEP P_{CE}^1 in the section $X = 0.95$ to the critical point P_C^A in the section $X = 6.6$. Figure 33 gives a schematic representation of this

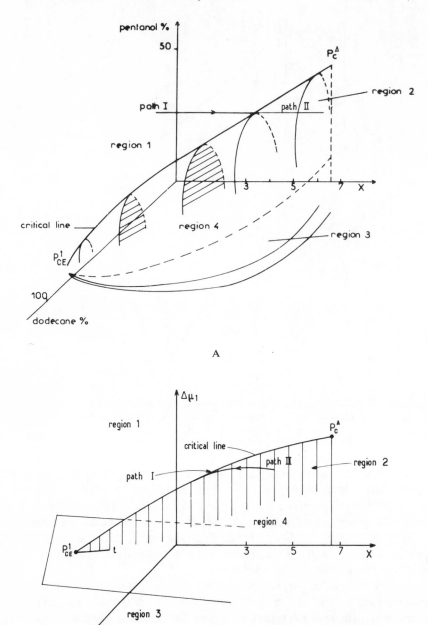

FIGURE 33.   Schematic representations of the phase diagram of the water-dodecane-pentanol-SDS system at fixed temperature. (A) In the mixed density-field space consisting of the pentanol concentration - dodecane concentration - X. (B) In the field space: μ_1, μ_2, X; μ_1 and μ_2 are two field variables related to the chemical potentials of pentanol and dodecane. Paths I and II are the routes followed, respectively, in the single and two-phase regions to approach several points of the critical line P_{CE}^l - P_C^A. Region 1: microemulsion (L_2); region 2: two-phase region (d_1); region 3: lamellar liquid crystalline phase (D); region 4: polyphasic region. The line t is a line of triple points L_2 D L_2 (equilibria t_4).

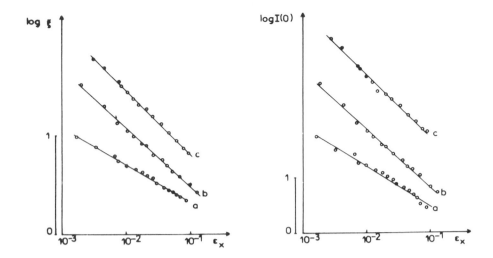

FIGURE 34. Log-log plots of the correlation length ξ and of the total intensity $I(o)$ vs. the reduced variable $\xi_x = (X_C - X)/X_C$ for the three following critical points: a: $X_C = 1.207$; b: $X_C = 1.55$; c: $X_C = 3.424$. These three critical points were approached along path I in the single phase region at fixed temperature (see Table 3). For clarity the data relative to each critical sample have been shifted along the vertical axis.

line of critical points in the field space μ_1, μ_2, X, and the mixed space C_A C_O X where μ_1 and μ_2 are two variables related to the chemical potentials of oil and alcohol and C_A and C_O are the pentanol and dodecane concentrations, respectively.

Figure 33 shows two possible paths of approaching at fixed temperature any critical point P_C^X of the line P_C^l. Paths I and II are lying within the single phase and the two-phase regions, respectively. Along both paths, the variable of approach is the field X, the alcohol and oil concentrations being kept constant to their critical values. This special direction of approach in a quaternary mixture is equivalent to that of a ternary mixture where the Fisher renormalization applies. Therefore, the values of the critical exponents measured along paths I and II are expected to be 10% stronger than the Ising values.

In order to determine whether the critical behavior is the same at several distinct positions of the critical line, we have carried out light scattering and refractive index measurements for several critical samples. The exponents ν and γ have been measured along path I and the exponent β along path II. The three critical points defined by $X_c = 1.207$, 1.55, and 3.424 were approached along path I. The experimental study of each critical point requires the preparation of several samples. Each sample has the critical alcohol and oil concentrations; they differ in the value of X. Both ξ and $I(o)$ increase with X. They are found to follow power laws.

$$I(o) = I_o\, \epsilon_x^{-\gamma_x} \quad \text{and} \quad \xi = \xi\, \epsilon_x^{-\nu_x} \tag{2}$$

The x index indicates that the critical point is approached by varying X. ϵ_x is the reduced variable X: $\epsilon_x = (X_c - X)/X_c$ where X_c is the value of X at the critical point. Figure 34 gives log-log plots of $I(o)$ and ξ vs. ϵ_x. The values of ν_x and γ_x obtained by a least squares fitting of the data for the three critical points studied are given in Table 3. Their values depend on the critical point investigated. They are found to decrease with X. This special behavior will be discussed later in this paper.

Four critical points defined by X = 1.034, 1.207, 1.372, and 1.552 have been approached along path II. For each critical point, about 15 two-phase samples have been prepared. After equilibrium of the mixtures at 21°C for 1 week, the refractive index of each phase has been

Table 3
CRITICAL TEMPERATURE T_c AND CRITICAL RATIO FOR THE POINTS MEASURED ALONG PATHS I AND II

Path I

X_c		1.207	1.550	3.424
T_c (°C)		21.84	25.40	21.80
ν_x		0.38 ± 0.03	0.57 ± 0.03	0.64 ± 0.03
γ_x		0.78 ± 0.06	1.10 ± 0.06	1.20 ± 0.06

Path II

X_c	1.034	1.207	1.372	1.552
T_c (°C)	21.0	21.0	21.0	21.0
β_x	0.40 ± 0.04	0.38 ± 0.04	0.46 ± 0.04	0.42 ± 0.04

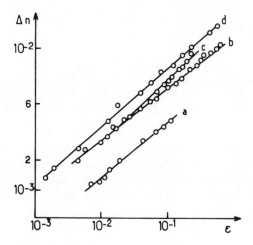

FIGURE 35.   Variation in log-log plot of the refractive index difference Δn against the reduced variable ξ_x for the four following critical points, a: $X_C = 1.034$; b: $X_C = 1.207$; c: $X_C = 1.379$; d: $X_C = 1.552$. The critical points were approached along path II shown in Figure 33 (Table 3) (T = 20.4°C).

measured at 20.4°C. Lowering of temperature has been carried out in order to ensure stability of the phase during the measurement. The refractive index difference Δn between the two coexisting phases is plotted against the reduced variable $\epsilon_x = (X - X_c)/X_c$ in a log-log plot (Figure 35). For all the critical systems studied, the experimental data are well fitted by the power law:

$$\Delta n = N_o \, \epsilon_x^{\beta_x} \tag{3}$$

Values of β_x and X_c obtained by a least squares fit are given in Table 3. Taking into account, the experimental accuracy, the value of β_x is found the same for all the critical points investigated. Besides, as expected, this value is close to the renormalized critical exponent ($\beta/1 - \alpha = 0.365$). This finding contrasts with the behavior of ν and γ found in the single phase region.

2. Critical Behavior at Constant Composition

If one takes into account the temperature variation, the system has one additional degree of freedom so that the critical line P_C^1 becomes a critical surface S_C^1. The section at constant

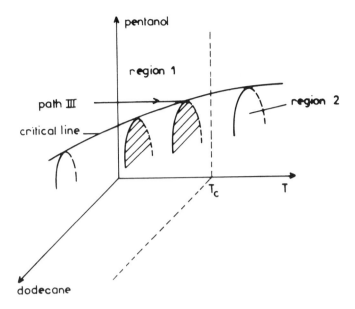

FIGURE 36. Schematic representation of the phase diagram of the water-dodecane-pentanol-SDS system at X constant in the dodecane concentration-alcohol concentration-temperature space. Several critical points were approached in the single phase region along path III.

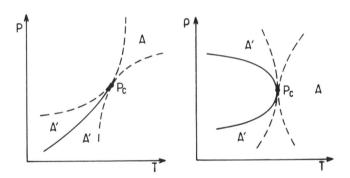

FIGURE 37. Liquid-vapor coexistence curve (solid line) and critical point for a pure fluid in (a) the temperature-pressure space and (b) in the density-temperature space. The dotted lines are an aid in visualizing how regions in the PT plane near the critical point are mapped into corresponding regions in the ρT plane.[35]

X for this surface gives a critical line. Figure 36 shows a scheme of the phase diagram in the temperature, pentanol, dodecane concentrations space for fixed X. In this space any critical point can be approached in the single-phase region by varying temperature (path III), the overall composition of all the components being kept constant with their critical values. Such a path which imposes a triple density constraint does not correspond to any of the approaches discussed by Griffith and Wheeler. Due to the geometrical effect associated with the Legendre transformation of variables, a path asymptotically parallel to the coexistence surface (APCS) in a field space can be easily achieved in practice in a density-temperature space as shown for a pure compound in Figure 37. In this figure the regions denoted A, A' which include all the approaches — A, P, C, S — are narrow in the field diagram and rather large in the density one.

Outside regions A and A' the exponents are renormalized. In the case of multicomponent mixture it is difficult to have one idea of the extent of regions A and A'. In some cases, measurements made as a function of temperature with constant overall composition have given the unrenormalized exponents.[26] On Figure 38a are displayed the experimental boundaries found between regions 1 and 2 in the plane X = 1.55 at several temperatures and the projection of the critical line. As temperature increases, the coexistence curve moves toward the higher alcohol content. A similar pattern of phase behavior is observed for the other values of x except for the lowest value (X = 1.034). In this case, it appears that the coexistence line is less sensitive to temperature, at least in the temperature range investigated, 15 to 35°C (Figure 38b). One can see in Figure 39 that the angle between the path followed and the coexisting surface depends on the critical point considered.

Six critical samples defined by X = 1.034, 1.207, 1.372, 1.552, 3.448, and 5.172 were prepared. Both ξ and I(o) increase as the temperature is raised and diverge at the critical point. Figure 40 shows log-log plots of I(o) and ξ vs. the reduced temperature $\epsilon = (T_c - T)/T_c$ for the six critical mixtures. The data are well described by the power laws:

$$I(o) = I_o \, \epsilon_t^{-\gamma_t} \quad \text{and} \quad \xi = \xi_o \, \epsilon_t^{-\nu_t} \tag{4}$$

The t index indicates that the critical point is approached by varying temperature. The values obtained by a least squares fitting procedure for T_c, γ_t, ν_t, and ξ_o are given in Table 3. The value T_c deduced from the fitting of the experimental data is slightly different from the experimentally found temperature T_d at which the mixture separates. The difference found between T_c and T_d (0.1°C to 0.15°C) is of the order of magnitude of the experimental uncertainty on the T_d measurement. The values of the exponents ν and γ determined in using T_d are very close to those obtained when temperature is set free in the fit of the experimental data. The uncertainties given in Table 4 correspond to the variations found when the critical temperature considered is either T_c or T_d. The γ_t:ν_t ratio is found to be constant and is close to 2. A continuous decrease of the exponents ν_t and γ_t is observed as the CEP P_C^A is approached. As X_c decreases, the values of ν_t and γ_t show a trend similar to that found for ν_x and γ_x (Figure 41). For a given critical point, the values of ν and γ obtained along paths II and III are very close. The decrease of ν_t and γ_t is accompanied by an increase of ξ_o from 30 Å to more than 300 Å and also an increase of I_o (Figure 41). The ratio I_o:ξ_o^2 is not found constant. New experiments are in progress in order to obtain more accurate values of the constant I_o.

In order to examine to what extent the critical exponents are depending on the determination of the critical composition, we have prepared two samples d_1 and d_2 located in the plane X = 1.55 near the critical concentration (d). The dodecane concentration difference between the samples d_1 and d_2 (1.8%) is large in comparison with the uncertainty on the critical concentration (0.5%). The exponent ν measured for the three cases has the same value (d_1: $\nu = 0.55 \pm 0.03$; d: $\nu = 0.53 \pm 0.03$; d_2: $\nu = 0.54 \pm 0.03$). These results suggest that the observed behavior cannot be related to errors in the determination of the critical point.

C. Discussion

The results presented in this section confirm the field character of the water to surfactant ratio established from phase composition analysis. Moreover, they show that the variable X can be used similarly with temperature to approach a critical point. Critical exponents obtained from measurements in the multiphase regions are consistent with the theoretical predictions. In contrast, those determined in the single phase region show a singular behavior. In this last approach, the two independent fields used, X and T, give the same evolution of the critical behavior as the critical point considered approaches the CEP P_{CE}^1.

The continuous variation of the effective critical exponents ν and γ that we have evidenced

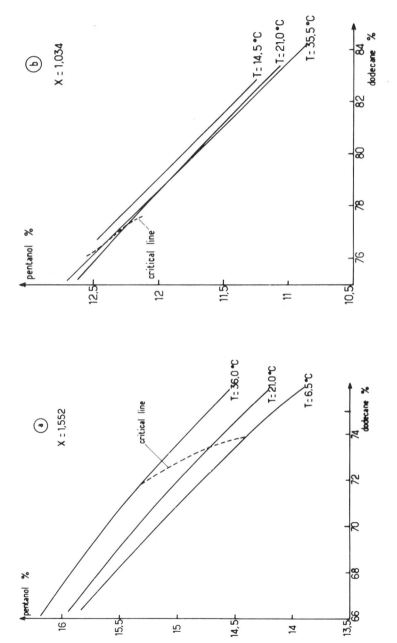

FIGURE 38.   Partial experimental pseudoternary phase diagrams (expressed in wt %) of the pentanol-dodecane-water-SDS system at several temperatures, (a) X = 1.552; (b) X = 1.034. In both figures, the full lines represent the boundary between regions 1 and 2 (region 1 lies above the coexistence curve), the dashed lines represent the critical line.

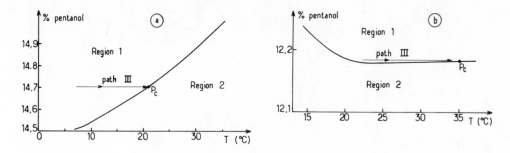

FIGURE 39.   Partial experimental pseudoternary phase diagrams of the pentanol-dodecane-water-SDS system. Sections at constant X and constant dodecane concentration represented in the plane pentanol concentration-temperature (a) X = 1.55, dodecane concentration = 73.5%; (b) X = 1.034, dodecane concentration = 76.5%. In both figures, the full lines represent the boundary between the microemulsion region L_2 (region 1) and the two-phase region d_1 (region 2).

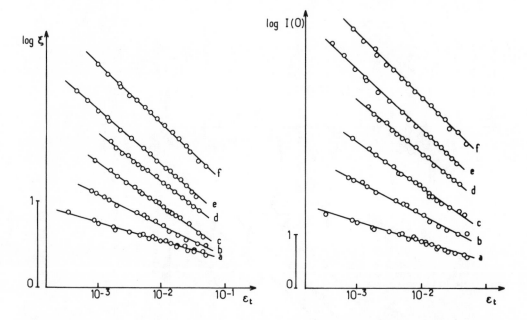

FIGURE 40.   Log-log plots of the correlation length ξ and the total intensity I(o) vs. the reduced temperature for the six following critical mixtures. a: $X_C = 1.034$; b: $X_C = 1.207$; c: $X_C = 1.372$; d: $X_C = 1.552$; e: $X_C = 3.448$; f: $X_C = 5.172$. Each critical point has been approached in the single phase region along path III (Table 4).

seems very similar to that previously reported by Corti et al.[31] for nonionic aqueous micellar solutions as the surfactant is varied. Indeed the values of the critical exponents found for the aqueous solutions of the polyoxyethylene amphiphiles C_6E_3, C_8E_4, $C_{12}E_8$, and $C_{14}E_7$ are ranging between 0.63 and 0.44 for ν, and 1.25 and 0.87 for γ while the ξ_o factor varies between 3.4 and 17.5 Å. The critical phenomena found in these micellar solutions are not of the simple type described by the Ising model.[9] Corti et al.[31] have suggested that a new theoretical model is needed to describe the critical behavior of micellar solutions. However, it is worth mentioning that in some microemulsion and micellar systems such a variation in critical exponents is not observed.[26,27] The water-AOT-decane system presents in the temperature-composition space a line of critical points ended by a CEP. The critical exponents ν and γ measured in several distinct points of this line have almost the same values, even in the vicinity of the CEP. They are found to be in good agreement with the Ising indexes.

Table 4
CRITICAL TEMPERATURE T_c AND CRITICAL RATIO X_c FOR THE SIX POINTS MEASURED ALONG PATH III

X_c	1.034	1.207	1.372	1.552	3.448	5.172
T_c (°C)	36.35	37.54	34.52	32.25	33.43	35.47
ν_t	0.21 ± 0.02	0.34 ± 0.03	0.47 ± 0.03	0.53 ± 0.03	0.59 ± 0.03	0.64 ± 0.03
γ_t	0.40 ± 0.04	0.63 ± 0.06	0.86 ± 0.06	1.01 ± 0.06	1.14 ± 0.06	1.21 ± 0.06
ϵ_0(Å)	331 ± 20	141 ± 10	64.6 ± 6	46.8 ± 5	27.5 ± 3	30.9 ± 3
I_0^*	0.407	0.144	0.051	0.035	0.017	0.047

* Arbitrary unit.

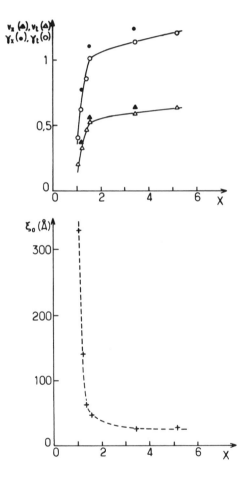

FIGURE 41. Variation of the critical exponents ν_t, γ_t, ν_x, γ_x and of the scale factor ξ_0 vs. X.

Although the complex behavior observed in the quaternary mixture studied in this paper is not fully understood, several possible explanations can be considered. As previously emphasized, it is known that in fluids (pure or multicomponent mixtures) the exponents depend on the path followed in approaching the critical point.[35] This possibility complicates the analysis of the results. However, the identical variation observed in the single phase by varying two different fields (I and II) suggests that the observed behavior is not only due

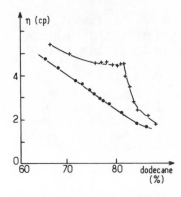

FIGURE 42.   Viscosity of microemulsions
(21°C) vs. the dodecane volume fraction for samples located in two X planes: + X = 1.02; o
X = 1.55.

to the choice of the specific path followed. In both cases, the deviation from Ising values
increases significantly in the vicinity of the CEP.

The exponent variation could be due to a crossover between two sets of exponents, one
of them being the Ising indexes; these indexes are obtained for the X_c = 3.4 and 5.172
samples. In this case, the second set of exponents does not correspond to the mean field
values since the exponents measured for the lowest X values are much less than the mean
field indexes (ν = 0.5; γ = 1). The second set of exponents should be more likely associated
with a "special" CEP in relation to the existence of the liquid crystalline phase D. With
regard to this explanation, it should be noted that we have observed in the vicinity of the
CEP an anomalous behavior in viscosity similar to that previously reported for microemulsions near a critical point.[55] We have measured, at fixed temperature, the viscosity of mixtures
located along the coexistence surface. Measurements have been carried out along paths
defined by a fixed value of X; the variable along each path is the dodecane and alcohol
concentration. Figure 42 shows the variation of the viscosity along the two paths considered
(X = 1.02 and X = 1.55). The data obtained for the X = 1.02 plane show an anomalous
behavior in the oil concentration range around the critical point (volumic critical dodecane
concentration 82%). This anomaly disappears as X increases. Indeed in the plane defined
by X = 1.55, viscosity continuously decreases as the oil concentration increases; in this
case, no particular feature is observed at the critical point (volumic critical dodecane concentration 77%). Hence, the anomalous behavior observed for the X planes close to the CEP
P_{CE}^l (low values of X) is not due only to the presence of a critical point. This behavior is
more likely related to the existence of the lamellar phase and could be due to precursor
fluctuations of the smectic phase. Let us recall that in the vicinity of the CEP, the critical
microemulsions and the birefringent phase have very similar compositions. In contrast, in
the AOT-water-decane system in which we observe neither an anomalous behavior of viscosity nor an evolution of critical exponents, the liquid crystalline phase is far in composition
from the critical line.

VI. CONCLUSION

The results presented in this chapter give, for the first time, an overall picture of the phase
diagram of a quaternary mixture. The diagram obtained shows many common features with
those of simpler surfactant systems containing only two or three components. In addition
to the very wide microemulsion region, two lyotropic phases occur in the diagram of the
water-dodecane-pentanol-SDS system. One of them, the lamellar phase, exists over a wide

range of water and oil concentrations. A great variety of phase equilibria involving two, three, and even four phases is observed in the oil-rich part of the diagram. Besides, in this last region, the water over surfactant ratio is found to behave as a field variable. This very important result has permitted proposition of a complete description of the phase diagram. In this description, which accounts for the continuity of the multiphase regions, some three-phase equilibria are generated by CEPs, others by indifferent states. Recent results suggest that the occurrence of such states and also that of azeotropic-like states are quite frequent in surfactant multicomponent mixtures.

Although the overall picture of the diagram is known, further experimental work is necessary to precise details in some parts of the diagram, particularly in the water-rich part of the diagram. Moreover, several critical lines and CEPs predicted in the proposed description still have to be located. Several important problems remain to be investigated. One of these is the structure of the dilute water lamellar phase. Another one is the temperature dependence of the equilibria found in the oil-rich part of the diagram. Both areas certainly deserve further research studies.

This chapter has left untouched the theoretical aspects relative to phase stability. Several models of microemulsion have been proposed.[18,56-62] Recent experimental studies have shown the importance of interactions between water in oil micelles on the stability of oil-rich microemulsions.[42,63,64] In particular, the existence of the critical line P_C^l found in the system under study can be interpreted as a liquid-gas transition due to interactions between inverse micelles.[42,63]

The experimental results of the critical behavior at several critical points or CEPs of the quaternary mixture water-dodecane-pentanol-SDS call attention to many unsolved problems. Although some data are successfully interpreted by the theory for mixtures of simple molecules, the variation of critical exponents with the position of the point along the critical line is an unanswered question. The large discrepancy found between theory and experiment in the vicinity of the CEP P_{CE}^l could be due to the occurrence of the oil-rich lamellar phase. Along this line of thought, the knowledge of the structure of these particular mesophases should be very useful for the understanding of the phase transition.

ACKNOWLEDGMENTS

Part of this work was done in collaboration with P. Honorat and G. Guerin and with the technical assistance of M. Maugey and O. Babagbeto.

It is with pleasure that the authors acknowledge comments and criticism from B. Widom (Cornell University), V. Degiorgio (Milan University), and J. Prost (Bordeaux). They are also grateful to P. Bothorel for fruitful discussions in the course of this work.

REFERENCES

1. **Ekwall, P.,** Composition, properties and structures of liquid crystalline phases in systems of amphiphilic compounds in *Advances in Liquid Crystals,* Vol. 1, Brown, G. H., Ed., Academic Press, New York, 1975, 1.
2. **Winsor, P. A.,** Binary and multicomponent solutions of amphiphilic compounds. Solubilization and the formation and theoretical significance of liquid crystalline solutions, *Chem. Rev.,* 68, 1, 1968.
3. **Fontell, K.,** *Liquid Crystals and Plastic Crystals,* Vol. 2, Gray, G. W. and Winsor, P. A., Eds., Ellis Howood, Chichester, 1974, 80.
4. **Tiddy, G. J. T.,** Surfactant-water liquid crystal phases, *Phys. Rep.,* 57, 1, 1980.
5. **Tiddy, G. J. T. and Walsh, M. F.,** Lyotropic liquid crystals, in *Aggregation Process in Solution,* Vol. 28, Wyn Jones, E. and Gornally, J., Eds., Elsevier, Amsterdam, 1983, 151.

6. **Charvolin, J.,** Agrégats de molécules amphiphiles en solution, in *Colloides et Interfaces,* Editions de Physique, Paris, 1983, 33; **Charvolin, J.,** Polymorphism of interfaces, *J. Chim. Phys., 80,* 15, 1983.

7. **Radley, K. and Reeves, L. W.,** Studies of ternary nematic phases by NMR. Alkali metal decylsulfates-decanol-D_2O, *Can. J. Chem., 53,* 2998, 1975.

8. **Radley, K., Reeves, L. W., and Tracey, A. S.,** Effect of counterion substitution on the type and nature of nematic lyotropic phases from NMR studies, *J. Phys. Chem., 80,* 174, 1976.

9. **Yu, L. J. and Saupe, A.,** Observation of a biaxial nematic phase in potassium laurate-1-decanol-water mixtures, *Phys. Rev. Lett., 45,* 1000, 1980.

10. **Hendricks, Y. and Charvolin, J.,** Structural relations between lyotropic phases in the vicinity of the nematic phases, *J. Phys., 42,* 1427, 1981.

11. **Forrest, B. J. and Reeves, L. W.,** New lyotropic liquid crystals composed of finite nonspherical micelles, *Chem. Rev., 81,* 1, 1981.

12. **Ekwall, P., Mandell, L., and Fontell, K., Eds.,** Solubilization in micelles and mesophases and the transition from normal to reversed structures, in *Molecular Crystals and Liquid Crystals,* Vol. 8, Gordon & Breach Science Publ., New York, 1969, 157.

13. **Gillberg, G., Lehtinen, H., and Friberg, S.,** NMR and IR investigation of the conditions determining the stability of microemulsions, *J. Colloid Interface Sci., 33,* 40, 1970.

14. **Ahmad, S. E., Shinoda, K., and Friberg, S.,** Microemulsions and phase equilibria, *J. Colloid Interface Sci., 47,* 32, 1974.

15. **Friberg, S. and Burasczenska, I.,** Microemulsions containing ionic surfactants, in *Micellization, Solubilization, and Microemulsions,* Vol. 2, Mittal, K. L., Ed., Plenum Press, New York, 1977, 791.

16. **Healy, R. N., Reed, R. L., and Stenmark, D. G.,** Multiphase microemulsion systems, *Soc. Pet. Eng. J., 16,* 147, 1976.

17. **Shah, D. O. and Schecter, R. S., Eds.,** *Improved Oil Recovery by Surfactant and Polymer Flooding,* Academic Press, New York, 1977.

18. **Bellocq, A. M., Biais, J., Bothorel, P., Clin, B., Fourche, G., Lalanne, P., Lemaire, B., Lemanceau, B., and Roux, D.,** Microemulsions, *Adv. Colloid Interface Sci., 20,* 167, 1984.

19. **Bellocq, A. M., Bourbon, D., and Lemanceau, B.,** Three-dimensional phase diagram and molecular and interfacial properties of the system water-dodecane-pentanol-sodium octylbenzene sulfonate, *J. Colloid Interface Sci., 79,* 419, 1981.

20. **Lang, J. C.,** Chemical and thermal development of nonionic surfactant phase equilibria, in *Physics of Amphiphiles, Micelles, Vesicles and Microemulsions,* Degiorgio, V. and Corti, M., Eds., North Holland, Amsterdam, 1985, 336.

21. **Kahlweit, M. and Strey, R.,** Phase behavior of ternary systems of the type H_2O-oil-nonionic surfactant, *Angew. Chem., 24,* 654, 1985.

22. **Lang, J. C. and Morgan, R. D.,** Nonionic surfactant mixtures. I. Phase equilibria in $C_{10}E_4$-H_2O and closed-loop coexistence, *J. Chem. Phys., 73,* 5849, 1980.

23. **Huang, J. S. and Kim, M. W.,** Critical behavior of a microemulsion, *Phys. Rev. Lett., 47,* 1462, 1981.

24. **Kotlarchyk, M., Chen, S. H., and Huang, J. S.,** Critical behavior of a microemulsion studied by small angle neutron scattering, *Phys. Rev., A28,* 508, 1983.

25. **Roux, D. and Bellocq, A. M.,** Experimental evidence for an apparent field variable in a critical microemulsion system, *Phys. Rev. Lett., 52,* 1895, 1984.

26. **Abillon, O., Chatenay, D., Langevin, D., and Meunier, J.,** Light scattering study of a lower critical consolute point in a micellar system, *J. Phys. Lett., 45,* L223, 1984.

27. **Honorat, P., Roux, D., and Bellocq, A. M.,** Light scattering study of the critical behavior in a ternary microemulsion system, *J. Phys. Lett., 45,* L961, 1984.

28. **Bellocq, A. M., Bourbon, D., Lemanceau, B., and Fourche, G.,** Thermodynamic interfacial and structural properties of polyphasic microemulsion systems, *J. Colloid Interface Sci., 89,* 427, 1982.

29. **Cazabat, A. M., Langevin, D., Meunier, J., and Pouchelon, A.,** Critical behavior in microemulsions, *Adv. Colloid Interface Sci., 16,* 175, 1982; *J. Phys. Lett., 43,* L89, 1982.

30. **Corti, M., Degiorgio, V., and Zulauf, M.,** Nonuniversal critical behavior of micellar solutions, *Phys. Rev. Lett., 48,* 1617, 1982.

31. **Corti, M., Minero, C., and Degiorgio, V.,** Cloud point transition in nonionic micellar solutions, *J. Phys. Chem., 88,* 309, 1984.

32. **Dorshow, R., de Buzzaccarini, F., Bunton, C. A., and Nicoli, D. F.,** Critical-like behavior observed for a five-component microemulsion, *Phys. Rev. Lett., 47,* 1336, 1981.

33. **Stanley, M. E.,** *Introduction to Phase Transition and Critical Phenomena,* Oxford University Press, New York, 1971.

34. **Beysens, D.,** Status of the experimental situation in critical binary fluids, *Nato Adv. Study Ser., 82,* 25, 1982.

35. **Griffiths, R. B. and Wheeler, J. C.,** Critical points in multicomponent systems, *Phys. Rev. A, 2,* 1047, 1970.

36. **Skoulios, A.,** Amphiphiles: organisation et diagrammes de phases, *Ann. Phys.,* 3, 421, 1978.
37. **Bellocq, A. M., Biais, J., Clin, B., Gélot, A., Lalanne, R., and Lemanceau, B.,** Three-dimensional phase diagram of the brine toluene-butanol-sodium dodecyl-sulfate system, *J. Colloid Interface Sci.,* 74, 311, 1980.
38. **Fontell, K.,** Liquid crystallinity in lipid water systems, *Mol. Cryst. Liq. Cryst.,* 63, 59, 1981.
39. **Prigogine, I. and Defay, R.,** *Thermodynamique Chimique,* Liège, Belgium, 1950.
40. **Roux, D. and Bellocq, A. M.,** Critical behavior in a microemulsion system. I. Experimental evidence for a field like variable, in *Surfactants in Solution,* Mittal, K. L. and Bothorel, P., Eds., Plenum Press, New York, in press.
41. **Dvolaitzky, M., Dimiglio, J. M., Ober, R., and Taupin, C.,** Defects and curvature in the interfacial film of birefringent microemulsions, *J. Phys. Lett.,* 44, L229, 1983.
42. **Roux, D. and Bellocq, A. M.,** Phase diagrams and interactions in oil rich microemulsions, in *Physics of Amphiphiles, Micelles, Vesicles and Microemulsions,* Degiorgio, V. and Corti, M., Eds., North Holland, Amsterdam, 1985, 842.
43. **Levelt Sengers, A., Hocken, R., and Sengers, J. V.,** *Phys. Today,* 42, 1977.
44. **Kumar, A., Krishnamurthy, H. R., and Gopal, E. S. R.,** Equilibrium critical phenomena in binary liquid mixtures, *Phys. Rep.,* 98, 57, 1983.
45. **Beysens, D., Bourgou, A., and Calmettes, P.,** Experimental determinations of universal amplitudes combinations for binary fluids. I. Statics, *Phys. Rev. A,* 26, 3589, 1982.
46. **Fisher, M. E.,** Renormalization of critical exponents by hidden variables, *Phys. Rev.,* 176, 257, 1968.
47. **Zollweg, J. A.,** Shape of the coexistence curve near a plait point in a three-component system, *J. Chem. Phys.,* 55, 1430, 1971.
48. **Ohbayashi, K. and Chu, B.,** Light scattering near the plait point of a ternary liquid mixture: ethanol-water-chloroform, *J. Chem. Phys.,* 68, 5066, 1978.
49. **Wold, L. E., Pruit, G. J., and Morrisson, G.,** The shape of the coexistence curve of ternary liquid mixtures near the plait point, *J. Phys. Chem.,* 77, 1572, 1973.
50. **Goldburg, W. I. and Pusey, P. N.,** Observation of critical parameter renormalization in a three-component liquid mixture, *J. Phys.,* 33, 105, 1972.
51. **Bloemen, E., Thoen, J., and Van Dael, W.,** The specific heat anomaly in some ternary liquid mixtures near a critical solution point, *J. Chem. Phys.,* 75, 1488, 1981.
52. **Johnston, R. G., Clark, N. A., Wiltzius, P., and Cannell, D. S.,** Critical behavior near a vanishing miscibility gap, *Phys. Rev. Lett.,* 54, 49, 1985.
53. **Bellocq, A. M., Honorat, P., and Roux, D.,** Critical behavior in a microemulsion system. II. Experimental evidence for a continuous variation of critical exponents, in *Surfactants in Solution,* Mittal, K. L. and Bothorel, P., Eds., Plenum Press, New York, in press.
54. **Bellocq, A. M., Honorat, P., and Roux, D.,** Experimental evidence for a continuous variation of effective critical exponents in a microemulsion system, *J. Phys.,* 45, 743, 1985.
55. **Cazabat, A. M., Langevin, D., and Sorba, O.,** Anomalous viscosity of microemulsions near a critical point, *J. Phys. Lett.,* 43, L505, 1982.
56. **Ruckenstein, E.,** Stability of microemulsions, *J. Chem. Soc. Faraday Trans. 2,* 71, 1690, 1975.
57. **Talmon, Y. and Prager, S.,** Statistical thermodynamics of phase equilibria in microemulsions, *J. Chem. Phys.,* 69, 2984, 1978.
58. **Mukerjee, S., Miller, C. A., and Forb, T.,** Theory of drop size and phase continuity in microemulsions, *J. Colloid Interface Sci.,* 1982.
59. **Jouffroy, J., Levinson, P., and de Gennes, P. G.,** Phase equilibria involving microemulsions (remarks on the Talmon-Prager model), *J. Phys.,* 43, 1241, 1982.
60. **Safran, S. A. and Turkevich, L. A.,** Phase diagrams for microemulsions, *Phys. Rev. Lett.,* 50, 1930, 1983.
61. **Safran, S. A., Turkevich, L. A., and Pincus, P.,** Cylindrical microemulsions: a polymer-like phase?, *J. Phys. (Paris) Lett.,* 45, L69, 1984.
62. **Widom, B.,** A model microemulsion, *J. Chem. Phys.,* 81, 1030, 1984.
63. **Roux, D., Bellocq, A. M., and Leblanc, M. S.,** An interpretation of the phase diagrams of microemulsions, *Chem. Phys. Lett.,* 94, 156, 1983.
64. **Kotlarchyk, M., Chen, S. H., Huang, J. S., and Kim, M. W.,** Structure of three-component microemulsions in the critical region determined by small-angle neutron scattering, *Phys. Rev. A,* 29, 2054, 1984.

Chapter 3

NONAQUEOUS MICROEMULSIONS

Stig E. Friberg and Yuh-Chirn Liang

TABLE OF CONTENTS

The concept of microemulsions recently has been advanced into the area of nonaqueous systems. In these systems, water is replaced by a polar organic compound such as ethylene glycol, glycerol, or acetamide. Some of the progress in the research in nonaqueous micellar solutions, lamellar liquid crystals, microemulsions, and critical solutions will be reviewed.

I. INTRODUCTION

Microemulsions are, at present, the focus of extensive research efforts and the progress in the field has been reported regularly in review articles.[1,2] Their definition is still open to debate[3,4] and probably will be in the future.

Microemulsions are transparent vehicles containing large amounts of both water and hydrocarbon. They are colloidally dispersed systems and hence show an essential distinction from molecular solutions of hydrocarbons and water. The latter kind may be exemplified by the system water, benzene, and isopropanol (Figure 1).

The diagram makes it obvious that the efficacy of isopropanol to ensure mutual solubility of hydrocarbon and water is not pronounced. A solution of equal amount of hydrocarbon and isopropanol will dissolve only 12% water. The system may be described as essentially an isopropanol-hydrocarbon solution with a small amount of water dissolved.

In a microemulsion, the hydrocarbon/water colloidal solution is commonly stabilized by a combination of an ionic surfactant and a medium chain length alcohol such as pentanol. A comparison of their solubilization power with that of isopropanol is useful in order to understand the specifics and advantages of the microemulsion over the molecular solution.

The surfactant or the pentanol are *individually* not superior to the isopropanol; on the contrary, the pentanol alone is even less efficient than the isopropanol in dissolving water into the hydrocarbon and the surfactant has too small a solubility region in water to be useful by itself. Instead, the outstanding properties of the surfactant and pentanol are found with their combination. According to Figure 1, the high solubility of water at low hydrocarbon content is maintained proportionally with added hydrocarbon to the pentanol/surfactant mixture. The difference at 50% hydrocarbon is striking; with isopropanol water may be dissolved to one fourth of the remaining half of the vehicle; in the microemulsion three quarters of the remaining space will be filled by water.

The water/hydrocarbon microemulsions have found a great variety of uses[5] and they are today not only scientifically but also technologically well established. Against this background and against the future need to use more polar compounds in liquid fuels, an interest has begun to surface on the application of the microemulsion concept to systems of hydrocarbons combined with more polar substances, which do not show pronounced solubility in the hydrocarbon. The research efforts in this area have been initiated in recent years and the volume, so far, is modest. In this chapter, some of the progress for micellar solutions, lyotropic liquid crystals, and microemulsions will be reviewed.

II. MICELLAR SOLUTIONS

Investigations of micellization in nonaqueous polar solvents were reported during the 1960s with the excellent contributions by Reinsborough et al.[6-10] using pyridimium chloride at high temperature as solvent and with investigations in ethylene glycol[11,12] and in amides and sulfoxides.[12,13]

Recent investigations by Evans and collaborators[14-18] have given excellent information not only on micellization in nonaqueous polar media but have also provided essential information on the process in water.

Illustrative information was obtained from a comparison of micellization in hydrazine[18] and in water at high temperatures.[15]

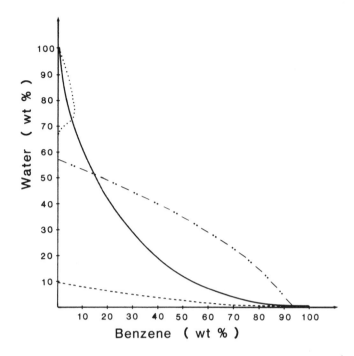

FIGURE 1.   The concurrent solubility of benzene and water in isopropanol
(——) is not pronounced. The water solubility in pentanol plus hydrocarbon
is even less (- - -). The solubility of hydrocarbon in aqueous solutions of
surfactant (· · ·) is greater than that in isopropanol (—··—), but the limited
solubility of the surfactant in water makes the formation of microemulsions
impossible. First, the combination of pentanol and surfactant in a W/O microemulsion gives the high water solubility experienced in a microemulsion.

Evans' comparison of the properties of water and hydrazine (Table 1[18]) demonstrated the
similarities and differences between the two media. The similarities are found in properties,
which do not depend on the tendency of water to form a three-dimensional hydrogen-bonded
network such as density, boiling and melting points, and enthalpy of evaporation. On the
other hand, the hydrogen bond network of water causes pronounced differences between
the properties of the two compounds in the room temperature range. There hydrazine shows
none of the characteristic properties of water such as volume increase at freezing, temperature
of maximum density, and compressability maximum.

These differences are reflected in the relation between the two solvents and nonpolar
molecules. For example, the Gibbs' free energy for transfer of a methyl group from hydrazine
to water[16] varies from -660 to $+60$ cal/mol over the temperature range 10 to 45°C. The
more structured water at low temperature is evidently favorable for the solubility of nonpolar
substances, a conclusion reached by Shinoda more than 15 years ago.[19]

With this follows positive entropy changes with micellization in water as a result of a
large positive entropy change from the removal of ordered water around the hydrocarbon
chains incompletely compensated by the negative entropy change due to the association of
the amphiphiles into a micelle. The entropy change for micellization in hydrazine is negative
as is the case in water at high temperatures.[18]

In summary, the ordered structure of water at low temperatures has an effect on the
thermodynamics of micellization, but it is not a decisive influence. The order causes the
entropy change for micellization to be positive at low temperatures and the micellization is
mainly entropy driven. At higher temperatures this contribution is lost, but with the loss of
the ordered structure, the enthalpy change has become negative and now provides the

Table 1
COMPARISON OF HYDRAZINE AND WATER

Property	N_2H_4	H_2O
m.p., °C	1.69	0.0
b.p., °C	113.5	100.0
T_c, °C	380	374.2
P_c, atm	145	218.3
Density at 25°C, g/cm^{-3}	10,036	0.9971
η at 25°C, P	0.00905	0.008904
γ at 25°C, deg/cm	66.67	72.0
ΔH_f at m.p., cal/mol	3,025	1,440
ΔS_f, cal/mol/K	11	5.3
ΔH_v at b.p., cal/mol	9,760	9,720
ΔH_v at 25°C	10,700	10,500
ΔS_v at b.p., gibbs/mol	25.2	26.1
M (gas), D	1.83—1.90	1.85
ϵ at 25°C	51.7	78.3
η_D	1.4644	1.3325
sp conductance at 25°C, mho/cm	3×10^{-6}	5×10^{-8}
Ion product, mol^2/cm^{-6}	2×10^{-25}	10^{-14}
C_p(liquid), cal/mol/K	23.62 (298 K)	17.98 (298 K)
C_p(solid), cal/mol/K	15.3 (274.7 K)	8.9 (273 K)

necessary energy for the process. These latter conditions are the governing ones for the polar organic solvents also at low temperatures.

III. LYOTROPIC LIQUID CRYSTALS

The earliest contributions to the area of lyotropic liquid crystals without water as a solvent came from Winsor,[20] who pointed out the fact that some amphiphilic compounds, such as dioctylsulfosuccinate, are in the liquid crystalline state at room temperature. Attempts by Winsor to find lyotropic liquid crystals with polar solvents replacing water failed.[20] Such systems, in which the water is replaced by a polar solvent, were introduced much later when Moucharafieh[21] published a comparison between the geometric dimensions of lamellar liquid crystals of lecithin with ethylene glycol or with water as solvent.

This publication was followed by a whole series of publications which described use of solvents such as alkane diols[22] and oligomers of ethylene glycol.[23] The success with the oligomers of ethylene glycol encouraged attempts to use polyethylene glycols as solvents. The success was limited;[24] the molecular weight of the polymer 540 makes its size comparable to the common amphiphilic substances. The results showed diols with larger alkane chains and higher oligomers of ethylene glycol to penetrate the space between the lecithin molecules while the solvent molecules of lower molecular weight remained between the lecithin polar group layers to a higher degree.

A comparison of the dynamics of these solvent molecules in the liquid crystalline structure with that of water in the same structure was obviously an important investigation. Larsen and collaborators[25,26] used the quadrupolar splitting from deuterated ethylene glycol to describe the motion and bonding of the molecules by a three-state model (Figure 2). The bonded state giving a split of 2240 Hz was modeled as an ethylene glycol molecule hydrogen bonded to the phosphate group (Figure 3). Literature values for the bond lengths and angles were used to calculate the splitting for the bonded state with the spherical harmonic addition theorem[27] to show a simple expression for the order parameter (Figure 3):

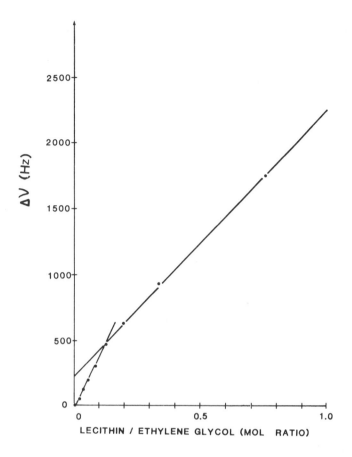

FIGURE 2. The quadrupolar splitting of deuterons on the methylene groups on ethylene glycol vs. molecular ratio. (From Larsen, D. W., Friberg, S. E., and Christenson, H., *J. Am. Chem. Soc.*, 102, 6565, 1980. With permission.)

$$S = \frac{1}{8} (3 \cos^2\alpha - 1)(3 \cos^2\beta - 1)(3 \cos^2\gamma - 1)$$

The calculation gave a value of 2.0 kHz supporting this simple model for the bound ethylene glycol.

The "free" state had no quadrupolar splitting and the intermediate state showed a split of 230 Hz. The results indicated only one ethylene glycol molecule to be strongly bonded per lecithin molecule. In comparison, it should be observed that the water system shows seven water molecules to display strong bonds to the lecithin molecule.[28]

Evans and collaborators[29] used ethylammonium nitrate as a solvent to obtain lamellar liquid crystals with lecithin as an amphiphile. Independent of these investigations, vesicles in glycerol were investigated by McIntosh and collaborators.[30] This investigation should be viewed against the biological effects of glycerol on the cell walls.

An interesting case of a nonaqueous lyotropic liquid crystal is the combination of the amphiphiles lecithin and polyethylene glycol dodecyl ether.[31] In this case, with only two amphiphiles present there is no organic solvent in the formulation. Instead, it is assumed that the system uses the polar part of the nonionic surfactant as the solvent. With this assumption, a direct parallel is found to the combination of lecithin with oligomers of ethylene glycol[23] but with the case of the nonionic surfactant, the polyether is anchored to a hydrocarbon chain, which penetrates into the lecithin layer.

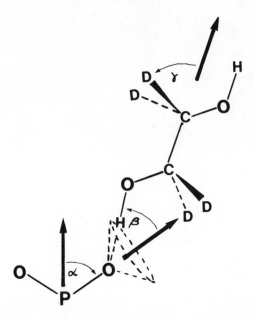

FIGURE 3.   Model for calculations of deuteron splitting of ethylene glycol hydrogen bonded to the phosphate group of lecithin. (From Larsen, D. W., Rananavare, S. B., El Nokaly, M. A., and Friberg, S. E., *Finn. Chem. Lett.*, 96, 1982. With permission.)

This system gave a lamellar liquid crystal first when the concentration of the ether exceeded a certain value. Above that limit the interlayer distance was *reduced* with increasing amount of the ether (Figure 4). At ether concentrations below the limit for the liquid crystalline phase, the measured interlayer spacings were considerably greater and strongly *increasing* with the amount of ether. The results were interpreted as follows. For concentrations less than the limit for the liquid crystalline phase to appear, the added ether was localized between the polar group layers of the solid lecithin. This localization leads to a strong increase of the interlayer distance, an increase proportional to the added amount.

The presence of the surfactant between the polar layers causes a perturbation of the lecithin molecules in their close-packed layers. At a certain concentration, the perturbation has reached such a level that the hydrocarbon chains of the surfactant begin penetration between the lecithin hydrocarbon chains. Once the penetration has begun, the perturbation is enhanced and all the nonionic surfactant molecules are now found with their hydrocarbon chains between the hydrocarbon chain of the lecithin molecules. The reduction of interlayer spacing with further addition of nonionic surfactant is easily understood from the shorter hydrocarbon chain length of the nonionic surfactant.

Recently, lamellar liquid crystals with ionic surfactant/long chain alcohol amphiphile combination[32] have been found with glycerol as a solvent instead of water. These structures are fascinating against their stability toward high solvent contents. A content in excess of 90% by weight means an interlayer distance well in excess of 150 Å. The question of the stability of such a structure is certainly not trivial.

IV. MICROEMULSIONS

The concept of microemulsions of hydrocarbons and polar organic substances such as ethylene glycol, glycerol formamide, and similar compounds was recently introduced by three groups independent of each other. All were published in 1984.[33-35] None of the groups

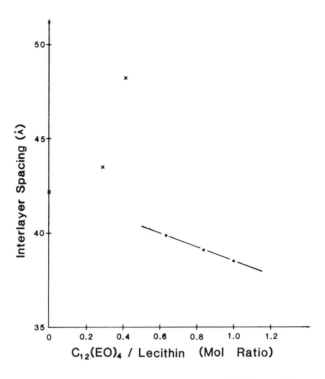

FIGURE 4.   Interlayer spacing in the system lecithin/tetraethylene glycol vs. their mole ratio. ($\times$) Before liquid crystal is formed; ($\bullet$) liquid crystalline range.

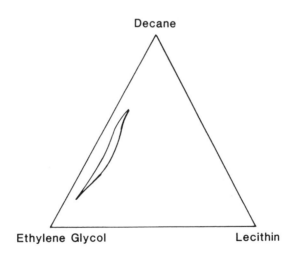

FIGURE 5.   The region for the nonaqueous microemulsion in the system lecithin, decane, and ethylene glycol.

had knowledge of the other and it is obvious that all articles are to be considered as truly original introductory publications on the subject.

Friberg and Podzimek[33] found microemulsions in the system ethylene glycol, lecithin, and decane. The microemulsion region was narrow with approximately equal ethylene glycol and decane content, the surfactant requirement being approximately 10% by weight (Figure 5). Lattes'[34] system contained formamide as the polar solvent, cyclohexane as the hydro-

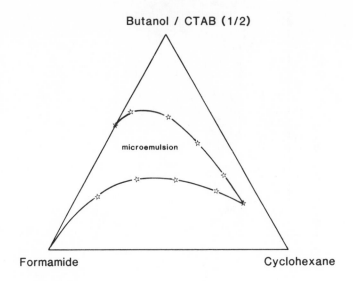

FIGURE 6.   The microemulsion region in the system formamide, cyclohexane, butanol, and cetyltrimethylammonium bromide.[34]

carbon, and an ionic surfactant/butanol combination as stabilizer. Typical results showed
total surfactant requirements of 40% for equal amounts of formamide and hydrocarbon
(Figure 6). As is the case for aqueous microemulsions, the stabilizer to solvent ratio to obtain
initial solubilization was extremely high both for formamide and hydrocarbon.

Robinson et al.[35] investigated dilute glycerol in heptane microemulsions stabilized by
Aerosol OT. They found spherical droplets with attractive interactions between them as the
stability limit was approached.

Rico and Lattes[36] also used the formamide system to obtain microemulsions with perfluorinated hydrocarbons. A combination of a fluorinated alcohol and fluorinated potassium
soap gave excellent microemulsions with modest surfactant requirements.

An interesting choice of surfactant is the substituted phosphonium salts.[37] These are not
soluble in water but readily so in formamide, in which solvent the sudden change in surface
tension was judged as indicating a critical micellization concentration.

The aqueous W/O microemulsions emanate directly from the well-known inverse micellar
solutions in Ekwall's water/soap/alcohol systems.[38] A comparison of the phase regions in
these aqueous systems with the corresponding combinations in which water is replaced by
glycerol recently has been made by Friberg and Liang.[39,40] The results showed similarities
and differences. An example of these is the fact that the aqueous system with butanol gave
a single isotropic solution region reaching from the water corner uninterrupted to the butanol
corner. A similar region was found with hexanol in the glycerol system (Figure 7).

The long chain alcohols also showed similarities as demonstrated by the decanol systems
(Figure 8). Both the water and the glycerol systems contain an alcohol isotropic solution,
a lamellar liquid crystalline phase, and an aqueous (glycerol) isotropic solution. The difference is found in the solubility in water and glycerol by the surfactant. The pronounced
solubility in water by the ionic surfactant gives both an isotropic aqueous solution with high
concentration of surfactant as well as a liquid crystalline phase of hexagonally packed
cylinders with approximately 50% surfactant. Both these features were absent in the glycerol
system. The features in "horizontal" direction simply were absent in the combination with
glycerol.

The analysis of microemulsion phenomena, so far, has been restricted to the glycerol-
SDS-hexanol system which is similar to the corresponding aqueous system with butanol.

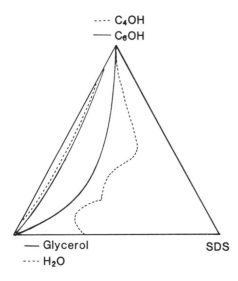

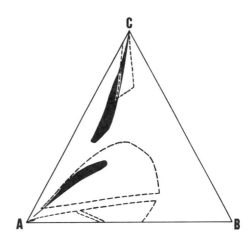

FIGURE 7. A comparison of the solubility region in the water/SDS/butanol system (- - -) with the one in the glycerol/SDS/hexanol system (——).

FIGURE 8. A comparison between the phases in the systems SDS/decanol with water (- - -) and with glycerol (black).

The addition of hydrocarbon gave solubility regions with pronounced glycerol solubilization into the hydrocarbon (Figures 9A through E).

The structure of these solutions was evaluated using light scattering and determination of equilibria. The results showed as expected that the nonhydrocarbon system contained a critical point at 52.63% glycerol and 43.06% hexanol with 4.31% surfactant. The light scattering results showed a maximum of scattering intensity at this point (Figure 10). The solutions with higher amounts of surfactant gave maxima, which emanated from the point moving toward slightly enhanced glycerol to hydrocarbon ratios with higher surfactant content. These results do not support the notion of a microemulsion base containing inverse micelles with longer lifetime than the normal fluctuations in a solution. Instead, the results indicate a critical solution to exist.

Addition of hydrocarbon gave solubility regions characterized by a critical point (Figures 9B through E). With increasing hydrocarbon content the critical point was not moved toward higher hydrocarbon to glycerol ratios. The tie-lines in the three-dimensional space showed only small deviations from the hydrocarbon to hexanol ratio at the critical point. Light scattering results (Figure 11) gave maxima close to the critical point.

With these limited results at hand, it is obvious that any conclusions in the form of "microemulsion" structure for these nonaqueous systems are premature. Instead, the results rather indicate these solubility areas to contain molecular solutions in which aggregates of a critical nature appear.

It is essential to realize that critical behavior does not, per se, prohibit a microemulsion structure. As pointed out earlier by Senatra and Giubilaro[41] dielectric properties of aqueous microemulsion follow a critical phenomenon pattern. Later investigations[42-51] have confirmed this conclusion and have also reached a reasonable consensus in relating the critical behavior to clustering of micelles while leaving the size of individual swollen micelles without drastic changes during critical separation.

It is equally essential to realize the fact that the interpretation of NMR data by Lindman and collaborators[52-54] show the micelles for short chain cosurfactants to be characterized by a diffuse and highly dynamic interface. It appears probable that a variation of the chain length of the cosurfactant may enable the formation of systems with properties spanning the

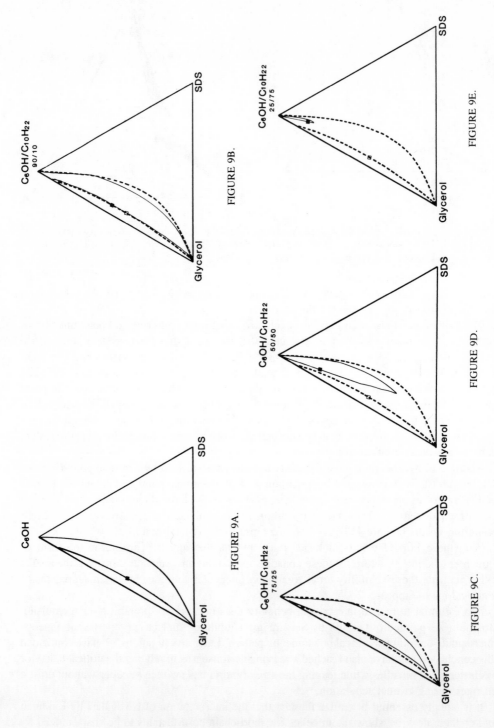

FIGURE 9A.

FIGURE 9B.

FIGURE 9C.

FIGURE 9D.

FIGURE 9E.

FIGURE 9.   The solubility area in systems with glycerol/SDS/(hexanol:hydrocarbon) solutions. Hexanol/decane weight ratio: Figure A — 100/0; Figure B — 90/10; Figure C — 75/25; Figure D — 50/50; Figure E — 25/75.

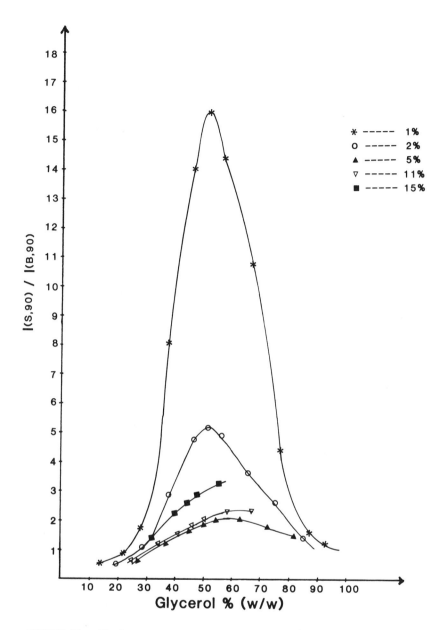

FIGURE 10.   The light scattering intensity from compositions with percentage numbers measured perpendicularly into the solubility region (Figure 9A) from the low surfactant solubility region gave maximum at the critical point. Farther away from the solubility limit, the intensity maximum became less accentuated.

entire range between well-defined microemulsion droplets and their clusters to molecular solutions.

The nonaqueous systems described here offer an interesting parallel to the aqueous systems and the future clarification of their structure and dynamics may also be useful in order better to understand the aqueous systems.

ACKNOWLEDGMENT

The financial support of this research from the Department of Energy, Office of Basic Sciences is gratefully acknowledged.

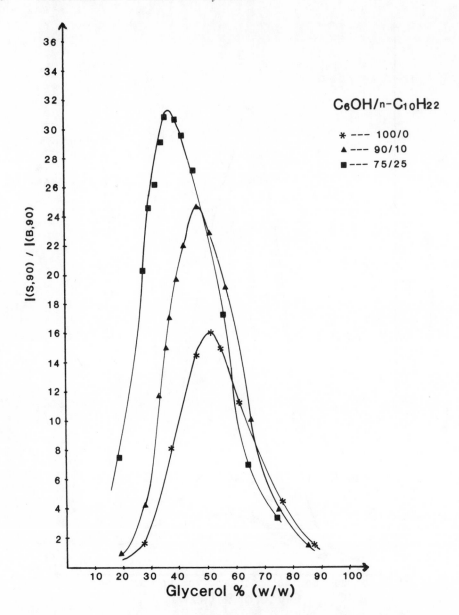

FIGURE 11.   The maximum of light scattering intensity from compositions 1% into
the solubility region of glycerol-in-decane ''microemulsions'' from the base system
(Figure 9A) were transferred toward lower glycerol contents with increased hydrocarbon content as was the critical points (Figure 9B and C).

REFERENCES

1. **Rand, R. P.,** *Food Emulsions,* Friberg, S. E., Ed., Marcel Dekker, New York, 1976, 278.
2. **Friberg, S. E. and Venable, R. L.,** *Encyclopedia of Emulsion Technology,* Vol. 1, Becher, P., Ed.,
   Marcel Dekker, New York, 1983, 287.
3. **Lindman, B. and Danielsson, I.,** *Colloids Surf.,* 3, 391, 1981.
4. **Friberg, S. E.,** *Colloids Surf.,* 4, 201, 1982.

5. **Langevin, D.,** *Reverse Micelles: Biological and Technological Relevance of Amphiphilic Structures in Apolar Media,* Luisi, P. L. and Staub, B. E., Eds., Plenum Press, New York, 1984, 287.
6. **Reinsborough, V. C. and Bloom, H.,** *Aust. J. Chem.,* 20, 2583, 1967.
7. **Reinsborough, V. C. and Bloom, H.,** *Aust. J. Chem.,* 21, 1525, 1968.
8. **Reinsborough, V. C. and Bloom, H.,** *Aust. J. Chem.,* 22, 519, 1969.
9. **Reinsborough, V. C.,** *Aust. J. Chem.,* 23, 1471, 1970.
10. **Reinsborough, V. C. and Valleau, J. P.,** *Aust. J. Chem.,* 21, 2905, 1068.
11. **Ionescue, L. G. and Fung, D. S.,** *J. Chem. Soc. Faraday Trans. 1,* 77, 2907, 1981.
12. **Singh, H. N., Saleem, S. M., and Singh, R. P.,** *J. Phys. Chem.,* 84, 2191, 1980.
13. **Sugden, S. and Wilkins, H.,** *Chem. Soc. J.,* p. 1291, 1929.
14. **Evans, D. F., Chen, S. H., Schriver, G. W., and Arnett, E. M.,** *J. Am. Chem. Soc.,* 103, 481, 1981.
15. **Evans, D. F., Yamauchi, A., Roman, R., and Casassa, E. Z.,** *J. Colloid Interface Sci.,* 88, 89, 1982.
16. **Evans, D. F. and Wightman, P. J.,** *J. Colloid Interface Sci.,* 86, 515, 1982.
17. **Evans, D. F. et al.,** *J. Phys. Chem.,* 87, 4538, 1983.
18. **Evans, D. F. and Ninham, B. W.,** *J. Phys. Chem.,* 87, 5025, 1983.
19. **Shinoda, K. and Fijuhira, M.,** *Bull. Chem. Soc. Jpn.,* 4, 2612, 1968.
20. **Winsor, P. A.,** *Liquid Crystals and Plastic Crystals,* Vol. 1, Gray, G. W. and Winsor, P. A., Eds., Ellis Horwood, Chichester, 1974, 199.
21. **Moucharafieh, N. and Friberg, S.E.,** *Mol. Cryst. Liq. Cryst.,* 49, 231, 1979.
22. **El Nokaly, M. A., Ford, L. D., Friberg, S. E., and Larsen, D. W.,** *J. Colloid Interface Sci.,* 84, 228, 1981.
23. **El Nokaly, M. A., Friberg, S. E., and Larsen, D. W.,** *Liq. Cryst. Ordered Fluids,* 4, 441, 1984.
24. **El Nokaly, M. A., Friberg, S. E., and Larsen, D. W.,** *J. Colloid Interface Sci.,* 98(1), 274, 1984.
25. **Larsen, D. W., Friberg, S. E., and Christenson, H.,** *J. Am. Chem. Soc.,* 102, 6565, 1980.
26. **Larsen, D. W., Rananavare, S. B., El Nokaly, M. A., and Friberg, S. E.,** *Finn. Chem. Lett.,* 96, 1982.
27. **Woessner, D. E.,** *J. Chem. Phys.,* 36, 1, 1962.
28. **Finer, E. G.,** *J. Chem. Soc. Faraday Trans. 2,* 69, 1590, 1973.
29. **Evans, D. F., Kaler, E. W., and Benton, W. J.,** *J. Phys. Chem.,* 87, 533, 1983.
30. **McDaniel, R. V., McIntosh, T. J., and Simon, S. A.,** *Biochim. Biophys. Acta,* 731, 97, 1983.
31. **Ganzuo, L., El Nokaly, M. A., and Friberg, S. E.,** *Mol. Cryst. Liq. Cryst.,* 72, 183, 1982.
32. **Friberg, S. E. and Liang, P.,** *Colloids Surf.,* in press.
33. **Friberg, S. E. and Podzimek, M.,** *Colloid Polym. Sci.,* 262, 252, 1984.
34. **Escoula, B., Hajjaji, N., Rico, I., and Lattes, A.,** *J. Chem. Sci. Chem. Commun.,* 12, 33, 1984.
35. **Fletcher, P. D. I., Galal, M. F., and Robinson, B. H.,** *J. Chem. Soc. Faraday Trans. 1,* 80, 3307, 1984.
36. **Rico, I. and Lattes, A.,** *Nouv. J. Chem.,* 8, 424, 1984.
37. **Rico, I. and Lattes, A.,** *J. Colloid Interface Sci.,* in press.
38. **Ekwall, P.,** *Advances in Liquid Crystals,* Brown, G. H., Ed., Academic Press, New York, 1975, 1.
39. **Friberg, S. E. and Liang, Y.-C.,** to be published.
40. **Friberg, S. E. and Liang, P.,** to be published.
41. **Senatra, D. and Giubilaro, G.,** *J. Colloid Interface Sci.,* 67, 448, 1978.
42. **Cazabat, A. M., Chatenay, D., Langevin, D., and Pouchelon, A.,** *J. Phys. (Paris) Lett.,* 41, L441, 1980.
43. **Cazabat, A. M., Langevin, D., and Sorba, O.,** *J. Phys. (Paris) Lett.,* 43, L505, 1982.
44. **Cazabat, A. M., Langevin, D., Meunier, J., and Pouchelon, A.,** *J. Phys. (Paris) Lett.,* 43, L89, 1982.
45. **Cazabat, A. M., Langevin, D., and Pouchelon, A.,** *J. Colloid Interface Sci.,* 73, 1, 1980.
46. **Fourche, G., Bellocq, A.-M., and Brunetti, S.,** Letters to the editors, *J. Colloid Interface Sci.,* 88, 302, 1982.
47. **Dorshow, R., de Buzzaccarini, F., Bunton, C. A., and Nicoli, D. F.,** *Phys. Rev. Lett.,* 47, 1336, 1981.
48. **Tabony, J., Drifford, M., and De Geyer, A.,** *Chem. Phys. Lett.,* 96, 119, 1983.
49. **Huang, J. S. and Kim, M. W.,** *Phys. Rev. Lett.,* 47, 1462, 1981.
50. **Kotlarchyk, M., Chen, S.-H., and Huang, J. S.,** *Phys. Rev. A,* 28, 508, 1983.
51. **Kotlarchyk, M., Chen, S.-H., Huang, J. S., and Kim, M. W.,** *Phys. Rev. A,* 29, 2054, 1984.
52. **Stilbs, P., Rapacki, K., and Lindman, B.,** *J. Colloid Interface Sci.,* 95, 583, 1983.
53. **Lindman, B., Ahlnas, T., Soderman, O., and Walderhaug, H.,** *Faraday Discuss. Chem. Soc.,* 96, 001, 1983.
54. **Lindman, B., Stilbs, P., and Moseley, M. E.,** *J. Colloid Interface Sci.,* 83, 569, 1981.

Chapter 4

NONIONICS

J. C. Ravey

TABLE OF CONTENTS

I. NONIONICS

The partial mutual miscibility of oil and water can be promoted by using a huge number of nonionic compounds: they range from the simple amphiphatic molecules like short alcohols to block copolymers, including, of course, the very surface active agents. According to the type and concentration of each component of such a ternary one-phase mixture, one could speak of mutually saturated liquids, cosolubilization, selective adsorption, etc. However, in systems of interest here, it will be much simpler to regard water and hydrocarbon as solubilized with the nonionic amphiphile acting as cosolvent,[1] and, for the structures, to use the wording of molecular aggregation, a microemulsion will be the ensemble of the aggregates due to the presence of surfactant molecules organized in the transparent, stable, and isotropic (pseudo-) ternary mixtures which contain substantial amounts of both oil and water. The very fundamental point will be the structural knowledge of the actual aggregates in relation to the phase behavior of these mixtures, as a function of the parameters of importance: chemical nature of oil and surfactant, temperature, and influence of eventual additives (mainly organic/inorganic salts).[1-8,31-39,48-60]

A. Trends in Recent Investigations

In spite of the considerable development of the research initiated in view of important technical applications,[9-13] the understanding of the correlation between structures and solubilization properties is still at a low level, at least as far as nonionic surfactants are concerned. This results from both the large number of systems of interest and the complexity of most of their ternary phase diagrams: generally, the pseudoternary diagrams of ionic systems (i.e., with a cosurfactant) often appear comparatively simpler.

The most important progress accomplished during these last years essentially concerns the knowledge of the phase behavior of a few series of compounds belonging to the class of surface-active ethyleneoxide adducts.[14-34,121] Concurrently, very extensive and systematic measurements of interfacial tension also have been carried out on multiphase systems with such types of surfactant, although generally restricted to commercial grade compounds and to a few overall compositions but in the presence of salt and alcohol.[12,35-37]

As far as the structural determinations are concerned, their number remains quite modest[14,38-64] and most of them also have been performed on systems with polyoxyethylene alcohols/phenols.

In many cases, these results are still open to discussion. For example, let us be reminded of the controversy about the structure of the "simple" aqueous micelles of some of these nonionic surfactants.[39,66-68] The difficulty in the interpretation of the experimental results is due mainly to the dense packing of the aggregates in all these systems, giving rise to important repulsive/attractive interparticle effects.[65] Moreover, it is often safe to consider that a given composition of the one-phase mixture corresponds to a certain structure which is modified by any oil or water dilution: this is quite obvious for systems whose composition corresponds to isolated domains in phase diagrams. In spite of their shortcomings, these structural investigations have contributed to emphasize the prime importance of a well-defined "interfacial" surfactant film (from the structural and dynamic points of view), confirming earlier thermodynamic theories.[1]

B. Models and Thermodynamic Aspects

Recent general models for the formation and stability of microemulsions have been proposed these last few years,[69-81] most of them being designed for both ionics and nonionics. However, specific theoretical studies of the influence of the water-oligo-oxyethylene chain interactions on the structure and phase behavior in ternary oil/water/nonionic surfactant systems do not seem to have been well developed. Treatments of such interactions have been initiated for binary aqueous mixtures and for block copolymers, making use of models peculiar to polymeric chains or lattice models.[25,70,71] Clearly, this problem is quite complex, since we have to consider the concomitant water-water, water-oxyethylene bondings together with the possible conformations of the hydrophilic chain. It is also quite obvious that this lack of knowledge constitutes a serious drawback in the full understanding of the behavior of the nonionic surfactant systems.

Perhaps the simplest theory for nonionic microemulsions made of oil or water spherical droplets is the one proposed by Robbins,[69] which could be extended to the case of fluctuating interfacial films[72,73] of mean curvature radius. Here, solubilized water (oil) was assumed to be located entirely at the center of the spherical globules, and the force balance was given by a formal variation of the lateral stress over the molecule in each separate part.

Talmon and Prager[86] developed a model which allows construction of phase diagrams which bear some resemblance to some of the simpler experimental ones. It is not clear whether or not such a model can represent true nonionic microemulsions termed "bicontinuous".

Other recent treatments have focused on factors which contribute to the Gibbs force energy,[75-85] including the interaction energy among droplets and the effect caused by the entropy of dispersion. Arguments are adduced which suggest that a planar regular structure cannot replace the globular shape since the former is unstable to external and thermal perturbations; "chaotic" movements of the surface may also occur, leading to, in some cases, a marked similarity with the dynamics present in nearly critical systems. An important conclusion emerging from these works is that while the free energy associated with the entropy of dispersion may be small, it plays a major role with regards to both the interfacial tension and phase behavior. The same conclusion also can be drawn for binary (aqueous and apolar) nonionic surfactant mixtures, mainly for compositions in the vicinity of their isotropic liquid-liquid coexistence curves.[71]

On the other hand, a mere inspection of the phase diagrams of nonionics shows that large parts of them are often constituted by liquid crystal domains. Hence, there is no reason to believe *a priori* that the molecular aggregates must be spherical particles everywhere in any isotropic phase for the sole reason that it is macroscopically isotropic. As a matter of fact, many theoretical treatments contain "ready-to-use" explanations for the existence of most of the various types of structures. But they are generally not yet able to predict them in relation to the phase behavior and wait for new experimental structural results in order to get confirmation of the calculations. Therefore, there is still much to be done to completely delineate the thermodynamics of nonionics; parallely, the recent extensive experimental investigations also have to be further developed.

C. The Nonionic Systems of Importance

Several ternary oil/water/alcohol mixtures can give rise to phase diagrams where it has been possible to identify regions of isotropic "microemulsions"[87-95] in addition to regions with small aggregates and normal ternary solutions. But independently of any structural aspect, and due to their relative simplicity, they have been proved to be of valuable utility for a basic understanding of the general phase behavior of many nonionic surfactant systems, as shown by the recent, beautifully simple papers of Kahlweit et al.[15,16] They are useful, too, when the apparently complex influence of added salts has to be analyzed. Although the interfacial tensions may be ultra-low for a few compositions in the very vicinity of critical points, these "simple" systems do not have the technical importance of true surface-active agents. Therefore, in this chapter we shall exclusively focus our attention on these last compounds.

We refer to the famous reviews of Schönfeldt and Schick[96,97] for an inventory of most of the nonionics used in industry, which are comprehensive treatises on organic, physical, biological, and analytical chemistry of the nonionic surfactants. Since as far as oil/water solubilization and microemulsions are concerned, the far most studied nonionic emulsifiants are polyoxyethylene alkylphenols and polyoxyethylene alcohols, this chapter will be restricted to the discussion of the properties of such compounds, without claiming exhaustiveness. We recognize that phosphate esters, amides, amines, polyols, or biological surfactants have been studied as well,[97,98] but they are too sparse to allow a systematization of their properties. Therefore, in the following, ''nonionic'' will be synonymous of oligo-oxyethylene alcohol/alkyl phenol.

This review will be extended to a novel class of nonionic microemulsion-forming emulsifiants: the fluorinated polyoxyethylene alcohols.[31-33] Indeed, the interest in investigations of such compounds is twofold. First, and foremost, is the comparative study with a homologous series of hydrogenated surfactants, in order to make clearer the influence of the hydrophobic tail on the formation of the molecular structures. Also, from a practical point of view, these microemulsions have potential applications in biotechnology as respiratory gas carriers.

In the following section of this chapter we shall summarize the current state of knowledge of the phase behavior of oil/water/polyoxyethylene alcohols in relation to the characteristics of the surfactant molecule (its hydrophilic-lipophilic balance or HLB), of its solutions in oil or in water (''cloud points''), and of its multiphase systems (phase inversion temperature or PIT). Effects of additive also will be tentatively quantified. Indeed, the conditions of existence of microemulsions are strongly correlated to the properties of binary (aqueous and apolar) surfactant solutions, and to their aptitude to solubilize some amount of oil or water, respectively. The last part of this chapter will be devoted to structural aspects, calling attention to the relation between structures in isotropic phases and the presence of liquid crystalline phases.

II. THE PHASE BEHAVIORS

A. General Characteristics of a Nonionic Surfactant

The low polarity of the elementary polar groups is compensated by their accumulation along a long hydrophilic chain. This introduces extra geometrical and configurational constraints as compared to ionic detergents: when aggregated, these molecules cannot accommodate too large a disorder between adjacent chains, favoring the formation of (mainly lamellar) liquid crystalline phases (when the temperature is not too high). We can then guess that the mean structure of the interfacial film in microemulsions will locally look like that of an anisotropic phase; curvature constraints should allow the formation of much larger oil in water globules than for ionics when the repulsion between oxyethylene chains is low (i.e., at lower temperatures). This also can be related to the fact that critical micelle concentrations (CMC) are generally much smaller than those of the corresponding ionic surfactants.

The two most important features of these nonionic systems are as follows:

1.   First, they can accommodate brines of salinity much higher than can classical ionics.[35] This is partly due to some kind of ion complexation along the oxyethylene chains.
2.   Secondly, the soft interactions between polar groups (belonging to the same or to different aggregates) are relatively sensitive to any change in temperature. Hence, given the delicate force balance which presides over the existence of the structures, we get phase diagrams whose outlook may be markedly temperature dependent.

In order to characterize the potential relative solubility of a surfactant in water and in oil, the concept of HLB has been introduced by Becher.[99] The (semi-) empirical definition proposed for ethoxylated nonionic surface-active agents has been proved to be particularly valuable in the formulation of emulsions, appearing as an intrinsic quantity, i.e., depending only on the chemistry of the molecule:

$$HLB = 20 \ M_H/(M_L + M_H) \tag{1}$$

where M_H and M_L are the molecular weights of the hydrophilic and lipophilic moieties.

In keeping with its intrinsic character, Equation 1 should be more or less modified when the chemical nature of the molecule is varied (e.g., the introduction of a benzyl group in the hydrophobic chain). Some authors also speak of "effective" HLB to express the fact that the relative surfactant solubility in oil/water changes with experimental conditions. There have been a few attempts for an HLB formulation in terms of surfactant structure, cohesive energy density, solubility parameter,[100-103] etc.; a survey of this problem can be found in the papers of Becher.[99] We believe it is better to turn back to intrinsic definitions like Equation 1 with its "chemical" aspect, and to describe the phenomenon of interest in terms separate from this HLB and of the intrinsic characteristics of the other components.

Indeed, there is no reason to believe *a priori* that the HLB is a universal parameter for any property of an oil and/or water surfactant mixture. According to Equation 1 applied to hydrogenated polyoxyethylene alcohols ($C_m \ EO_n$), the HLB remains constant if the addition of one oxyethylene group is counterbalanced by the increase in the hydrocarbon chain by three methylene groups. Close inspection of the data in literature relative to binary aqueous phase separation shows that one CH_2 group is as effective as four EO groups. But for the CMC, the influence of 1 CH_2 is stronger than 12 EO groups! The difference in behavior is still enhanced in the case of fluorinated compounds; while one CF_2 group is equivalent to 1.5 to 1.8 CH_2 groups for the CMC, one EO group in a fluorinated compound is as effective as two EO groups in the parent hydrogenated surfactant as far as the phase inversion temperature (PIT) is concerned. Nevertheless, the well-known general relation between HLB calculated according to Equation 1 and the type of emulsion still holds, the HLB range (8 to 10) corresponding to the possibility of obtaining both microemulsions and phase inversion (transformation of oil-in-water to water-in-oil emulsions), through isotropic multiphase systems. Therefore, once again we have to stress the necessity of a best thermodynamic representation of the relationship between the structure of the surfactants and their surface-chemical properties.

B. Mean Features of Aqueous Systems

A typical temperature-composition phase diagram of aqueous nonionic surface-active agents is shown in Figure 1. Several liquid crystalline phases are generally present for concentrations above 40%. A strong hydration of hydrophilic chains of higher EO content results in occurrence of a micellar phase L_1 and tends to reduce the predominance of the lamellar phase over the hexagonal phase.[26,27] Besides, the liquid crystal stability is low for alkyl chain lengths below C_{10}, and the corresponding phase diagram may reduce to only the L_1 phase, limited by an upper liquid-liquid coexistence curve, with a lower consolute critical point (T_w). This curve is called the cloud point curve (T_{cp}), and is typical for most water-soluble chain oligomers and polymers. This curve constitutes the keystone for the understanding of all the problems of oil solubilization, microemulsions, and isotropic multiphase systems. The presence of liquid crystals must be considered as a secondary occurrence, even when they considerably distort the cloud point curve, but give some indication about the probable structure of molecular aggregates in the isotropic binary and ternary mixtures. According to the terminology introduced by Shinoda and co-workers,[104-106] at low concentration the surfactants dissolve in water (the water phase), while at higher concentration water dissolves into the surfactant and forms the so-called "surfactant phase".

For the few systems where these L_1 phases do not exist (case of "long" hydrocarbon tail

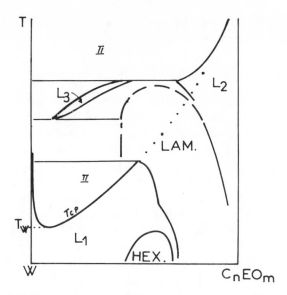

FIGURE 1.   A schematic typical phase diagram of an aqueous polyoxyethylene alcohol. According to the relative values of m and n its
upper or lower part may not exist in temperature range 0 to 100°C,
and the extent of the lamellar phase may be much more reduced. T_w
is the lower critical temperature, and T_{cp} stands for the cloud point
curve.[26]

with low HLB, like $C_{12}EO_3$, or the equivalent fluorinated compounds like $C_6F_{13}CH_2EO_5$,
etc.), we can assume that they would still exist for temperatures below 0°C, and the general
features of the isotropic multiphase equilibria would not change, except for a shift toward
lower temperatures.

Far from the lower critical temperatures, the oil solubility in the dilute L_1 phase is generally
low (typically one hydrocarbon chain for five surfactant molecules), inducing small changes
in the CMCs (generally, a decrease) and in the sizes of the aggregates. A few interpretations
of the mechanism of that limited solubilization have been proposed,[106-110] assuming different
locations and distributions of the solubilizate according its polarity: in analogy with ionic
emulsifiants, the saturation concentration decreases as the length of the alkyl chain increases.
Alcohols, acids, and amines seem to be solubilized much more strongly than hydrocarbons
of the same carbon number. Micellar weights also seem to increase with an elevation of
temperature.[2,69,106-110]

The cloud point curve is appreciably influenced by the presence of certain additives:

- Most of the electrolytes suppress the cloud point in proportion to their concentration.[105,109,111] An electrolyte with the lowest lyotropic number is the most effective.
  Moreover, there is a marked increase of oil solubilization which roughly parallels the
  depression of the cloud point, together with an increase of the size of the micellar
  aggregates.
- Shorter saturated hydrocarbons generally do not depress the cloud point very much,
  while longer ones somewhat enhance it. Aliphatic alcohols (the longer ones), fatty
  acids, or phenols depress the cloud point remarkably, the different effects being due
  to the different modes of incorporation of the solubilizate. As a matter of fact, this
  depression is generally restricted to the smallest oil contents, and this does not prevent
  a much larger solubilization to be achieved.

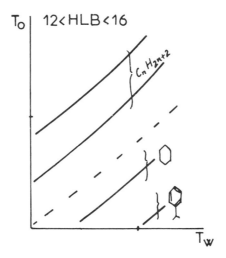

FIGURE 2. Example of correlation between the consolute temperatures of binary apolar (T_o) and aqueous (T_w) mixtures of nonionic surfactants for different types of hydrocarbons.[112]

These observations have to be understood more clearly by using whole ternary phase diagrams. The important result to point out here is the enhancement of hydrocarbon solubilization in the isotropic phase L_1 due to the proximity of the lower critical point, i.e., where the solubility of the surfactant in water tends to be much reduced.

C. Nonpolar Solutions

Like ordinary organic compounds, binary mixtures of oil and nonionic surfactants exhibit lower liquid-liquid solubility curves, with an upper consolute critical point.[110-112] The critical concentration is about 10 to 20%, while typical values are 1 to 5% in aqueous solutions for the same surface active agent. From this observation it can be inferred that the micellar aggregation must be low, the micelle formation being likely only if both the moieties of the surfactant are large.[110]

The consolute temperatures are very dependent on solvent, and they decrease in the order:

saturated n-hydrocarbons $> \ldots$ heptane $>$ (water) $>$ c-hydrocarbons $>$ aromatics

Moreover, they are found to increase roughly linearly with HLB (within the range 8 to 12), just as they do in aqueous solutions. As a result, the HLB appears as the right (intrinsic) parameter for a satisfactory description of the parallel behavior of both the consolute (aqueous and apolar) temperatures. In other words, there must be a strong correlation between the miscibility temperature (T_o) in hydrocarbon and the (purely aqueous) cloud point (T_w) of the surfactant, which must essentially depend on the chemical nature of the oil, as shown in Figure 2. Such a correlation has effectively been found by Shinoda[110] to be a result of prime importance: for longer saturated hydrocarbons, $T_o > T_w$, while for the more polar oils, $T_o < T_w$. Such a difference in behavior reflects the actual preferential location of the oil on the surfactant chains.

In oil dilute mixtures (when the EO groups are not blocked by polar molecules), some water solubilization can be achieved, probably due to interbinding with polyglycol groups. In contrast with the oil incorporation in aqueous solutions, water solubilization is not particularly enhanced by the proximity of the lower miscibility curve of the binary oil/surfactant mixture. On the contrary, the temperature has to be raised to a large extent (typically several tens of degrees) to a "critical" temperature (T_H) (and generally within a quite restricted

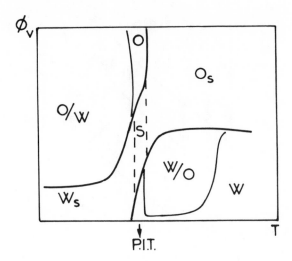

FIGURE 3.   Typical phase volume fractions of water (W), oil (O), surfactant/oil (O_s), surfactant/water (W_s), and surfactant phases, together with volumes of O/W and W/O emulsions as present 1 day after mixing.[117]

range of temperature) before a marked water solubilization may take place. Generally, the oil with the higher consolute temperature (T_o) is also the one with the higher T_H.[113] This property is fundamental for the phase inversion phenomenon and isotropic three phase equilibria since it generally ensures T_H values higher than the lower consolute temperature T_w (which corresponds to the enhancement in oil solubilization), even when T_w is higher than the upper consolute temperature of the oil-surfactant mixture. The structural implications by this type of behavior will be discussed later.

In total analogy with aqueous systems, the addition of salts generally depresses the T_H temperatures: here also the decrease of T_H is nearly proportional to the molarity of the salts, the effect being due mainly to the anions of the electrolytes.[109,114] Quite interestingly, the inspection of data in literature shows that the decreases of T_H and T_{cp} (aqueous solutions) are roughly parallel for a given anion, independent of the HLB. (Perhaps the salts seem a little less effective in the oil-rich than in aqueous systems.)

D. Emulsions and Phase Inversion Temperature

Studies of emulsion stability in systems with comparable amounts of oil and water and with low nonionic surfactant content have shown for a long time the strong correlation between the phase inversion phenomenon and the existence of three isotropic phase equilibria with low or ultra-low interfacial tensions.[115-119] Represented on the schematic diagram in Figure 3 are typical phase volume of the oil and water phases, of the third one termed surfactant phase (or middle phase) by Shinoda, and of "creams" which are still present a few days after mixing. The emulsion stability is maximum just before the surfactant phase exists, and just after it has disappeared, and is practically zero inbetween.

The surfactant phase is then the very microemulsion domain, which is the stable state of these oil/water mixtures in the presence of a small surfactant amount. For effective nonionic surface-active agents, that phase exists only for a rather narrow range of the experimental conditions. Thus, the temperature of inversion of phase (PIT) can also be defined as the narrow temperature range for which the surfactant phase exists which contains as much water as oil. This temperature must be somewhere between T_H (which enhances water solubilization in oil phase) and T_{cp}, above which the incorporation of hydrocarbon is strongly favored. A crude estimate could be

$$PIT \simeq \frac{1}{2} (T_{CP} + T_H)$$

Hence, from the lengthy previous discussions, we can now get a straightforward qualitative description of the behavior of the ''microemulsion phase''-temperature (PIT) according to the changes in characteristics of the various components: HLB, chemical nature of oil, effect of electrolytes, etc. We shall come back later to this description from a more qualitative point of view.

Obviously, for a more basic understanding of these peculiar surfactant systems, we have to consider the whole ternary and quaternary phase diagrams (case of a fourth component, or use of temperature as a fourth variable).

III. TERNARY AND QUATERNARY PHASE REPRESENTATIONS

A. Ternary Mixtures

One of the very first ternary phase diagrams illustrating solubilization by nonionics was published as early as in 1948 by McBain.[120] Although incomplete, it clearly indicated the complexity of multicomponent surfactant systems like Triton® X 100/benzene/water. So far, only a few groups have been interested in the equilibria and the exact delineation of the various domains in a whole ternary diagram with nonionic surface-active agents. Some diagrams with pure, commercial grade and mixtures of surfactants were studied, resulting in a patchwork of (pseudo-) ternary representations.[14-34,121]

In order to get a better understanding of the relevance of nonionic emulsifiants in technical applications, Shinoda and the Japanese school have developed a formidable body of systematic studies.[103-106,122] They have published series of temperature-composition diagrams which are vertical sections at constant surfactant weight fraction of the prismatic representations oil/water/surfactant/temperature. These investigations have concerned systems with relatively low surfactant content, and they have served as a basis for the development of straightforward theoretical treatments of phase behavior according to the well-known Winsor's terminology of the phase equilibria.[1,69]

About 10 years ago, Friberg and the Scandinavian school initiated the important work on full determination of the isotropic phase domains in systems with highly pure compounds, mainly the dodecyl polyoxyethylene alcohols.[19,20] From these invaluable sets of data, they have been able to make clear the systematic evolution of the one phase realms when temperature, HLB, and nature of the oil are varied. These results tally with Shinoda's findings, which also concern long-hydrocarbon-chain surfactants, i.e., the most relevant for technical applications. It is worth noting that these *real* diagrams bear some resemblance to the (Winsor) simple theoretical ones only at higher temperatures.

Since then, other diagrams for other systems have been determined which exhibit a behavior quite similar to that stressed by Friberg, these systems including hydrogenated as well as fluorinated compounds.[21-25,28-33]

However, full studies of the phase equilibria including plait-points and tie-lines for eventual anisotropic phases have been published very lately and remain very few. A particularly impressive one done by Kunieda and Shinoda[17] is for $C_{14}H_{30}/C_{12}EO_5$. Its PIT is 47.8°, characterized by the presence of one isolated isotropic phase domain with an oil/water ratio equal to one, which may be in equilibrium with almost pure oil and water. This feature seems characteristic for most of the longer hydrocarbon-chain systems, as apparent from other data from Shinoda.[103-106,122]

Other particularly well-studied ternary mixtures from Lang[25] are for $C_{10}EO_4/C_{16}H_{34}$ and $C_{10}EO_5/C_{16}H_{34}$. Lamellar phases are still present, the diagram keeping a high degree of complexity particularly in the water corner (multiequilibria between two microemulsion phases: water and the lamellar phase).

Much simpler phase behavior has been obtained by Kunieda and Friberg,[121] using the short chain-surfactant C_8EO_3 and decane which cannot stabilize a liquid crystalline phase. This behavior appears much like that of ordinary alcohols,[88-92] as already discussed by Kahlweit[15] and Fox:[90] in a temperature range between two critical end points (CEP), there is a three-phase region due to the superposition of three miscibility gaps (water + surfactant-, surfactant + oil-, and oil + water phases).

At this point, it is worth noting that classical ionic surfactant/cosurfactant/brine/oil systems give phase equilibria quite similar to this simple last one, but the effect of temperature is generally the reverse.[121]

Then whatever the system, for compositions with low surfactant content and comparable amounts of oil and water, there is a range of temperature where the classical succession of equilibria water/surfactant-, water/surfactant/oil-, surfactant/oil phases take place when temperature increases, independent of the presence of liquid crystals for larger amounts of surfactant (of course, when the emulsions have disappeared).

This can be understood from Gibbs phase rules[15,16] for shorter nonionics,[88-92] as a consequence of the following peculiarity: when the CEP of the aqueous-(oil + surfactant) mixture touches the oil/water solubility gap (at $T = T_L$), the plait-point of this gap still faces the oil-surfactant basis. For more involved systems, it could merely be noted that, for values of T in the neighborhood of T_L, the part of the phase diagram with larger oil/water ratios (at least with low surfactant content) is almost insensitive to temperature, reflecting then a higher stability of the intermolecular interactions, enhanced by the presence of a small amount of bounded water (let us recall the apparently general inequation: $T_{cp} < T_H$).

For comparison, typical ternary diagrams for short and long chain surfactants are schematically represented in Figures 4 and 5, in relation to the kind of phase diagram extensively investigated by Shinoda and co-workers. We have chosen the case where the aqueous critical temperature is enhanced by the oil (a longer chain hydrocarbon); the presence of a lamellar phase depressed the cloud point of the more concentrated solutions. Clearly the main difference between long and short surfactant systems lies in the extent and the fugitive aspect of the surfactant phase domains (microemulsions). However, a limited stability with temperature and composition does not prevent a cosolubilization of comparable amounts of oil and water by using a very small surfactant concentration. As an apparent rule, the less stable the microemulsion the smaller the needed emulsifiant amount. For example, the respective couples of minimum weight fraction and temperature range for the microemulsions in [C_8EO_3 + C_{10}], [$C_{12}EO_5$ + C_{14}], and [$C_{12}EO_5$ + C_{10}] are as follows: [25%, 14 to 30°C], [12%, 44 to 52°C], and [7%, 35 to 39°C].[17,121] This trend is quite in line with the behavior of simple alcohol systems, characterized by larger multiphase region and moderate solubilizing power.

A simple picture for the microemulsion domain with longer nonionic surfactant is a slender bridge between the water-corner and the oil-surfactant basis, temperature ascending along the "massifs" constituted by the liquid crystal domains.

We have also to note that:

● Whether the oil is a short (saturated) hydrocarbon or not, the junction is favored toward the oil corner or the middle of the oil-surfactant basis.
● In isotropic three-phase systems, the surfactant phases are in equilibrium with almost pure water and with oil with a small content of surfactant. At the high temperature-CEP, the two oil and surfactant phases smoothly coalesce; but the coalescence of water and surfactant phases at a low temperature-CEP is not always quite clear, due to the possible multiphase equilibria between three isotropic and one anisotropic phases.[25]
● The oil/water solubilization is larger when the nonionic is monodisperse than when the distribution of hydrophilic chain lengths if broad,[106] in correlation with a larger

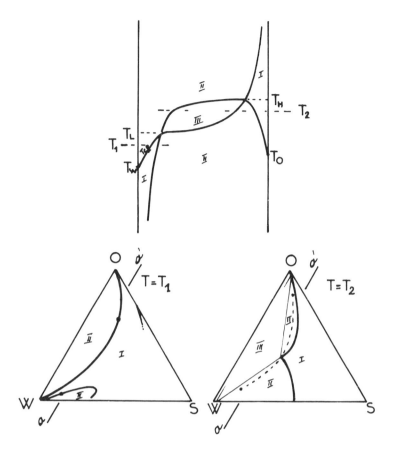

FIGURE 4.   Relation between ternary phase diagrams and temperature-composition representation for a given (a-a') low weight fraction of a shorter nonionic surfactant.[121]

extent of the three-phase domain. Nevertheless, the use of commercial grade surfactants does not change the general behavior.[24,28]

• In order to increase the amount of solubilization, longer hydrocarbon-chain surfactants are to be used.

• All the general features discussed so far are not exclusively those of hydrogenated compounds, but similar rules and phase behavior have also been obtained for fluorinated compounds. Indeed, Mathis and co-workers have shown that fluorocarbon/water miscibility can be still achieved by using polyoxyethylene alcohols, but with a perfluorinated hydrophobic chain, and that the similarity of both systems is quite striking. This similarity holds for phase diagrams and aggregate structures as well.[34,58]

To conclude that discussion on that part of the ternary systems with lower surfactant content, let us point out that three main types of ternary phase diagrams still may be defined for the longer nonionic surfactant, according to the location and outlook of the one isotropic phase realms (Figure 5).[28] They could be termed as cloud point, PIT, and "reverse" diagrams. In the first ones, two phase realms appear in the water corner, in connection with the cloud point curve of the aqueous surfactant. The second ones correspond to generally isolated realms, for the oil/water ratio equal to 1. They also correspond to a larger area of the triangular three-phase domain — from this point, the surfactant affinity for oil (in presence of water) prevails over its affinity for water (in presence of oil). In the last ones, there remain only phases with higher surfactant/water ratios: in all likelihood they will contain only

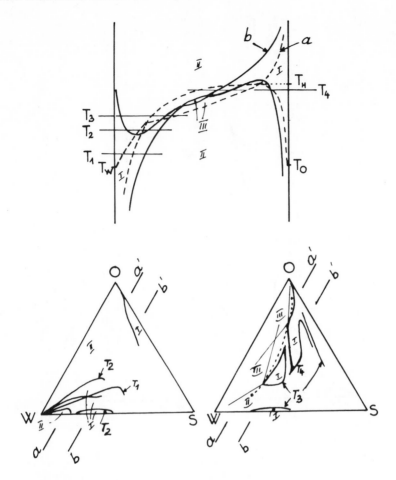

FIGURE 5.   Visualization of the relation between ternary phase diagrams and the temperature. Composition representations for two (a-a'; b-b') given low surfactant weight fractions: case of longer nonionic emulsificants.

definite water in oil structures, at least in their oil richer part. This classification, then, retains some resemblance to the so-called Winsor's diagrams.

B. Quaternary Systems: Effects of Additive

Although the adjunction of additives (mainly salts or alcohols) to these nonionic surfactant systems can be of prime importance in technical applications (e.g., in oil tertiary recovery,)[35-37] the exact determinations of their whole quaternary (and higher order) phase diagrams are very few at the present time.

Some oil/water/salt/alcohol mixtures have been investigated instead,[15,87-93] lending weight to the hypothesis that the patterns of phase behavior of many of these "ordinary" liquids mimic those of surfactant systems.[87] However, noteworthy differences in degree between the two types of compounds have been recognized: for alcohols there are no indications of island-like regions, and the isotropic multiphase domains are much larger than those found in many surfactant systems.

Another difficulty arises from the (geometrical) representation of multicomponent systems themselves. To get around this problem, some of the various species are grouped into "best" pseudocomponents.[89,123,124] In practically all former investigations concerning brine/oil solubilization, it has been assumed that the brine partitions as one component, although it has

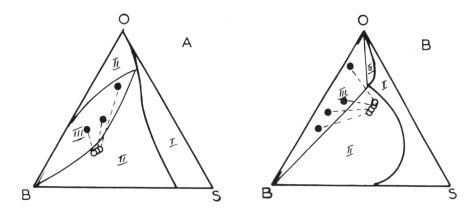

FIGURE 6. Examples of isotropic multiphase realms in the pseudoternary phase diagram of a commercial $C_{10}EO_6$ at T = 55°C, for two brine salinities (Na_2CO_3; A = 70 g/ℓ; B = 150 g/ℓ). Also shown is the composition (○) of the surfactant phase corresponding to various three phase systems (●).[30]

already been suggested that in a three-phase region small relative adjustments of salinity can produce large changes in the nature of the coexisting phases.[124]

To get a deeper insight in the concomitant influence of the temperature and the salinity in the phase behavior of nonionics, systematic determinations of pseudoternary diagrams have been initiated quite recently.[28-30,57] Until now they have concerned a commercial grade $C_{10}EO_6$ surfactant in the range of temperature 50 to 80°C, and with three different brines (NaCl, Na_2CO_3, $CaCl_2$) of ionic strengths between 0 and 5. We have clearly shown that temperature and salinity have quite similar effects on the evolution of these phase diagrams: the series of patterns obtained by varying the temperature at constant ionic strength and those obtained by varying the salinity at constant temperature are practically the same, in passing from cloud point to reverse diagrams. As a rule, the anisotropic phase realms and emulsion regions become smaller and smaller and die out above 60°C. Temperature and salinity also have a similar decreasing effect on the Bragg spacings of the lamellar phases.

Correlatively a three isotropic phase region appears, whose extent first increases, and then decreases. Most interesting is the way the three-phase region is just appearing in the pseudoternary representation: indeed, at earlier stages we get a small triangle exclusively confined in the water corner, characterized by a mean oil/surfactant ratio equal to 1, without contact with oil-water or surfactant-water bases. This realm then temporarily loses its triangular shape. Such configurations are apparently far from looking like idealized Winsor or Robbins diagrams (see Figure 6). Even if they do not, the point corresponding to the surfactant phase of a given three-phase system generally depends on the overall composition of the latter. And it does not even seem to belong to any one phase region in this pseudoternary representation. This clearly indicates that the salt component is not equally distributed in the "water component" of each of the three phases in equilibrium.

In fact, phase composition analyses show that:

- The water-phase practically contains only salt and water, and no oil at all.
- The oil-phase is a dilute solution of surfactant, with only traces of water and a very small quantity of salt.
- The surfactant phase contains most of the surfactant and some amount of oil and brine; the salinity of this brine is generally noticeably less than the overall brine salinity, increasing with the latter and with temperature. The water content has the opposite trend, paralleling the behaviors of surfactant and simple alcohol systems.

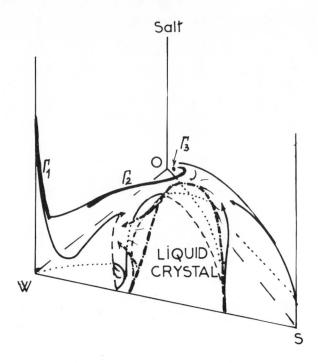

FIGURE 7.   Schematic three-dimensional representation of the surfactant phase domain as a function of the salt concentration (see text).[29]

From these series of apparently complex and unusual (pseudo-) ternary diagrams, it has
been possible to get a three-dimensional representation of the phase realms, as schematically
shown in Figure 7. The Γ_1, Γ_2, and Γ_3 curves represent the compositions of the water,
surfactant, and oil phases, respectively. Here too, the surfactant phase domain can be
described as a slender bridge between water corner and oil-surfactant basis, on which Γ_2 is
drawn in such a way that its distance to the oil-water axis is minimal. The Γ_1-Γ_2 curves do
not appear to form one smooth continuous curve, suggesting breaks due to the presence of
the liquid crystals which "distort" the boundary of the one isotropic phase in the low oil
content region. Interestingly, if we perform a transformation where every vertical coordinate
is expanded by a factor inversely proportional to the water concentration, we can obtain a
three-dimensional phase diagram having the same features as the temperature-composition
representation of the true ternary systems.

Similarly, if the fourth component is a long alcohol, the pseudoternary diagram may be
quite analogous to that of the ternary system (without alcohol) but at a higher temperature,[125]
or with another hydrocarbon.

In fact, gathering the data from the literature, it can be concluded that T_D, the temperature
at which a given type of (pseudoternary) diagram exists, is modified by water salinity (S),
equivalent alkane chain length (ACN), alcohol concentration (A), and HLB according to
the following simple relation (valid, of course, for a certain range of the parameter values):

$$T_D = -a + b.\,HLB + c.\,ACN + d.S + e.A$$

where (-b) is the temperature-HLB equivalence, etc., d and e depend on the chemical nature
of the electrolyte and the alcohol, and a and b depend on the type of diagram actually
concerned.

For example, for temperatures giving rise to PIT-diagrams, we got:

$$\text{PIT } (°C) \simeq -160 + 15.5 \text{ HLB} + 1.8 \text{ ACN} + d.S \qquad (2)$$

with $d = -0.68$ and $-0.25°$ ℓ/g for Na_2CO_3 and $NaCl$, respectively.

As far as salinity and ACN are concerned, these results are quite in agreement with findings of Bourrel,[35,37] although obtained from minimum interfacial tension investigations of multiphase systems. This emphasizes the strong correlation between phase behavior and physicochemical properties in surfactant systems, a correlation which should also be extended to the actual aggregation structures.

In Equation 2, derived for polyoxyethylene alcohols, HLB has been calculated according to Equation 1. However, we have also found that the same relation still holds for polyoxyethylene alkylphenols on the condition that their HLB calculation is just a little modified:[57] in the evaluation of M_L (Equation 1), the benzyl radical has to be replaced by only one methylene group. In view of the larger polarity of the benzyl group, it is quite satisfactory to find that the incorporation of that radical in the hydrophobic chain leads to a smaller intrinsic HLB than does the adjunction of a linear hexyl chain. A quite analogous and well-known result also holds for the ACN calculation of benzyl substituted hydrocarbons.

As already noted, Equation 2 is still roughly valid for the cloud point of binary aqueous surfactant solutions (in this case, $b \simeq 17$).

We shall also emphasize that the parameter d which describes the effect of the salt is practically independent of the phenomenon under consideration, whether it is the PIT, the cloud point, or even the CMC: the effect of the salt mainly concerns the water component by an alteration of its intrinsic properties. The same conclusion also seems valid when the additive is a longer alcohol. On the other hand, the use of ionic emulsifiants as cosurfactants seems to modify the phase behavior of the nonionics to a very large extent, but this will not be discussed in the present chapter.[126-128]

IV. STRUCTURAL ASPECTS

A. Local and Semilocal Investigations

A number of earlier investigations were concerned with the size and shape determinations of dilute micelles and (water) swollen micelles, by using classical techniques (like light scattering, viscosity, ultracentrifugation, etc.) as mentioned in previous discussions. Many attempts to apply these "semilocal" investigations to much more concentrated systems like the microemulsions have been made more recently, but with unequal success due to the difficulties introduced by interparticle effects. A matter for recent study includes other types of information like the location within an amphiphile solution of particular solutes, obtained from local methods like spectroscopy. For example, we can quote the studies of water-EO group interactions in aqueous systems,[98,129,130] or the competing effects in water- and benzene bondings on nonionic surfactants in apolar media.[60,130] However, measurements of thermodynamic quantities like partial molar volumes also gave valuable information on local structures in ternary aggregates.[131-134]

B. Relation to Liquid Crystalline Phases

At this point, we have to recall the earlier review by Winsor based on the "R theory", where general emphasis was placed on the importance of the formation and structure of liquid crystalline solutions in the understanding of the isotropic amphiphile solutions.[1]

Similarly, in recent papers, Friberg[131] has proposed that lyotropic liquid crystals should be considered as parts of a continuous solubility region analogous to the one for short-chain amphiphatic systems, indicating that the isotropic-anisotropic phase transition is obviously more related to packing constraints than entropy of formation.

We have to mention here the few X-ray diffraction studies of the hydrocarbon solubilization in liquid crystal phases,[1,135] determining the effective area per polar group and the thickness of paraffin layers or diameter of hydrocarbon cylindrical cores. However, we can note that no definite conclusions were stated about the exact penetration of water and oil into the surfactant film, nor to the chain conformations. Quite recent quadrupolar relaxation studies have indicated that the hydrocarbon chains of the oil are in liquid state, without marked penetration into the hydrophobic surfactant film.[136,137]

C. General Features

Apart from the gathering of punctual observations, the number of actual structural determinations in systematic relation to phase behavior of nonionics is still very limited. Moreover they reveal only one particular aspect of the structures, depending on the characteristics of the experimental technique which is used.

Classical NMR (relaxation times, lines widths, etc.) give information on thermal motion of particular grouping or atoms; they indicate, for example, that the molecular motion of a surfactant chain engaged in interfacial film is much more restricted than in liquid hydrocarbon, and that hydrocarbon cores of O/W globules are in a liquid-like state. Changes in chemical shifts reflect the nature of the external environment of the molecules giving rise to the peak. A good discussion of these types of investigation can be found elsewhere.[38]

As an example of recent results, we can quote ^{13}C shift analyses in fluorinated compounds,[59] which tend to show that a trans-conformation of the hydrophobic surfactant chain is more likely in the interfacial film of these O/W microemulsions. This is also found quite coherent with the paramagnetic relaxation measurements of ^{19}F in presence of oxygen solubilized in microemulsions. When combined with other semilocal investigation (like small-angle neutron scattering), these results allow an evaluation of the penetration rate of the (solubilized) fluorocarbon into the hydrophobic part of the interfacial film (typically one fluorocarbon for two to three surfactant molecules). Obtaining clear indications about the conformation of the EO chain in this film is much less evident.

As far as "inverse" structures are concerned, another example concerns the sites of water bonding on surfactants in hydrocarbon medium,[60] and the indication of the formation of a water-rich core only for water concentration in excess of a well-defined minimum water content. That condition of emergence of such "water pools" has been emphasized for various nonionics, by using most different experimental techniques, including conductivity, viscosity, light scattering, or fluorescence measurements.[61,62]

For O/W globules, the hydration phenomenon also significantly affects many transport properties of these systems, and particularly the dielectric relaxation; most of the results agree with the mean hydration rate of two water molecules per EO group.[14,45-48]

A swollen micelle-microemulsion transition has also been reported[43] for a system containing alcohol, reflected by a rather abrupt transition in properties at a certain value of hexadecane uptake: for higher oil/surfactant ratios, data have been interpreted in terms of globular particles, while for smaller values of that ratio particles seemed anisotropic in shape. It is not yet possible to state it is a general behavior.

In other light-scattering investigations (with alcohol also), a marked increase of the diameter of the particles has been noted when increasing the oil/surfactant ratio, a water dilution at constant size being assumed.[42]

As far as the (true) surfactant phase is concerned (oil/water ratio $\simeq 1$, very small surfactant content), strong experimental evidence of any detailed structure has not yet been proposed. There are only assumed pictures; Shinoda has considered a lamellar-type structure with both water and hydrocarbon layers, while according to Friberg, the geometries should be much more indefinite.

Given the lack of general consensus at the present time, partly due to the diversity of systems studied so far, we believe it could be much better to conclude this chapter by a

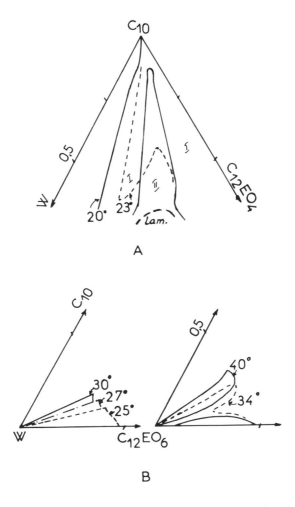

FIGURE 8. Phase diagrams of systems investigated by small angle neutron scattering for temperature on both sides of the PIT (A = decane/C$_{12}$EO$_4$; B = decane/C$_{12}$EO$_6$).[56,57]

separate presentation of a few recent interpretations of different investigations which have been particularly designed to correlate structures and phase behavior of pure C$_{12}$EO$_n$ surfactants.

D. NMR/Self-Diffusion Studies[38,39,44,49,50]

Results of pulsed NMR self-diffusion measurements have been reported for many compositions of the various isotropic regions of the systems C$_{12}$EO$_4$/C$_{10}$ and C$_{12}$EO$_4$/C$_{16}$, i.e., at various temperatures.

By comparing the self-diffusion coefficients of each component with those of the neat liquids, some indications about the rigidity of aggregate structures have been deduced.

While closed water droplets were not found unlikely in C$_{12}$EO$_4$/C$_{10}$ for samples in the microemulsion phase close to the decane corner (see Figure 8A), structures being either water or hydrocarbon discontinuous were ruled out for all surfactant phases. They could be bicontinuous, or very loose with rapid changes in aggregate size and shape.

The low temperature phases in the water corner were effectively recognized as O/W globules. Similarly, so called reversed (W/O) micelles with closed water cores were indicated but not clearly documented.

E. Dielectric Studies and Transport Properties[14,48]

These measurements were also intended in order to get clear distinction between oil-continuous, water-continuous, and "mixed"-continuity solutions, and to enable some conclusions to be drawn about the nature of the aggregates.

Extensive studies have concerned the $C_{12}EO_4/C_7$, C_{10}, C_{16} systems, at many temperatures. Here, also, the conductivity in the low-temperature (aqueous) regions (see Figure 8B) was that expected for a water-continuous phase. In the high temperature regions of the heptane systems (which join the alkane corner), the aggregates are of the W/O type, while in the hexadecane system, where the regions join near the middle of the phase diagram, the structures are assumed polyoxyethylene-oxide-water-continuous. Consequently, in the former case, the complete inversion in structures could be performed through a series of disordered bicontinuous arrangements. But for the longer hydrocarbons, this complete inversion process could be impossible, the system becoming so unstable that two phases are formed.

For heptane, in lower water-content phases, it has been suggested that there is an arrangement of large lamellar aggregates at the lower temperature end; at the high surfactant end, the evidence was that pools of water are involved. Thus, at some composition, the first unbound or free water must appear; when the hydrogen bonding breaks down and the entropy of mixing is insufficient, the water should become "encapsulated".

F. Neutron Scattering Studies[52,59]

Small angle neutron scattering measurements have been performed in order to study the structural changes induced by temperature variations in the vicinity of T_{cp} (splitting of the aqueous domain) and of T_H (coalescence of the isolated surfactant phase to the domain of high surfactant/water ratio).

1. Temperature Above the PIT

The system was $C_{12}EO_4/C_{10}$ at about 20°C (see Figure 8A). At lower temperatures (17°C), two distinct one-phase domains exist. The coalescence is complete above 25°C.

a. Low Temperature and Low Water Content

There are very small "hank-like" aggregates in the binary decane-surfactant mixture (the aggregation number N is about 10), for concentrations above 5%. (In the hank-like structure, the EO chains of each layer of the bilayer interpenetrate each other). Series of results concerning N and the eventual apparent CMC have been discussed in terms of the oil nature, HLB, and temperature.[52] Progressive addition of water promotes the formation of larger and larger aggregates. In the case of decane, all the water molecules are bound to the oxyethylene sites, and the "swollen micelles" change into lamellas (bilayer with separate layers).

Further addition of water induces a demixing, one of the phases being a lamellar liquid crystal. This occurs when the number of water molecules bound to one EO_4 chain is $\alpha \simeq 10$. The area per polar group σ is 42 Å², and the EO chain is in the extended conformation. Quite interestingly, everywhere in the lamellar phase, σ is constant and also equal to 40 to 42 Å². The limiting factor in aggregate growth could be mainly due to the presence of a thin "belt" of nonhydrated EO groups, all around the hydrophilic core.

Then, at low temperature, all the systems are characterized by a constant value of σ, and from a limiting value of the surfactant/water ratio ($\alpha = 10$), the only means to get a central film of water is the formation of infinite aggregates. Yet, a water core can exist in swollen micelles or microemulsions if the conditions are changed.

b. Low Temperature and Higher Water Content

This microemulsion phase is made of (mainly oblate?) water in oil globules; σ increases from 43 to 60 Å², and α from 7 to 12 when water is incorporated (N ~ 2000) until the second demixing occurs (in fact we get emulsions). The EO chain is in the extended configuration, probably unlike the hydrophobic chain. Typical dimensions are 250 Å.

c. Higher Temperature and Lower Water Content

At 21 to 22°C, the two-phase systems become a one isotropic phase system, the aggregates keep a mean lamellar shape, with a central water film: the temperature change is too small to induce in the surfactant packing the disorder necessary for a globular shape, given the relatively low water content of the sample. It is important to note that now the EO hydration rate is only α = 6, instead of α ≃ 10 to 11 if there were no water film. The core formation results from a sudden partial water release from the EO sites (at $T = T_H$), and not only from the excess of water incorporated to micelles with the previous maximum hydration rate ($\alpha_m = 10$).

These structural results put in light the correlation between microemulsions and (nonspherical) swollen micelles: whatever the type of aggregates, the state of the surfactant film must be compatible with the packing constraints mainly due to temperature, and, to a lesser degree, to the oil-hydrocarbon-surfactant chain interactions.

d. The Surfactant Rich Phase

The isotropic phase with higher surfactant/water ratios and containing more than 60 to 70% surfactant can be considered as a disordered dispersion of small multilamellar domains. Moreover, we have shown that there is absolutely no discontinuity in Bragg spacings when passing from the isotropic realm to its joined lamellar phase. As an important result, all these systems seem characterized by a nearly constant area per polar group, which seems to result from a coiled conformation of the hydrophobic chain of the surfactant (as in W/O microemulsions).

A minimum water content (about 10%) is necessary to promote the formation and stability of these domains, with a typical dimension of 200 Å; then the "continuous phase" of that peculiar concentrated dispersion, filling the voids, is constituted by some kind of molecular solution of mainly nonaggregated oil, water, and surfactant molecules.

Increasing the overall water content seems to induce the conformation change meander → zig-zag of the EO chain before the free water film exists. Let us recall that this constancy of σ was already pointed out by Skoulios some years ago, in another type of polyoxyethylene alkyl phenol system.[1,135]

2. Temperature Below the PIT

The system is $C_{12}EO_6/C_{10}$ in the temperature range 20 to 40°C. With an elevation of temperature, the maximum decane solubilization also increases (to 40°C). At about 30°C, however, this domain splits into one sector-like realm (still touching the water corner) and another small region, isolated along the water-surfactant axis. Above that temperature, swollen miceles do not form any longer O/W microemulsions by continuous incorporation of decane (see Figure 8B).

a. True Swollen Micelles (T < 30°C)

The higher the temperature, the larger the aggregation number of surfactant (N), but the smaller the anisometry of the particle. In fact, demixing occurs when particles are spherical. The whole solubilization process between 25 and 30°C is carried out at constant area per polar head and with a small decrease of N (σ = 50 Å²). The rate of penetration of decane into the hydrophobic part of the surfactant film increases with the oil/surfactant ratio (typically from 5 to 2 surfactant molecules per decane molecule), but does not depend here on temperature. At a given temperature, the EO chain, initially in the "meander" conformation, is a little stretched by the oil incorporation. Consequently, the higher the temperature, the larger the aggregation number of the binary aqueous rod-like micelles; the longer these micelles, the larger the amount of oil which can be solubilized, within some limits, of course.

b. The O/W Microemulsion Structures

In contrast to this above behavior, between 30 and 40°C the aggregation number in the microemulsion domain markedly increases (from 600 to 1500) with the oil/surfactant ratio.

Correlatively, the oil penetration ratio still increases (but more moderately), together with a probable small variation of σ. These small modifications of that "membrane" are sufficient to allow a further oil solubilization and a marked increase in size of the O/W globules; here, these globules keep a nearly constant main shape, with the apparent main ellipticity p $\approx$ 3. These results are quite analogous to those already found for W/O microemulsions, as previously discussed. It is also worth noting that O/W fluorinated microemulsions exhibit quite similar structural properties.[58,59]

V. CONCLUSION

Each part of the previous discussions about experimental results affords us a few pieces to the puzzle of the knowledge of aggregate structures in nonionics systems. Although many of them are still fuzzy or not quite understandable, the consistency of many others allows us to discern a clear picture of some aspects of the problem. This is particularly true when conditions favor the existence of O/W or W/O globules. Some light has also been shed onto the dual question of swollen micelles vs. microemulsions. Besides, something more than a hypothesis has been proposed to describe the structures in some surfactant-rich regions. Of course, everything cannot be directly generalized to any nonionic system, but extrapolations to other nonionics should be straightforward.

On the other hand, these structural features, although not comprehensive, make clearer the qualitative descriptions of many aspects in phase behavior. For example, we have some insight into the experimental conditions which would promote the rather sudden oil- or water-solubilization enhancement, in relation to a release of the local stiffness in surfactant film due to hydration forces.

Further advances surely will be achieved in the near future, but the most effective ones will probably come from systematic studies of series of compounds. An example of quite a promising investigation would be a further comparative study between fluorinated and hydrogenated surface-active agents.

REFERENCES

1. **Winsor, P. A.,** Binary and multicomponent solutions of amphiphilic compounds. Solubilization and the formation, structure and theoretical significance of liquid crystalline solutions, *Chem. Rev.*, 68, 1, 1968.
2. **Shah, D. O., Bansal, V. K., Chan, K., and Hsieh, W. C.,** The structure, formation and phase-inversion of microemulsions, in *Improved Oil Recovery by Surfactant and Polymer Flooding,* Shah, D. O. and Schechter, R. S., Eds., Academic Press, New York, 1977, 293.
3. **McBain, M. E. L. and Hutchinson, E.,** *Solubilization and Related Phenomena,* Academic Press, New York, 1955, 66.
4. **Schulman, J. H., Matalon, R., and Cohen, M.,** X-Ray and optical properties of spherical and cylindrical aggregates in long chain hydrocarbon polyoxyethylene oxide systems, *Faraday Soc. Discuss.*, 11, 117, 1951.
5. **Ahmed, S. I., Shinoda, K., and Friberg, S.,** Microemulsions and phase equilibria, *J. Colloid Interface Sci.*, 47, 32, 1974.

6. **Prince, L. M.,** Microemulsions versus micelles, *J. Colloid Interface Sci.,* 52, 182, 1975.
7. **Shah, D. O., Walker, R. D., Hsieh, W. C., Shah, N. J., Dwivedi, S., Nelander, J., Pepinsky, R., and Deamer, D. W.,** Some structural aspects of microemulsions and cosolubilized systems, SPE 5815 presenced at Improved Oil Recovery Symp., Tulsa, Okla., 1976.
8. **Jönsson, B., Nilsson, P. G., Lindman, B., Guldbrand, L., and Wennerström, H.,** Principles of phase equilibria in surfactant-water systems, in *Surfactants in Solution,* Vol. 1, Mittal, K. L. and Lindman, B., Eds., Plenum Press, New York, 1984, 3.
9. **Healy, R. N. and Reed, R. L.,** Physicochemical aspects of microemulsions floodings, *Soc. Pet. Eng. J.,* 14, 441, 1974.
10. *Improved Oil Recovery by Surfactant and Polymer Flooding,* Shah, D. O. and Schechter, R. S., Eds., Academic Press, New York, 1977.
11. **Johansen, R. T. and Berg, R. L., Eds.,** *Chemistry of Oil Recovery,* ACS Symp. Ser. No. 91, American Chemical Society, Washington, D.C., 1979.
12. **Salager, J. L. and Anton, R. E.,** Physicochemical characterization of a surfactant: a quick and precise method, *J. Dispersion Sci. Technol.,* 4, 253, 1983.
13. **Shinoda, K. and Kunieda, H.,** Conditions to produce so-called microemulsions: factors to increase the mutual solubility of oil and water by solubilizer, *J. Colloid Interface Sci.,* 42, 381, 1973.
14. **Bostock, T. A., McDonald, M. P., Tiddy, G. J. T., and Waring, L.,** Phase diagrams and transport properties in nonionic surfactant/paraffin/water systems, SCI Chemical Society Symp. on Surface Active Agents, Nottingham, England, 1979.
15. **Kahlweit, M.,** The phase behavior of systems of the type H_2O-oil-nonionic surfactant-electrolyte, *J. Colloid Interface Sci.,* 90, 197, 1982.
16. **Kahlweit, M., Lessner, E., and Strey, R.,** Influence of the properties of the oil and the surfactant on the phase behavior of systems of the H_2O-oil-nonionic surfactant, *J. Phys. Chem.,* 87, 5032, 1983; 88, 1937, 1984.
17. **Kunieda, H. and Shinoda, K.,** Phase behavior in systems of nonionic surfactant/water/oil around the hydrophile-lipophile-balance-temperature, *J. Dispersion Sci. Technol.,* 3, 233, 1982.
18. **Mulley, B. A. and Metcalf, A. D.,** Nonionic surface active agents. VI. Phase equilibria in binary and ternary systems containing nonionic surface-active agents, *J. Colloid Sci.,* 19, 501, 1964.
19. **Friberg, S. and Lapczynska, I.,** Microemulsions and solubilization by nonionic surfactants, *Prog. Colloid Polym. Sci.,* 56, 16, 1975.
20. **Friberg, S., Buraczewska, I., and Ravey, J. C.,** Solubilization by nonionic surfactants in the HLB-temperature range, in *Micellization, Solubilization, and Microemulsions,* Vol. 2, Mittal, K. L., Ed., Plenum Press, New York, 1977, 901.
21. **Harusawa, F., Nakamura, S., and Mitsui, T.,** Phase equilibria in the water-dodecane-pentaoxyethylene dodecyl ether system, *Colloid Polym. Sci.,* 252, 613, 1974.
22. **Lo, I., Florence, A. T., Treguier, J. P., Seiller, M., and Puisieux, F.,** The influence of surfactant HLB and the nature of the oil phase on the phase diagrams of nonionic surfactant-oil-water systems, *J. Colloid Interface Sci.,* 39, 319, 1977.
23. **Lo, I., Madsen, F., Florence, A. T., Treguier, J. P., Seiller, M., and Puisieux, F.,** Mixed nonionic detergent systems in aqueous and non-aqueous solvents, in *Micellization, Solubilization, and Microemulsions,* Vol. 1, Mittal, K. L., Ed., Plenum Press, New York, 1977, 455.
24. **Lissant, K. J. and Bradley, G. M.,** Phase diagrams of surfactant systems. The system water-hydrocarbon-oxyethylated-n-decanol, in *Solution Chemistry of Surfactant,* Vol. 2, Mittal, K. L., Ed., Plenum Press, New York, 1979, 919.
25. **Lang, J. C.,** Phase equilibria in and lattice models for nonionic surfactant water mixtures, in *Surfactants in Solution,* Vol. 1, Mittal, K. L. and Lindman, B., Eds., Plenum Press, New York, 1984, 35.
26. **Mitchell, D. J., Tiddy, G. J. T., Waring, L., Bostock, T., and McDonald, M. P.,** The plan behavior of polyoxyethylene surfactants with water: mesophase structure and partial miscibility, *J. Chem. Soc. Faraday Trans. 1,* 79, 975, 1983.
27. **Tiddy, G. J. T.,** Surfactant-water liquid crystal phases, *Physics Rep.,* 57, 2, 1980.
28. **Buzier, M. and Ravey, J. C.,** Solubilization properties on nonionic surfactants. I. Evolution of the ternary phase diagrams with temperature, salinity, HLB and ACN, *J. Colloid Interface Sci.,* 91, 20, 1983.
29. **Buzier, M. and Ravey, J. C.,** Solubilization properties on nonionic surfactants. II. 3-D representation of the three phase domains, *J. Colloid Interface Sci.,* 103, 594, 1985.
30. **Buzier, M. and Ravey, J. C.,** Three dimension phase diagram of nonionic surfactants: effect of the salinity and the temperature, in *Surfactants in Solution,* Bothorel, P. and Mittal, K. L., Eds., Plenum Press, New York, 1985.
31. **Mathis, G., Leempoel, P., Ravey, J. C., Selve, C., and Delpuech, J. J.,** A novel class of nonionic microemulsions. Fluorocarbons in aqueous solutions of fluorinated poly(oxyethylene) surfactants, *J. Am. Chem. Soc.,* 106, 6162, 1984.
32. **Stebe, M. J.,** Unpublished work.

33. **Robert, A. and Tondre, C.,** Solubilization of water in binary mixtures of fluorocarbons and nonionic fluorinated surfactants: existence domains of reverse microemulsions, *J. Colloid Interface Sci.,* 98, 515, 1984.

34. **Mathis, G., Ravey, J. C., and Buzier, M.,** Structural and solubilization properties of nonionic surfactants: a comparative study of two homologous series, in *Microemulsions,* Robb, I. D., Ed., Plenum Press, New York, 1982, 85.

35. **Bourrel, M., Salager, J. L., Schechter, R. S., and Wade, W. H.,** A correlation for phase behavior of nonionic surfactants, *J. Colloid Interface Sci.,* 75, 451, 1980.

36. **Graciaa, A., Baraket, Y., Schechter, R. S., Wade, W. H., and Yiv, S.,** Emulsion stability and phase behavior for ethoxylated nonyl phenol surfactants, *J. Colloid Interface Sci.,* 89, 217, 1982.

37. **Bourrel, M. and Chambu, C.,** The rules for achieving high solubilization of brine and oil by amphiphilic molecules, SPE/DOE 10676 Symp. on Enhanced Oil Recovery, Tulsa, Okla., 1982.

38. **Lindman, B. and Wennerström, H.,** Structure and dynamics of micelles and dynamics of micelles and microemulsions, in *Solution Chemistry of Surfactants — Theoretical and Applied Aspects,* Fendler, E. J. and Mittal, K. L., Eds., Plenum Press, New York, 1982; **Nilsson, P. G. and Lindman, B.,** Solution structure of nonionic surfactant microemulsions from NMR self diffusion studies, *J. Phys. Chem.,* 86, 271, 1982.

39. **Nilsson, P. G., Wennerström, H., and Lindman, B.,** Structure of micellar solutions of nonionic surfactants. NMR self-diffusion studies of poly(ethylene oxide) alkyls ethers, *J. Phys. Chem.,* 87, 1377, 1983.

40. **Eshuis, A. and Mellema, J.,** Viscoelasticity and microstructure of nonionic microemulsions, *Colloid Polym. Sci.,* 262, 159, 1984.

41. **Nakagawa, T., Kuriyama, K., and Inoue, H.,** Micellar weights of nonionic surfactants in the presence of n-decane or n-decanol, *J. Colloid Sci.,* 15, 268, 1960.

42. **Hermansky, C. and Mackay, R. A.,** Light scattering measurements in nonionic microemulsion, *J. Colloid Interface Sci.,* 73, 324, 1980.

43. **Siano, D. B.,** The swollen micelle-microemulsion transition, *J. Colloid Interface Sci.,* 93, 1, 1983.

44. **Lindman, B., Stilbs, P., and Moseley, M. E.,** Fourier transform NMR self diffusion and microemulsion structure, *J. Colloid Interface Sci.,* 83, 569, 1981.

45. **Epstein, B. R., Foster, K. R., and Mackay, R. A.,** Microwave dielectric properties of ionic and nonionic microemulsions, *J. Colloid Interface Sci.,* 95, 218, 1983.

46. **Foster, K. R., Epstein, B. R., Jenin, P. C., and Mackay, R. A.,** Dielectric studies of nonionic microemulsions, *J. Colloid Interface Sci.,* 88, 233, 1982.

47. **Mackay, R. A. and Agarwal, R.,** Conductivity measurements in nonionic microemulsions, *J. Colloid Interface Sci.,* 65, 225, 1978.

48. **Boyle, M. H., McDonald, M. P., Rosi, P., and Wood, R. M.,** Dielectric studies of a nonionic surfactant-alkane-water system at low water content, in *Microemulsions,* Robb, D. I., Ed., Plenum Press, New York, 1982, 103.

49. **Lindman, B., Kamenka, N., Brun, B., and Nilsson, P. G.,** On the structure and dynamics of microemulsions self-diffusion studies, in *Microemulsions,* Robb, D. I., Ed., Plenum Press, New York, 1982, 115.

50. **Lindman, B., Kamenka, N., Kathopoulis, T. M., Brun, B., and Nilsson, P. G.,** Translational diffusion and solution structure of microemulsions, *J. Phys. Chem.,* 84, 2485, 1980.

51. **Bostock, T. A., McDonald, M. P., and Tiddy, G. J. T.,** Phase studies and conductivity measurements in microemulsion-forming systems containing a nonionic surfactant, in *Surfactants in Solution,* Vol. 3, Mittal, K. L. and Lindman, B., Eds., Plenum Press, New York, 1984, 1805.

52. **Ravey, J. C., Buzier, M., and Picot, C.,** Micellar structures of nonionic surfactants in apolar media, *J. Colloid Interface Sci.,* 97, 9, 1984.

53. **Ravey, J. C. and Buzier, M.,** Small angle neutron scattering by micellar nonionic surfactants in apolar media, in *Reverse Micelles: Biological and Technological Relevance of Amphiphilic Structures in Apolar Media,* Luisi, P. L. and Staub, B. E., Eds., Plenum Press, New York, 1984, 195.

54. **Ravey, J. C. and Buzier, M.,** Structure of nonionic microemulsions by small angle neutron scattering, in *Surfactants in Solution,* Vol. 3, Mittal, K. L. and Lindman, B., Eds., Plenum Press, New York, 1984, 1759.

55. **Friberg, S., Ravey, J. C., and Buzier, M.,** to be published.

56. **Ravey, J. C. and Buzier, M.,** Structures of inverse nonionic micelles and microemulsions: influence of the temperature, in *Emulsion and Microemulsions,* Shah, D. O., Ed., ACS Symp. Ser., American Chemical Society, Washington, 1984.

57. **Buzier, M.,** Phase Diagrams and Structures of Systems with Nonionic Surfactants. Influence of Salts and of the Temperature, Thèse d'Etat, Université de Nancy, France, 1984; **Ravey, J. C. and Buzier, M.,** Unpublished results.

58. **Ravey, J. C., Stebe, M. J., and Oberthür, R.,** Structure of a fluorinated nonionic O/W microemulsion, in *Surfactants in Solution,* Bothorel, P. and Mittal, K. L., Eds., Plenum Press, New York, 1985.

59. **Stebe, M. J., Serratrice, G., Ravey, J. C., and Delpuech, J. J.,** NMR as a complementary technique of the small angle neutron scattering: study of nonionic aqueous microemulsions of perfluorocarbons, in *Surfactants in Solution,* Bothorel, P. and Mittal, K. L., Eds., Plenum Press, New York, 1985.

60. **Christenson, H., Friberg, S. E., and Larsen, D. W.,** NMR investigation of aggregation of nonionic surfactants in a hydrocarbon medium, *J. Phys. Chem.,* 84, 3633, 1980.

61. **Kumar, C. and Balasubramanian, D.,** Studies on the Triton X 100-alcohol-water reverse micelles in cyclohexane, *J. Colloid Interface Sci.,* 69, 271, 1979.

62. **Kumar, C. and Balasubramanian, D.,** Spectroscopic studies on the microemulsions and lamellar phases of the system Triton X 100-hexanol-water in cyclohexane, *J. Colloid Interface Sci.,* 74, 64, 1980.

63. **Ribeiro, A.,** Aggregation states and dynamics of nonionic polyoxyethylene surfactants, in *Reverse Micelles: Biological and Technological Relevance of Amphiphilic Structures in Apolar Media,* Luisi, P. L. and Staub, B. E., Eds., Plenum Press, New York, 1984, 113.

64. **Magid, L.,** Solvent effects on amphiphilic aggregation (and references within), in *Solution Chemistry of Surfactants,* Mittal, K. L., Ed., Plenum Press, New York, 1979, 427.

65. Concentrated Colloidal Dispersions, *Faraday Discuss. Chem. Soc.,* No. 76, 1983.

66. **Degiorgio, V. and Corti, M.,** Laser light scattering study of nonionic micelles in aqueous solution, in *Surfactants in Solution,* Vol. 1, Mittal, K. L. and Lindman, B., Eds., Plenum Press, New York, 1984, 471.

67. **Ravey, J. C.,** Lower consolute curve related to micellar structure of nonionic surfactants, *J. Colloid Interface Sci.,* 94, 289, 1983.

68. **Hayter, J. B. and Zulauf, M.,** Attractive interactions in critical scattering from nonionic micelles, *Colloid Polym. Sci.,* 260, 1023, 1982.

69. **Robbins, M. L.,** Theory for the phase behavior of microemulsions, in *Micellization, Solubilization, and Microemulsions,* Mittal, K. L., Ed., Plenum Press, New York, 1977, 713.

70. **Kjellander, R. and Florin, E.,** Water structure and changes in thermal stability of the system poly(ethylene oxide)-water, *J. Chem. Soc. Faraday Trans. 1,* 77, 2053, 1981.

71. **Kjellander, R.,** Phase separation of nonionic surfactant solutions, *J. Chem. Soc. Faraday Trans. 2,* 78, 2025, 1982.

72. **Safran, S. A.,** Fluctuations of spherical microemulsions, *J. Chem. Phys.,* 78, 2073, 1983.

73. **Degennes, P. G. and Taupin, C.,** Microemulsions and the flexibility of oil/water interfaces, *J. Phys. Chem.,* 86, 2294, 1982.

74. **Huh, C.,** Interfacial tension and solubilizing ability of a microemulsion phase that coexists with oil and brine, *J. Colloid Interface Sci.,* 71, 408, 1979.

75. **Ruckenstein, E. and Krishnan, R.,** Swollen micellar models for solubilization, *J. Colloid Interface Sci.,* 71, 321, 1979.

76. **Ruckenstein, E. and Krishnan, R.,** Swollen micellar models for solubilization, *J. Colloid Interface Sci.,* 71, 321, 1979.

76. **Ruckenstein, E. and Nagarajan, R.,** Aggregation of amphiphiles in nonaqueous media, *J. Phys. Chem.,* 84, 1349, 1980.

77. **Ruckenstein, E.,** An explanation for the unusual phase behavior of microemulsions, *Chem. Phys. Lett.,* 98, 573, 1983.

78. **Mallikarjun, R. and Dadyburjor, D. B.,** Thermodynamics of solubilization, *J. Colloid Interface Sci.,* 84, 73, 1981.

79. **Mukherjee, S., Miller, C., and Fort, T.,** Theory of drop size and phase continuity in microemulsions, *J. Colloid Interface Sci.,* 91, 223, 1983.

80. **Ruckenstein, E.,** On the thermodynamic stability of microemulsions, *J. Colloid Interface Sci.,* 66, 369, 1978.

81. **Mitchell, D. J. and Ninham, B. W.,** Micelles, vesicles and microemulsions, *J. Chem. Soc. Faraday Trans. 2,* 77, 601, 1981.

82. **Rosen, M. J.,** Comparative effects of chemical structure and environment on adsorption of surfactants at the L/A interface and on micellization, in *Solution Chemistry of Surfactants,* Vol. 1, Mittal, K. L., Ed., Plenum Press, New York, 1979, 45.

83. **Gelbart, W. M., Ben Shaul, A., McMuller, W. E., and Masters, A.,** Micellar growth due to interaggregate interactions, *J. Phys. Chem.,* 88, 861, 1984.

84. **McMuller, W. E., Ben Shaul, A., and Gelbart, W. M.,** Rod/disk coexistence in dilute soaps solutions, *J. Colloid Interface Sci.,* 98, 523, 1984.

85. **Tanford, C., Nozaki, Y., and Rohde, M. F.,** Size and shape of globular micelles formed in aqueous solution by n-alkyl polyoxyethylene ethers, *J. Phys. Chem.,* 81, 1555, 1977.

86. **Talmon, Y. and Prager, S.,** Statistical thermodynamics of phase equilibria in microemulsions, *J. Chem. Phys.,* 69, 2984, 1978.

87. **Knickerbocker, B. M., Pesheck, C. V., Scriven, L. E., and Davis, H. T.,** Phase behavior of alcohol-hydrocarbon-brines mixtures, *J. Phys. Chem.,* 83, 1984, 1979.
88. **Bellocq, A. M., Bourbon, D., and Lemanceau, B.,** Thermodynamic, interfacial and structural properties of alcohol-brine-hydrocarbon systems, *J. Dispersion Sci. Technol.,* 2, 27, 1981.
89. **Fleming, P. D., III and Vinatieri, J. E.,** Phase behavior of multicomponent fluids, *J. Chem. Phys.,* 66, 3147, 1977.
90. **Fox, J. R.,** Tricritical phenomena in ternary and quaternary fluid mixtures, *J. Chem. Phys.,* 69, 2231, 1978.
91. **Ho, C. P. and Bender, T. M.,** Phase composition of aqueous/hydrocarbon systems containing organic salts, alcohol and sodium chloride, *J. Phys. Chem.,* 87, 2614, 1983.
92. **Smith, G. D., Donelan, C. E., and Barden, R. E.,** Oil-continuous microemulsions composed of hexane, water and 2-propanol, *J. Colloid Interface Sci.,* 60, 488, 1977.
93. **Keiser, B. A., Varie, D., Barden, R. E., and Holt, S. L.,** Detergentless water/oil microemulsions composed of hexane, water and 2-propanol. 2-NMR studies, effect of added NaCl, *J. Phys. Chem.,* 83, 1276, 1979.
94. **Lang, J. C. and Widom, B.,** Equilibrium of three liquid phases and approach to the tricritical point in benzene-ethanol-water-ammonium sulfate mixtures, *Physica,* 18A, 190, 1975.
95. **Wheeler, J. C.,** Geometric constraints at triple points, *J. Chem. Phys.,* 61, 4474, 1974.
96. **Schönfeldt, N.,** *Surface Active Ethylene Oxide Adducts,* Pergamon Press, London, 1969.
97. **Schick, M. J., Ed.,** *Nonionic Surfactants,* Marcel Dekker, New York, 1966.
98. **La Force, G. and Sarthz, B.,** A study of the isotropic phase in the ternary system. n-octylamine-p-xylene-water using NMR and light scattering, *J. Colloid Interface Sci.,* 37, 254, 1971.
99. **Becher, P.,** HLB — a survey, in *Surfactants in Solution,* Vol. 3, Mittal, K. L. and Lindman, B., Eds., Plenum Press, New York, 1984, 925.
100. **Lin, I. J. and Marszall, L.,** Partition coefficient, HLB and effective chain length of surface active agents, *Prog. Colloid Polym. Sci.,* 63, 99, 1978.
101. **Little, R. C.,** Correlation of surfactant HLB with solubility parameter, *J. Colloid Interface Sci.,* 65, 587, 1978.
102. **Marszall, L.,** The effective hydrophile-lipophile balance of nonionic surfactants in the presence of additives, *J. Colloid Interface Sci.,* 65, 589, 1978.
103. **Shinoda, K. and Takeda, H.,** The effect of added salts in water on the HLB of nonionic surfactants: the effect of added salts on the PIT of emulsions, *J. Colloid Interface Sci.,* 32, 642, 1970.
104. **Saito, H. and Shinoda, K.,** The solubilization of hydro-carbons in aqueous solutions of nonionic surfactants, *J. Colloid Interface Sci.,* 24, 10, 1967.
105. **Shinoda, K.,** Solvent properties of nonionic surfactants in aqueous solutions, in *Solvent Properties of Surfactant Solutions,* Shinoda, K., Ed., Marcel Dekker, New York, 1968, 27.
106. **Shinoda, K. and Kunedia, H.,** How to formulate microemulsions with less surfactant, in *Microemulsions: Theory and Practice,* Prince, L. M., Ed., Academic Press, 1977, 57.
107. **Mukerjee, P.,** Solubilization in micellar systems, *Pure Appl. Chem.,* 52, 1317, 1980.
108. **Mukerjee, P.,** Solubilization in aqueous micellar systems, in *Solution Chemistry of Surfactants,* Vol. 1, Mittal, K. L., Ed., Plenum Press, New York, 1979, 153.
109. **Nakagawa, T.,** Solubilization, in *Nonionic Surfactants,* Schick, M. J., Ed., Marcel Dekker, New York, 1966, 558.
110. **Shinoda, K.,** Criteria of micellar dissolution, *J. Phys. Chem.,* 85, 3311, 1981.
111. **Maclay, W. N.,** Factors affecting the solubility of nonionics emulsifiers, *J. Colloid Interface Sci.,* 11, 272, 1956.
112. **Shinoda, K. and Arai, H.,** Solubility of nonionic surface-active agents in hydrocarbon, *J. Colloid Sci.,* 20, 93, 1965.
113. **Shinoda, K. and Ogawa, T.,** The solubilization of water in nonaqueous solutions of nonionic surfactants, *J. Colloid Interface Sci.,* 24, 56, 1967.
114. **Kon-No, K. and Kitahara, A.,** Solubilization of water and secondary solubilization of electrolytes by oil-soluble surfactant solutions in nonaqueous media, *J. Colloid Interface Sci.,* 34, 221, 1970.
115. **Shinoda, K. and Arai, H.,** The correlation between PIT in emulsions and cloud point in solution of nonionic emulsifier, *J. Phys. Chem.,* 68, 3485, 1964.
116. **Shinoda, K.,** The correlation between the dissolution state of nonionic surfactant and the type of dispersion stabilized with the surfactant, *J. Colloid Interface Sci.,* 24, 4, 1967.
117. **Shinoda, K. and Sagitani, H.,** Emulsifier selection in water/oil type emulsions by the HLB-temperature system, *J. Colloid Interface Sci.,* 64, 68, 1978.
118. **Elworthy, P. H., Florence, A. T., and Rogers, J. A.,** Stabilization of oil-in-water emulsions by nonionic detergents, *J. Colloid Interface Sci.,* 35, 23, 1971.
119. **Bourrel, M., Graciaa, A., Schechter, R. S., and Wade, W. H.,** The relation of emulsion stability to phase behavior and interfacial tension of surfactant systems, *J. Colloid Interface Sci.,* 72, 161, 1979.

120. **Stenius, P.,** Micelles and reversed micelles: a historical overview, in *Reverse Micelles: Biological and Technological Relevance of Amphiphilic Structures in Apolar Media,* Luisi, P. L. and Staub, B. E., Eds., Plenum Press, New York, 1984, 1.

121. **Kunedia, H. and Friberg, S. E.,** Critical phenomena in a surfactant/water/oil system. Basic study on the correlation between solubilization, microemulsion and ultra low interfacial tensions, *Bull. Chem. Soc. Jpn.,* 54, 1010, 1981.

122. **Shinoda, K. and Saito, H.,** The effect of temperature on the phase equilibria and the types of dispersions of the ternary system composed of water, cyclohexane and nonionic surfactant, *J. Colloid Interface Sci.,* 26, 70, 1968.

123. **Salter, S. J.,** Selection of pseudo-components in surfactant-oil-brine-alcohols systems, SPE 7056, Symp. Improved Methods for Oil Recovery, Tulsa, Okla., 1978.

124. **Vinatieri, J. E. and Fleming, P. D.,** Use of pseudo-components in the representation of phase behavior of surfactant systems, SPE 7057, Symp. Improved Methods for Oil Recovery, Tulsa, Okla., 1978.

125. **Tondre, C., Xenakis, A., Robert, A., and Serratrice, G.,** Evidences of structural changes in reverse microemulsion systems formulated with nonionic surfactants, in *Surfactants in Solution,* Bothorel, P. and Mittal, K. L., Eds., Plenum Press, New York, 1985.

126. **Gillberg, G., Eriksson, L., and Friberg, S. E.,** New microemulsions using nonionic surfactants and small amounts of ionic surfactants, personal communication.

127. **Sagitani, H. and Friberg, S. E.,** Microemulsion systems with a nonionic cosurfactant, *J. Dispersion Sci. Technol.,* 1, 151, 1980.

128. **Shinoda, K., Dunedia, H., and Obi, N.,** Similarity in phase diagrams between ionic and nonionic surfactant solutions at constant temperature, *J. Colloid Interface Sci.,* 80, 304, 1981.

129. **Klason, T. and Henriksson, U.,** The interaction between water and ethylene oxide groups in oligo (ethylene glycol) dodecyl ethers as studied by ^{2}H NMR in liquid crystalline phase, in *Surfactants in Solution,* Vol. 1, Mittal, K. L. and Lindman, B., Eds., Plenum Press, New York, 1984, 93.

130. **Ribeiro, A.,** Aggregation states and dynamics of nonionic polyoxyethylene surfactants, in *Reverse Micelles: Biological and Technological Relevance of Amphiphilic Structures in Apolar Media,* Luisi, P. L. and Staub, B. E., Eds., Plenum Press, New York, 1984, 113.

131. **Friberg, S. E. and Flaim, T.,** Surfactant association structures, in *Inorganic Reactions in Organized Media,* Holt, S. H., Ed., ACS Symp. ser. No. 177, American Chemical Society, Washington, 1982.

132. **Güveli, D. E., Davis, S. S., and Kayes, J. B.,** Partial molal volumes and light scattering studies on certain polyoxyethylene monohexadecyl ethers and the effect of added aromatic solutes, *J. Colloid Interface Sci.,* 86, 213, 1982.

133. **Funasaki, N., Hada, S., and Neya, S.,** Volumetric study of solubilization of hydrophobic liquids in nonionic micelles, *J. Phys. Chem.,* 88, 1243, 1984.

134. **Nakamura, M., Bertrand, G. L., and Friberg, S. E.,** Partial molar enthalpies of benzene and water in tetraethylene glycol dodecyl ether-decane solutions, *J. Colloid Interface Sci.,* 91, 516, 1983.

135. **François, J. and Skoulios, A.,** Properties of aqueous amphiphilic gels, *Kolloid Z.Z. Polym.,* 219, 144, 1967.

136. **Moucharafieh, N. and Friberg, S. E.,** Effects of solubilized hydrocarbons on the structure and the dislocation pattern in a lamellar liquid crystal, *Mol. Cryst. Liq. Cryst.,* 53, 189, 1979.

137. **Friberg, S. E., Larsen, D. W., and Ward, A. J. I.,** Structural consequence in lamellar phases containing extremely large amounts of solubilized hydrocarbon, in *Macro and Microemulsions,* Shah, D. O., Ed., ACS Symp. Ser., American Chemical Society, Washington, D.C., 1984.

Chapter 5

MOLECULAR DIFFUSION IN MICROEMULSIONS

Björn Lindman and Peter Stilbs

TABLE OF CONTENTS

I. INTRODUCTION

The interest in studies of diffusional processes in microemulsions has grown rapidly during the past few years. The reasons are an increased number of available techniques providing easily interpretable data and the realization that self-diffusion data give unique information on the molecular organization and structure of microemulsions. Furthermore, many current and potential applications of microemulsions are dependent on molecular transport over macroscopic distances.

Diffusion rates of macromolecular or colloidal units in solution are necessarily much lower than those of small molecules in the same system; typically two to three orders of magnitude. The diffusion rate depends *inter alia* on macromolecule size and shape and on macromolecule-solvent interactions. To first approximation, the diffusion rate decreases in proportion to the linear extension of the macromolecule. For spherical macromolecules at low concentrations, the familiar Stokes-Einstein equation $D = kT/(6\pi\eta r)$ is generally an excellent approximation. D represents the self-diffusion coefficient, η the viscosity of the medium, and r the macromolecular radius. For partly supramolecularly organized constituents, the proportionally reduced diffusion rates reflect the degree of organization in an easily interpretable way. The underlying arguments and the structural implications deducible therefrom are elaborated upon in Sections II and IV.

A. Diffusion

When heating a liquid system, the energy added increases its internal kinetic energy, leading to overall increased rates of molecular and particle motion. Apart from macroscopic convection and convection-like phenomena (e.g., caused by thermal or density gradients in the system), one usually considers motional partitioning into internal molecular motions (rotation about bonds and vibration), and overall reorientation and translational (or lateral) diffusion of molecules and aggregates. Translational diffusion, further subdivided into mutual diffusion and self-diffusion, is the quantity focused upon in this chapter.

The theory of diffusion and molecular transport in solution is highly developed and extensive. We will only briefly outline the subject. There exist several excellent monographs on diffusion processes, to which the reader is referred for a more complete presentation.[1-5]

1. Self-Diffusion

Self-diffusion is the result of the thermal motion-induced random-walk process experienced by particles or molecules in solution. The distribution function of molecules with regard to their original positions in an infinitely large system with regard to an arbitrary reference time is a Gaussian one, increasingly growing in width as time increases.* In an isotropic system, without thermal or concentration gradients, the average displacement in all three directions is zero; the mean square displacement is nonzero, however, and is equal to:

---

* The situation differs for obvious reasons under certain circumstances. Statistical considerations must be taken into account if the system has very few particles or if the observation time is short. These factors are rarely of significance in the types of investigations considered in this chapter. Different manifestations of restricted diffusion (wall effects and effects of diffusion through semipermeable barriers) are often significant in micro- and macroheterogeneous systems and anisotropic diffusion rates (corresponding to different D values in different directions) are a characteristic feature of many liquid crystalline systems, for example.

$$< x^2 > = 2Dt \qquad (1)$$

in one dimension. The overall particle/molecule displacement in an isotropic three-dimensional space ($< r^2 >$) becomes proportionally larger;

$$< r^2 > = 6Dt \qquad (2)$$

In these equations, the brackets symbolize a time average over the time t and D, the self-diffusion coefficient. There are as many self-diffusion coefficients in a system as there are distinct components during the time of observation (the observation time may depend on the method used, however). For NaCl in water there are, in principle, four; one each for Na^+, Cl^-, hydrogen, and oxygen. Although water is a distinct species, the latter two self-diffusion coefficients differ, in principle, due to partial dissociation of water. (The actual values for proton and oxygen self-diffusion are closely the same, however.[6] Proton migration mechanisms in water have been thoroughly discussed in two papers by Halle and Karlström.[7])

Typical self-diffusion coefficients in liquid systems at room temperature range from a few times 10^{-9} m²/sec (small molecules in nonviscous solution), to 10^{-11} m²/sec. As a reference example, pertinent to the typical problems discussed in the present context, one can note that a typical micellar diffusion coefficient in dilute aqueous solution is one to five times the latter value at room temperature. With regard to the actual magnitudes of displacement, the figure 10^{-9} m²/sec corresponds to a root mean square displacement during 1 sec in the three-dimensional space of $7.7 \cdot 10^{-5}$ m (slightly less than 0.1 mm). For a 100 times lower self-diffusion coefficient, the displacement is consequently 10 times smaller. Self-diffusion coefficients are not dramatically temperature- or concentration-dependent; the temperature coefficient of diffusion in water, for example, is approximately 3% per degree at 25°C, a rather typical value.

Formally, the magnitude of the diffusion coefficient is determined by

$$D = k_B T/f \qquad (3)$$

where T represents the absolute temperature, k_B the Boltzmann's constant, and f the so-called frictional factor. For a sphere of radius r in a continuous medium of viscosity η, f is given by the familiar Stokes equation:

$$f = 6\pi\eta r \qquad (4)$$

which, when combined with Equation 3, leads to the Stokes-Einstein relation mentioned in the Introduction. For other geometries, and when the diffusing particle is of similar size as the solvent molecules, more complex theories and equations describe f (see, e.g., References 3 and 4).

2. Mutual Diffusion

In a nonequilibrium two-component system (Figure 1), e.g., solvent/solute layered upon (or below, depending on the relative densities) pure solvent, the mutual diffusion coefficient characterizes the relaxation of concentration gradients in the system according to Fick's law:

$$J = -D'(dC/dx) \qquad (5)$$

where dC/dx represents the solute concentration gradient, D' the mutual diffusion coefficient, and J the flow of solute molecules per unit cross-section area. The net molecular motion,

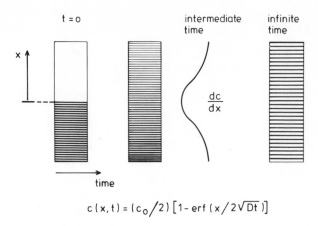

$$c(x,t) = (c_0/2)[1 - \text{erf}(x/2\sqrt{Dt})]$$

FIGURE 1.   A schematic visualization of the equilibration of the two layered solutions of different concentrations.

of course, originates from the same thermal motion that causes self-diffusion. As time passes, the once sharp boundary becomes blurred and eventually the concentration difference evens out and a single homogeneous solution (of lower concentration) remains.

In a two-component system, there is only one mutual diffusion coefficient.* Depending on the composition, it may approach either of the self-diffusion coefficients of the components. In an infinitely dilute "two-component" solution the (single) mutual diffusion coefficient is equal to the self-diffusion coefficient of the solute and not directly related to that of the solvent. At intermediate concentration ranges, the distinction between the two is important.[1-5]

3. Multicomponent Diffusion in Nonequilibrium Systems

This is, of course, the most common situation in nature and it is also the technically most important situation in the present context. The formal treatment of multicomponent diffusion is made in the framework of irreversible thermodynamics, and the reader is referred to References 3 or 8 for an introduction to the subject. In general, there are $(n - 1)^2$ different diffusion coefficients characterizing the relative fluxes of species in an n-component situation. Unlike the multicomponent self-diffusion coefficients measured by tracer of NMR techniques *(vide supra)*, multicomponent diffusion coefficients of that number and kind are difficult to estimate, measure, or interpret physically. Experimental values only exist to any appreciable extent for rather simple three-component systems (see Reference 3 for a partial but recent compilation).

The quantitative treatment of nonequilibrium multicomponent diffusion in heterogeneous or partly supramolecularly organized systems is, of course, even more complex than that of homogeneous solution phases (see, e.g., Reference 9 and further papers in that volume for an introduction to the subject).

---

*   With regard to the NaCl/water self-diffusion example just given, one should also note that although individual chloride ions diffuse approximately 50% faster than individual sodium ions in water, their net transport rates equal those of the sodium ions for sake of electroneutrality; the overall chloride ion transport is thus, to some extent, rate-determined by the net transport of sodium ions, the net transport rate of which is somewhat enhanced by the parallel transport of chloride ions.

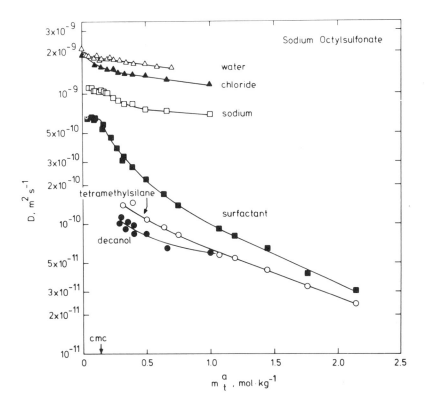

FIGURE 2A. Self-diffusion behavior of sodium octanesulfonate solutions. Water ($\triangle$), (small amount of added) chloride ($\blacktriangle$), sodium ($\square$), surfactant ($\blacksquare$), solubilized tetramethylsilane ($\circ$), and decanol ($\bullet$) self-diffusion coefficients (D in m^2/sec) are shown, as a function of surfactant concentration (m_t^a). (From Lindman, B. et al., *J. Phys. Chem.*, 88, 5048, 1984. With permission.)

II. PRINCIPLES ILLUSTRATED FROM THE SELF-DIFFUSION BEHAVIOR IN SIMPLE SURFACTANT SYSTEMS

A. Surfactant Micellization

The characteristic constituent self-diffusion behavior of an ionic surfactant upon micellization is illustrated in Figure 2A. To a good approximation all individual component (ion and neutral solubilizate) diffusion data obey a (free-micellarly bound) two-site model:

$$D_{obs} = p\, D_{micelle} + (1 - p)\, D_{free} \qquad (6)$$

in this concentration range and directly provide complete information on the aggregation behavior of the surfactant. Figure 2B illustrates the evaluated partitioning of the amphiphile and counterions, as well as the degree of counterion binding.

B. Solubilization Equilibria

Intrinsic diffusion rates of small to medium-size molecules in solution differ by one to two magnitudes from those of micelles and other supramolecular or macromolecular species in the same solution. The incorporation or binding of small molecules to these slowly diffusing units thus can be conveniently monitored due to the large effect of the binding on the self-diffusion coefficient of the substrate, according to the relation given in Equation 6. In a distinct water-micelle situation, a completely solubilized compound assumes a self-

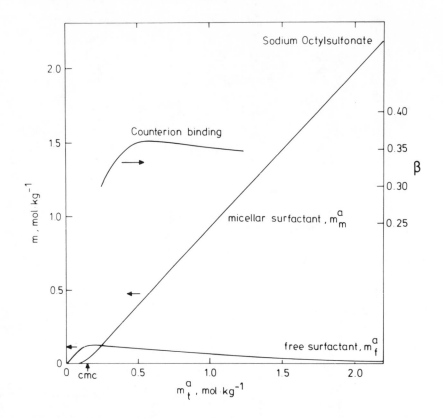

FIGURE 2B.   Self-association in solutions of sodium octanesulfonate. Concentrations of free (m_{21}^2) and micellar (m_m^2) surfactant and the degree of counterion binding (β) as a function of the total surfactant concentration. (From Lindman, B. et al., *J. Phys. Chem.*, 88, 5048, 1984. With permission.)

diffusion rate equal to that of the micelle itself. Partial solubilization, on the other hand, will manifest itself in self-diffusion coefficients of the solubilizate in between those of the micelle and the free molecule (dissolved in water), providing a new and convenient method for the quantification of solubilization processes and similar phenomena.[11-13]

A two-site model for the time-averaged molecular transport in solution makes possible a direct evaluation of experimental self-diffusion coefficients in terms of the degree of binding, p (cf., Equation 6). This quantity has a direct physicochemical meaning as such, but also can be transformed into a partition equilibrium constant of some desired form. Figure 3 illustrates a typical experimental set of spectra, recorded in about 10 min, leading to direct information on the solubilizate partitioning.

C. Micellar Breakdown by Solubilizates

Solubilizates necessarily perturb the basic micellar structure and may, at elevated concentrations, ultimately induce phase transitions. Lamellar mesophases typically form in water-surfactant systems upon addition of medium to long-chain alcohols to a solution of surfactant in water.

In a recent paper, we demonstrated that a dramatically enhanced self-diffusion rate is observed for trace amounts of solubilized hydrocarbon and for the surfactant itself, upon addition of moderate amounts of alcohols to a micellar surfactant-water solution.[14] This indicates that a breakdown of the micelle/water structure has occurred. Butanol was found to be the most effective "structure-breaker". As regards a comparison with the shorter-

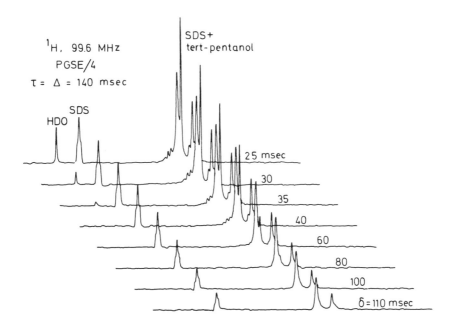

FIGURE 3. A sequence of 99.6 MHz pulsed-gradient spin-echo proton NMR spectra on a dilute sodium dodecylsulfate (SDS)/D₂O/tert-pentanol/tetramethylsilane (TMS) sample as a function of the magnetic field gradient duration, δ. Peak heights (A(i)) directly reflect self-diffusion coefficients of individual molecular types according to the relation $A(i) \propto \exp(-(\gamma G\delta)^2 D(i)(\Delta - \delta/3))$, where δ, G, and Δ are constants of the experiment. See Subsection III.B.1 for an overview of the experimental procedures.

chain alcohols methanol-propanol, the lower degree of solubilization accounts for the relatively lower destabilization effect of these compounds.*

D. Supramolecular Organization and Structure in Cubic Liquid Crystals

Data on molecular diffusion in lyotropic liquid crystals nicely demonstrate relations between diffusional rates and phase structure.[15,16] As in micelle/water systems, diffusional constraints arise as a result of barriers at the internal hydrophilic-hydrophobic interfaces. Since the structural elements of crystalline systems are immobile on the typical timescale of self-diffusion monitoring, effects on molecular self-diffusion become more pronounced.

One can deduce[16] that for unrestricted diffusion in two dimensions (and completely restricted diffusion in the third, like on a surface), the effective diffusion rate in any direction in a microcrystalline sample will be reduced to $^2/_3$ of the intrinsic value in the structural element to which molecular confinement occurs. Similarly, for a rod-like packing, the diffusion along the single continuous extension of the structural element averages out to $^1/_3$ of the intrinsic D value. In the case of structural confinement to spherical units, the resultant diffusion rate becomes essentially zero.

Static line-broadening effects normally necessitate specialized experimental procedures for NMR investigations of solids or liquid crystals. In liquid crystals of cubic symmetry, however, the static effects do not occur and their NMR spectroscopy is similar to that of

* The actual effect causing the "micellar breakdown" upon the incorporation of alcohols into the solution (most probably into the micelle-water interface) is the reduction of the barrier for micellar deformation and disruption. Alcohols of this short- to medium-chain type are rather evenly distributed between polar water domains and the micellar pseudophase (as indicated by the solubilizate partitioning studies mentioned in the preceding subsection and independent investigations by other methods), and the effect behind the altered interfacial properties of the micelle-water region is easy to visualize.

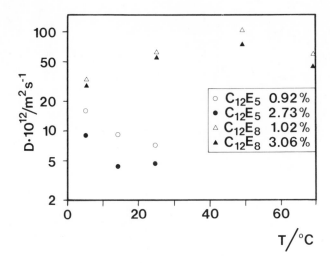

FIGURE 4.   The self-diffusion coefficients of $C_{12}E_5$ and $C_{12}E_8$ depend very differently on temperature and concentration (% by weight). Since under the experimental conditions $D_{surfactant} = D_{micelle}$ it can be deduced that $C_{12}E_5$ micelles grow strongly with increasing temperature and concentration while $C_{12}E_8$ micelles show only small growth effects. (From Nilsson, P. G., Wennerström, H., and Lindman, B., *Chem. Scr.*, 25, 67, 1985. With permission.)

normal liquids. Despite the characteristic macroscopic rigidity of cubic lyotropic liquid crystals (their physical appearance resembles that of Plexiglas or Lucite), one indeed finds experimentally that water diffusion rates are only a factor two or three lower than those of pure water, indicating that the systems are water-continuous. Observed surfactant diffusion rates lie one magnitude below these values, reflecting the combined structural/intrinsic molecular diffusion rate factors.[15,16]

E. Micelle Size, Shape, Hydration, and Intermicellar Interactions

Aqueous nonionic surfactants of the oligo (ethyleneoxide) type have been studied rather extensively to provide information on micelle size, shape, hydration, and intermicellar interactions.[17] Since the cmc's are very low, the second term of Equation 6 can be neglected and $D_{surfactant} = D_{mic}$. As exemplified in Figure 4, temperature, concentration, and surfactant chemical structure have important influences on D_{mic}. Estimating the micelle hydrodynamic radius from the Stokes-Einstein equation it can be demonstrated that for $C_{12}E_5$ (in C_xE_{yl}, C_x denotes a *n*-alkyl chain with x carbon atoms and y is the number of EO groups) the micelles grow both with increasing temperature and with increasing concentration. For $C_{12}E_8$ micelles, any growth effects are quite small.

Besides micelle size, the D_{mic} values are influenced by intermicellar interactions. At low temperatures and concentrations, D_{mic} decreases with concentration due to intermicellar repulsions. In the neighborhood of the cloud points, one observes increases in D_{mic} which are caused by intermicellar exchange of surfactant molecules made possible by intermicellar attractions under these conditions.

Water self-diffusion is retarded due to the diffusion of part of the water molecules with the micelles. Allowing for micelle obstruction effects it is possible to quantify micelle hydration. For the EO surfactants hydration decreases with temperature and increases with the number of EO groups. A striking observation is that hydration seems, at a given temperature, to a good approximation only to be given by the concentration of EO groups, which is illustrated in Figure 5.

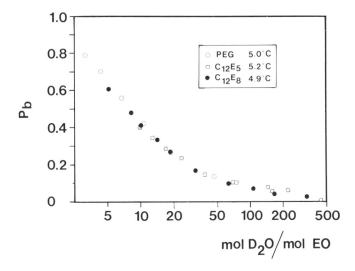

FIGURE 5.   At a fixed temperature, hydration of ethyleneoxide-containing systems depends mainly on EO-group concentration while it is rather insensitive to the physical entity where the EO groups appear. The figure gives the fraction of bound water vs. the molar ratio of water (D_2O) to EO groups for poly(ethyleneglycol), rod micelles of $C_{12}E_5$, and spherical micelles of $C_{12}E_8$. (From Nilsson, P. G., Wennerström, H., and Lindman, B., *Chem. Scr.*, 25, 67, 1985. With permission.)

From the water self-diffusion results it was also possible to demonstrate the occurrence of disc and rod micelles in different phases; the underlying principles will be further commented on later.

III. METHODS FOR MEASURING DIFFUSION

A. Self-Diffusion

1. The Nuclear Spin-Echo (SE) Method — Basic Principles

In a classic early paper by Hahn,[18] published only a few years after the first successful NMR experiments by the Purcell and Bloch groups, effects of multiple radiofrequency pulses in the NMR induction experiment were explored. Several types of effects related to so-called spin-echoes were noted and correctly interpreted. Unlike the situation only 10 years ago (when most NMR spectroscopy was made on continuous wave instruments in the swept mode), the majority of NMR instrumentation now works in the pulsed mode, as did the instrument used by Hahn. The basic pulsed NMR experiment is illustrated in Figure 6.

After an initial 90° rf pulse, spins at different precession rates dephase in the x′y′ plane. The origin of different precession frequencies may differ: chemical shift effects, effects of spin-spin couplings, and effects of an inhomogeneous magnetic field. The general useful feature of multipulse NMR experiments is some form or another of refocusing effects, i.e., the individual magnetic spin vectors regain phase coherence, to some extent, at some time after the second or higher numbered pulses.* The original experiments by Hahn were made with two or more 90° pulses, the effects of which are somewhat difficult to visualize. In the most common form today, so called spin-echo experiments utilize a 90° pulse, followed

---

* These experiments, of course, do not refocus the normal random spin dispersion due to transverse (T_2) spin relaxation processes, nor do they completely refocus effects related to homonuclear spin-spin couplings. In spin-echo experiments, the latter, in its simplest form, causes a cosinusoidal phase modulation of individual components of the echo.[19] In spin-echo experiments on strongly spin-coupled systems, the net phase-modulation effect is similar to the result of an efficient transverse relaxation process.

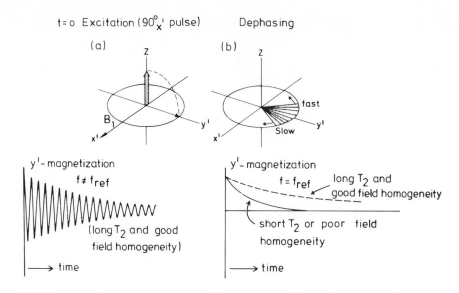

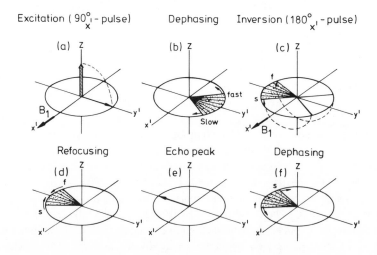

FIGURE 6. The basic pulsed NMR experiment. The signal appears in the form of transient, exponentially decaying, magnetization in the x'y'-plane; different signals precess in that plane at different rates with regard to a reference frequency and the y'-component is typically the quantity detected experimentally. Through Fourier transformation of that magnetization as a function of time one can reconstruct a frequency domain spectrum, with the reference frequency at the origin. Linewidths are inversely related to the respective time constants for magnetization decay in the x'y' plane.

FIGURE 7.   The 90°-τ-180°-pulse spin-echo experiment. For pulses of the same phase (and in the absence of transverse spin relaxation processes) spin vectors refocus after a time 2τ in the negative y'-direction.

by a single 180° pulse of the same phase (or shifted 90° with regard to the first). The effects of that pulse sequence are particularly simple to visualize if both pulses have the same phase.

a. The Static Gradient Method

The same experiment, done in an inhomogeneous magnetic field (in quantitative diffusion measurements, generally linearly inhomogeneous and inhomogeneous in one direction only), provides the basis for NMR self-diffusion experiments. The concept is simple; the 90 to 180° spin-echo experiment refocuses spins of any precession frequency, but only if the precession frequency is constant during the experiment. Therefore, if the nuclei move in an inhomogeneous magnetic field, their frequencies vary as well. For coherent motion in the direction of the gradient, the spins just refocus (ideally completely) in another direction, leading to a phase shift in the observed echo. (This provides a basis for measuring flow and similar phenomena by NMR.) From incoherent random motions like those in self-diffusion, a random phase shift of individual spin vectors occurs, leading to more or less incomplete refocusing at the time of the echo. The echo attenuation effect is quantitatively related to the self-diffusion coefficient of the molecule in which the nuclei are situated, according to the equation:[18]

$$A(2\tau)/A(O) = \exp(-2\tau/T_2 - 2\gamma^2 DG^2\tau^3/3) \tag{7}$$

where γ represents the gyromagnetic ratio of the nuclei, G the strength of the linear magnetic field gradient, and τ the interval between the two radiofrequency pulses. T_2 is the (natural) transverse spin relaxation time of the nuclei.

The basic restriction in all spin-echo diffusion experiments is evident in that equation; the irreversible effect from transverse spin relaxation always competes with the field-gradient/ diffusion-induced echo attenuation effect. Diffusion experiments are thus very much facilitated if the transverse relaxation times of the nuclei in question are long and their gyromagnetic ratio is high. Excessively rapid transverse spin relaxation actually makes diffusion measurements difficult or impossible for many nuclei. Although the underlying effects are not directly related, there is also a negative correlation between slow diffusion and rapid spin relaxation. In other words, in systems where the diffusion effect on the echo attenuation is particularly small, the transverse spin relaxation effect is often large. In the static gradient method, the only remedy is to shorten the rf pulse interval and further increase the strength of the field gradient. One should also note that it is not entirely straightforward to experimentally isolate the transverse relaxation effect from the diffusion attenuation effect in the static-gradient spin-echo experiment.

The majority of NMR diffusion measurements to date have been made on protons; the most abundant "NMR nucleus" and also the most facile nucleus for the purpose. Two alternative, "favorable" nuclei are ^{19}F and ^{7}Li. Other nuclei have more or less unfavorable spin-echo characteristics. A few examples of "possible spin-echo nuclei" and the origin of their negative characteristics are deuterium (in isotopically enriched samples): relatively rapid spin relaxation, low gyromagnetic ratio; ^{13}C: low natural abundance (expensive and often difficult chemical enrichment procedures), low gyromagnetic ratio; and ^{23}Na: rapid spin relaxation.

b. The Pulsed Field-Gradient Spin-Echo (PGSE) Method

The static-gradient experiment just described was significantly improved by Stejskal and Tanner in the mid-1960s,[20] in the form of the pulsed-gradient modification (Figure 8). Here the basic magnetic field is (essentially) homogeneous throughout the experiment. The effective dispersion/refocusing of the spins occurs in two identical (or as nearly identical as possible) field gradient pulses. The latter are almost always (but need not be) rectangular ones of the same sign, separated in time so as to fit into the chosen rf pulse interval.

The advantages over the static gradient method are twofold. The echo attenuation effect from diffusion can be separated from the transverse relaxation effect by performing the experiment at a fixed rf pulse interval (τ) (not eliminating, but keeping T_2-effects constant) and varying either the gradient pulse interval (Δ), the gradient strength (G), or the duration of the gradient pulses (δ). Also, the detection of the spin-echo is made in a relatively

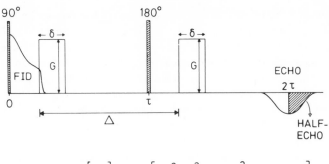

$$I(2\tau) = I(0) \cdot \exp\left[\frac{-2\tau}{T_{2i}}\right] \cdot \exp\left[-\gamma^2 \cdot G^2 \cdot D_i \cdot \delta^2 (\Delta - \delta/3)\right]$$

FIGURE 8.   The pulsed-gradient spin-echo experiment.

homogeneous magnetic field, eliminating the need for very rapid, broadband, electronic circuitry in the spectrometer.*

In a perfectly homogeneous background field, the relation between the echo attenuation and the pulse parameters in a single-component situation becomes:[20,21]

$$A(2\tau) = A(O) \exp(-2\tau/T_2 - (\gamma G \, \delta)^2 D(\Delta - \delta/3)) \qquad (8)$$

In the presence of significant background gradients, the relation (already in a strictly linear and orthogonal two-dimensional case) is considerably more complicated. In practice, it then becomes necessary to increase the pulsed gradient strengths one or two magnitudes above those of the background for adequate and correctable conditions of measurement.

It would thus appear highly desirable to reduce the background gradients as much as possible. Traditionally, however, the actual measurements are still made in an inhomogeneous field to high-resolution NMR standards. One underlying factor is instrumental. For practical reasons the normal shim coils are usually replaced by a pair of gradient coils; the resultant inherent inhomogeneity may exceed 1 G/cm. The other reason is related to the actual detection of the echo; for higher background gradients the echo becomes sharper and more easily observable. In common practice, the actual pulsed magnetic field gradient strength is of the order of 100 G/cm, corresponding to several tens of amperes through the field gradient coils. Rather specialized electronic circuitry is needed to provide clean and truly rectangular gradient pulses of this magnitude and duration (typically 0.1 to 50 msec).

The timescale (the actual period during which motion is monitored) of pulsed-gradient spin-echo experiments is equal to $(\Delta - \delta/3)$, which is typically in the 1- to 100-msec range. The actual value can be changed at will, provided that other experimental conditions (diffusion rates, gradient strength, etc.) provide large enough effects so as to allow an adequate determination of the actual echo attenuation effect from diffusion processes.

One should note that even macromolecular displacements over a millisecond time span correspond to distances that are magnitudes larger than typical molecular, supramolecular, or macromolecular radii (see Equations 1 and 2).

c. The Fourier Transform PGSE Method

Although straightforward as a concept in Fourier transform pulsed NMR[22] and demonstrated in principle in 1973,[23] the Fourier transform (FT) modification of the PGSE experiment

---

* For strong static gradients one must detect what is essentially a "spike" in the baseline, increasingly sharp as the strength of the gradient increases. Broadband electronic circuits are inherently noisy to NMR standards and have an adverse effect on the signal-to-noise ratio in the detection of the echo.

only recently has been developed into a practical tool, and extensively applied to physico-chemical problems (References 12 and 24 through 26 provide a deeper background to the actual practical experimental procedures). The basic idea is that the second half of the echo can be Fourier transformed, to separate individual contributions to the echo in the frequency domain, very much like in the normal basic pulsed FT-NMR experiment. For the detection of well-resolved spectra to be feasible in practice, the background magnetic field gradient must be of "high-resolution NMR magnitude" on the nonspinning sample (corresponding to a few Hz to at most a 10-Hz line width in the majority of applications.) Under adequate field-homogeneity conditions, it is straightforward to resolve the contributions to the combined spin-echo. Apart from "phase-distortion" effects originating from homonuclear spin-spin-couplings (so-called J-modulation effects[19]) and "amplitude distortion" effects originating from the fact that spin-echo spectra are partially relaxed and that different signals have different transverse relaxation rates (amplitudes in the spectrum thus deviate significantly from estimates based on the relative numbers of nuclei), spin-echo spectra look almost like normal NMR spectra and the interpretation is not much more difficult (see Figure 3).

2. The Neutron Spin-Echo Method

This technique[27] is conceptually similar to that of the nuclear spin-echo experiment just outlined. (A compact introduction can be found in Reference 28.) Although the procedures involve rather expensive and specialized equipment, the technique is particularly interesting in the present context, since the timescale is in the nanosecond range (rather than the millisecond range of NMR measurements), which turns out to correspond to molecular displacements of the same order of magnitude as the typical micellar radius, for example. One can foresee interesting contributions from neutron spin-echo measurements with regard to the investigation of diffusional processes in microemulsions in the near future.

3. Tracer Techniques

Labeling of molecules at a low level and monitoring the changes in label molecule concentration at different positions and/or times can be made in a number of ways to obtain self-diffusion coefficients. The two common techniques are the capillary tube method[29] and the diaphragm-cell method;[30] the former will be briefly described later in the chapter. For ease of detection and sensitivity it is most common to use radioactive labels. Radioactive isotopes of suitable lifetime are available for many elements of interest. Isotopes frequently used for surfactant systems are ^{3}H (β), ^{14}C (β), ^{22}Na (γ), ^{35}S (β), ^{36}Cl (β), and ^{45}Ca (β). If radioactive labels are not available, one may use mass spectrometric or optical detection methods; as an example, mass spectrometry has been used for ^{18}O in *inter alia* water self-diffusion studies.

The capillary tube method for studying self-diffusion involves filling a capillary cell with a solution containing the labeled molecules and inserting the capillary in a large amount of solution with identical composition except for the labeling. As a result of diffusion, the concentration of labelled molecules in the capillary tube is reduced with time. After a certain time (normally one or a few days), the radioactivity content (C) of the capillary is measured. This amount depends on the initial content (C_o), the capillary length (l), the diffusion time (t), and the self-diffusion coefficient in question. From an appropriate integration of Fick's second law, an expression can be derived which permits the calculation of the self-diffusion coefficient. For long diffusion times ($Dt/l^2 > 0.2$), the following simple relation applies with good approximation:

$$ D = \frac{4l^2}{\pi^2 t} \log \frac{8}{\pi^2} \cdot \frac{C_o}{C} $$

An advantage of tracer techniques is the high sensitivity often achievable permitting studies of a component at very low concentrations. Disadvantages are the long diffusion times, the laborious experimental procedures often required (normally one diffusion experiment only gives D of one constituent), and in particular for most surfactant systems the need for synthesizing radioactively labeled compounds. Also, one may encounter problems for surfactant systems due to stabilization of air bubbles and adsorption phenomena.

4. Transient Optical Grating Methods
a. Fluorescence Photobleaching Recovery (FPR)

The principles of FPR are conceptually very simple;[31] the species of interest contains (or is labeled chemically with) some suitable fluorophore. The solution is then locally subjected to an intense, brief (a millisecond or so) light pulse that irreversibly photobleaches a portion of fluorophores in that region. Subsequent monitoring of the same region with a low-intensity beam detects the re-emergence of fluorescence in the irradiated region, as bleached and unbleached species randomize their positions by diffusion. The detection period is typically of the order of a few seconds.

The technique usually requires the labeling of a specific molecule in the system under investigation by some suitable fluorophore. Due to the relatively large resultant local perturbation of the target molecule, the technique is often only of interest for the investigation of the diffusion of relatively large macromolecules.

The state-of-the-art FPR instrument[32,33] involves the use of a blue or green laser of the order of 1-W strength, a shutter, some optical mirrors and focusing lenses, and a fluorescence microscope with an integral photomultiplier and associated electronics, housing the actual sample. A more efficient detection scheme than focusing the entire bleaching pulse into one spot is to focus a moving image of a grating onto the sample, utilizing lock-in electronics to detect the modulation of the fluorescence intensity as the bleached pattern and the moving illumination fall in and out of phase.

b. Forced Rayleigh Scattering

This is essentially an extension of the quasi-elastic light scattering technique described in Section III.B.2. The concept is rather ingenious; the spontaneous, statistical fluctuations of normal solution systems are small, or at least smaller than desired from a measurement point of view. Several groups started, in the early 1970s, to investigate the possibility to photo-induce strong, coherent, spatially periodic temperature fluctuations in weakly absorbing media. The idea was to keep these fluctuations (of temperature, concentration, molecular orientation, etc.) much larger than the statistical ones, while still keeping the system close to thermal equilibrium. The name ''forced Rayleigh scattering'' was coined for this new technique. A good introduction can be found in Reference 34.

In essence, the typical experimental setup involves the use of two lasers of different wavelengths, illuminating the sample with crossed beams. One laser beam (high-power) is electronically pulsed (10- to 100-μsec light pulses), with millisecond to minute repetition times. A beamsplitter/mirror arrangement divides the beam into two parts, which recombine at the sample location, forming an interferometric fringe pattern in the sample. (An extreme mechanical stability of the optical setup is a necessity for adequate experiments by the technique.) In the fringe pattern (related to the perturbed system parameters, and appearing in the form of sinusoidal refractive-index variations in the sample) is then sampled by a second low-power laser, the beam of which is diffracted by the induced periodic refractive-index variations in the sample. (In measurements of self-diffusion, one utilizes instead photo-induced changes in photochromic dyes, labeling a molecule or particle type under investigation.) The detection electronics include photomultiplier units and a multichannel analyzer. Parameters accessible by the technique include thermal diffusivity (a combination of thermal conductivity, specific volume, and heat capacity), self-diffusion (utilizing photochromic dye labeling, as just mentioned), rotational diffusion (by specifically exciting molecules of a specific orientation by plane polarized light), and several others.

A recent investigation by these methods in the microemulsion field is presented in Reference 35, which also describes the experimental procedures in more detail.

B. Mutual Diffusion

1. Classic Concentration-Gradient Relaxation

Figure 1 has already illustrated the basic and rather obvious approach; the relaxation of the concentration gradient is monitored by some suitable technique, usually some optical one. The diffusion of colored species is easy to follow by simple spectrophotometric methods. Monitoring of concentration-related refractive-index changes, a more generally applicable technique, is usually made through interferometric techniques. Measurements of this kind are quite time-consuming (several hours to days), and require rather cumbersome postprocessing of acquired data (which usually appear in the form of photographic plates, scanned semimanually on photodensitometric equipment). The reader is referred to the literature[3,36,37] for a detailed description of these and alternative procedures.

2. Quasielastic Light Scattering

This is an elegant and relatively recent technique. Brief introductions to the procedures, as applied in the present field, can be found in References 38 to 40. In practice, one measures the autocorrelation function of the light intensity fluctuations of scattered light (at an angle θ) from a laser source, directed into a sample. The net intensity of the scattered light depends on the interference of the light from individual scatterers in the illuminated region. The phase and polarization of individual contributions to scattered light depend on the position and orientation of the source. As these vary as a result of Brownian motion, the net interference between different scatterers also fluctuates. The frequency of the scattered light fluctuates as well (due to the Doppler effect from the moving scattering centers), in a manner that also mimics the solution microdynamics. Different experimental designs for studying these effects exist; we will only briefly discuss the approach based on intensity fluctuation monitoring.

Intensity fluctuations of the scattered light (collected with a low-background photon-counting system) are studied through their autocorrelation function (obtained through processing of the intensity-related signal through a dedicated electronic autocorrelator). It can be demonstrated that, in the simplest cases, the autocorrelation function is exponential, with a time constant formally equal to

$$\tau = 1/(2q^2D) \tag{9}$$

Here q is the so-called scattering vector

$$q = (4\pi n/\lambda) \sin(\theta/2) \tag{10}$$

having the dimension length^{-1} (n represents the refractive index of the solution). The actual value of q can be changed to some extent by performing the experiment at different scattering angles. In dilute solution, for very small particles (in this context: [dimension · q] «1), or for spheres up to about 1 μm, D in Equation 9 approximates the particle self-diffusion coefficient. In more concentrated solutions, cross-correlations between different scattering centers contribute, and the actual quantity represented by the (average, or multiexponential) autocorrelation function from which D is evaluated is related to some kind of complex mutual diffusion coefficient, measured over a very short time interval, as compared to the macroscopic concentration-gradient relaxation method just mentioned. Most microemulsion systems are "concentrated" and have three or more constituents, so it is not altogether straightforward to interpret measured autocorrelation functions in terms of molecular or particle self-diffusion processes.

IV. SELF-DIFFUSION BEHAVIOR OF DIFFERENT MICROEMULSION SYSTEMS AND DEDUCED STRUCTURAL FEATURES

A. Principal Structural Models of Microemulsions and their Predicted Self-Diffusion Behavior

The principles behind using molecular self-diffusion coefficients, to get an insight into the structural organization of microemulsions, originate from the fact that small molecules diffusing as single entities diffuse much more rapidly than objects of macromolecular dimensions; for spherical macromolecules D is inversely proportional to the radius according to the Stokes-Einstein equation, which applies to dilute systems. Furthermore, if a small molecule is associated with or bound to a macromolecule or particle, its diffusion coefficient will be low. Therefore, if we have a disperse system with droplets of one liquid dispersed in another, the diffusion is low for a molecular species which is completely confined to the droplets and thus diffuses with the slowly moving droplets. This applies when diffusion is monitored over macroscopic distances as in the various NMR spin echo techniques and in tracer diffusion studies. (By the neutron spin-echo method and by NMR relaxation, short-range translational motions may be monitored.) In a droplet-type structure the molecules of the continuous medium, on the other hand, diffuse rapidly; except for a rather small obstruction effect, the diffusion will be as in a simple liquid or solution.

The interpretation of molecular self-diffusion coefficients is thus quite straightforward and it is possible, from a structural model of microemulsions, to predict its self-diffusion characteristics. Here we will consider, in particular, the following four clear-cut limiting cases:

1. *Water-in-oil droplet structure:* $D_{water} \ll D_{oil}$ and $D_{surfactant} \simeq D_{water} \simeq D_{droplet} \cdot D_{oil}$ will be of the same order of magnitude as D of neat oil or a simple solution.
2. *Oil-in-water droplet structure:* $D_{water} \gg D_{oil}$ and $D_{surfactant} \simeq D_{oil} \simeq D_{droplet} \cdot D_{water}$ will be of the same order of magnitude as D of neat water or a relevant aqueous solution.
3. *Bicontinuous structure:* D_{water} and D_{oil} both high while $D_{surfactant}$ is low. D_{water} and D_{oil} (eventually after applying an appropriate obstruction effect correction) will be of the order of magnitude of the D's of the neat liquids (alternatively relevant solutions). $D_{surfactant}$ is low because surfactant occurs in large aggregates but is predicted to be higher than in cases 1 and 2 because diffusion is unrestricted (although hindered).
4. *Molecule-disperse solution (nonstructured case):* D_{water}, D_{oil}, and $D_{surfactant}$ are all high (as in simple nonaggregated solutions in general).

We infer that, depending on the structural type, the ratios between the self-diffusion coefficients of the constituents can vary considerably. In particular, the quantity D_{water}/D_{oil} is a sensitive measure of microemulsion structure. In actual practice, we have observed variations in this quantity by as much as 10^5 between extreme cases of isotropic surfactant solutions. Since the individual self-diffusion coefficients are normally obtained to better than 5% precision, we have here a very sensitive measure of microemulsion structure.

The aforementioned structural examples are useful limiting cases but do not, of course, exhaust the possibilities. Although we have encountered examples closely corresponding to the limiting cases, intermediate-type structures are also important to discuss. Several possibilities exist:

1. **Attractive droplet structure** which, for part of the time, allows interdroplet exchange of molecules; this will increase the D values of the molecules in the droplets.
2. **Incomplete confinement into droplets:** The observed self-diffusion coefficient will then be a weighted average of the droplet and dispersed state D values. Since D_{free} » $D_{droplet}$ even a small fraction in the continuous medium can affect the observed D value considerably.
3. **Mixed structure type:** For the case of coexistence of two structural types again weighted average self-diffusion coefficients will be obtained.
4. **Partially organized molecule disperse solution:** Here the diffusion coefficients will be modified from case 4 toward the corresponding structural type.

Apart from the actual magnitudes of the self-diffusion coefficients of the constituents, structural information is obtainable from the variation of the self-diffusion coefficients with composition, for example, anticipated droplet volume fraction. In order to discuss this we have to consider briefly two phenomena; first, the *obstruction effect* of excluded volumes on the mobility of molecules in the medium, and secondly, the effect of *droplet-droplet* interactions on $D_{droplet}$.

A general treatise of the self-diffusion of small molecules in colloidal systems is provided by Jönsson et al.[41] using the cell model together with arguments from the thermodynamics of irreversible processes. This paper (which also reviews earlier work of relevance) contains many useful results for the problem considered here. One is that for spherical obstructing volumes the effective or observable self-diffusion coefficient is given by

$$D^{eff} = D^{o} \frac{1}{1 + \phi/2}$$

where D^{o} is the self-diffusion coefficient in the absence of the obstructing volumes (i.e., the self-diffusion coefficient of the molecules in question in the continuous medium) and ϕ the volume fraction of the spheres. For prolate or cylindrical obstructing volumes, the obstruction effect is also quite small up to high ϕ values while oblate aggregates with a high axial ratio give an obstruction factor of $2/3$ already at quite low volume fractions (Figure 9). This effect is useful for distinguishing between different aggregate shapes and has been applied to some simple nonionic surfactant systems[42] (Figure 10).

The theory of the concentration dependence of the self-diffusion of interacting colloidal particles is a field of much current activity.[43-46] In the case of hard spheres with no hydrodynamic interactions, the following relation for the self-diffusion coefficient has been derived:[44]

$$D = D_{o} (1 - 2\phi)$$

where D_{o} is the self-diffusion coefficient in the absence of interactions and ϕ the volume fraction of colloidal particles. This result has been supported by simulations.[45]

We note that for a droplet structure, molecules of the medium diffuse much more rapidly than molecules confined to the droplets. Furthermore, for such a case D_{medium} and $D_{droplet}$ decrease with increasing droplet volume fraction; for a hard sphere model, theories are available which predict the rate of decrease. Interactions between droplets decrease $D_{droplet}$. However, for attractive interactions where confinement to a single droplet is not maintained but interdroplet exchange is allowed (or fusion and fission processes occur), the diffusion of molecules of the droplets may occur more rapidly than diffusion of the droplets themselves. If this process is very efficient, the diffusion of droplet molecules may increase with increasing droplet volume fraction. (An example for the case of micelles is given in Reference 47.)

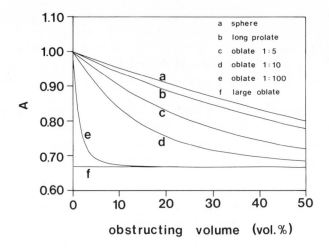

FIGURE 9. The theoretical obstruction term in solvent self-diffusion for some different particle shapes vs. the volume fraction of obstruction particles. (From Nilsson, P. G. and Lindman, B., *J. Phys. Chem.*, 88, 4764, 1984. With permission.)

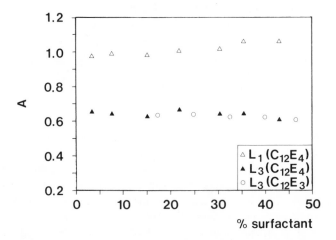

FIGURE 10. The experimental obstruction factor in water self-diffusion for solutions of nonionic surfactants with large micelles give evidence for rod micelles in the low temperature isotropic phase (L_1) and large disk micelles in the high temperature isotropic phase (L_3 or anomalous phase). (From Nilsson, P. G. and Lindman, B., *J. Phys. Chem.*, 88, 4764, 1984. With permission.)

B. Solubilized Micellar Solutions and Water-Rich Microemulsions

Self-diffusion coefficients of aqueous ionic surfactant solutions to which a solubilizate with low aqueous solubility has been added are exemplified in Figure 11. As observed for a wide range of systems the self-diffusion coefficients follow the sequence $D_{water} > D_{counterion} \gg D_{surfactant} > D_{solubilizate}$. In particular, we note that $D_{water} \gg D_{oil}$ and that D_{water} is very weakly concentration dependent; in fact, the concentration dependence of the water self-diffusion coefficient can be quantitatively rationalized in terms of hydration and an obstruction effect of spherical obstructing particles. For the ionic surfactant systems, $D_{micelle}$ decreases rapidly with concentration because of strong long-range electrostatic repulsions between micelles.

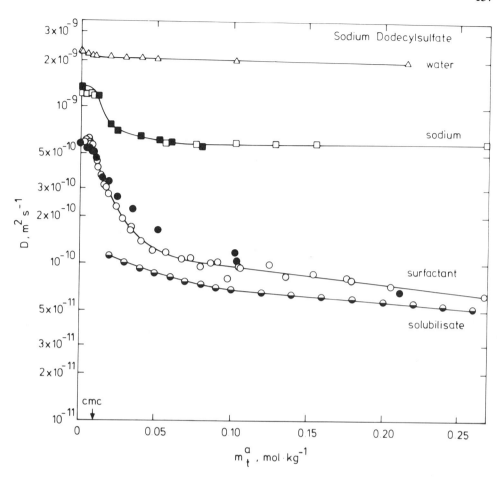

FIGURE 11. Self-diffusion behavior of aqueous solutions of SDS at 25°C. (From Lindman, B., Kamenka, N., Puyal, M. C., Rymdén, R., and Stilbs, P., *J. Phys. Chem.*, 88, 5048, 1984. With permission.)

For surfactants which are soluble to high concentrations (and perhaps also form rod micelles), the D_{water}/D_{oil} ratio can approach 1000 (see Figure 12).

Mackay et al.[48,49] have used a polarographic method to determine self-diffusion coefficients of inorganic ions (Cd^{2+}, $Fe(CN)_6^{3-}$, $Fe(CN)_6^{4-}$) and hydrophobic probes in various microemulsion systems. Water-rich (e.g., 60% water) microemulsions composed of $C_{16}N^+(CH_3)_3Br^-$, 1-butanol, hexadecane, KCl, and water or of $C_{16}SO_4^-Na^+$, 1-pentanol, mineral oil, KCl, and water gave low self-diffusion coefficients (of the order of $5 \cdot 10^{-11} m^2/$ sec) of 4-cyano-1-alkyl pyridinium ions and of alkyl viologens with long alkyl chains (>5 C) while the shorter probe molecules had a much faster diffusion. The results indicate the existence of distinct hydrophilic and hydrophobic regions and the confinement of the more hydrophobic probes to the oil domains.

C. Double-Chain Surfactant-Hydrocarbon-Water Systems

The most studied double-chain surfactant, besides the biological amphiphile lecithin, is Aerosol OT (sodium di-2-ethylhexylsulfosuccinate). Aerosol OT (AOT) is highly soluble in many hydrocarbons and these solutions can take up large amounts of water (up to approximately 60%, see Figure 13). AOT self-associates in hydrocarbons and these aggregates can take up ("solubilize") large amounts of water; often this type of system is considered as the counterpart to normal solubilized micellar solutions. The self-diffusion behavior[50-56]

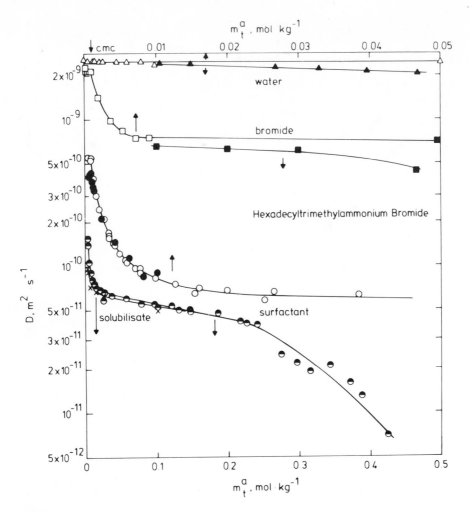

FIGURE 12.   Self-diffusion behavior of aqueous solutions of hexadecyltrimethylammonium bromide at 33°C. (Note upper and lower concentration scales and referring arrows.) (From Lindman, B. et al., *J. Phys. Chem.*, 88, 5048, 1984. With permission.)

of these systems is in fact as illustrated in Figure 13 effectively opposite to that described above for solubilized micellar solutions. Thus, $D_{oil} \gg D_{water} \simeq D_{surfactant} \simeq D_{droplet}$. This corresponds to a simple water-in-oil situation. It is striking that water diffusion stays slow up to the highest water contents studied. This suggests that interdroplet exchange of water is relatively insignificant.

Eicke's group has for a long period of time systematically applied a broad range of physicochemical approaches to study mainly the AOT-isooctane-water system. Recently, Eicke et al.[50,51] reported some temperature dependent water self-diffusion coefficients (^{1}H spin-echo NMR) for the system AOT-hexane-water. A rapid increase in D_{water} above 40°C was found and discussed in terms of percolation phenomena.

A number of cationic double-chain surfactant-hydrocarbon-water systems have recently been studied by Ninham, Evans, and co-workers[57-60] with interesting results. In the dido-decyldimethylammonium bromide-hydrocarbon-water system the surfactant is not soluble in hydrocarbon but there is an extensive isotropic solution region in the presence of water (for example, between approximately 15 and 60% water in the case of decane). The phase behavior is thus distinctly different from the AOT systems and so is the self-diffusion

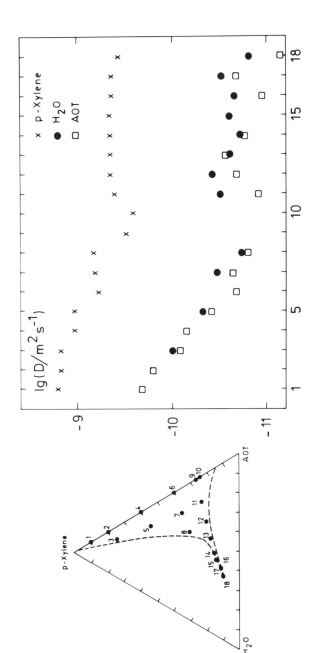

FIGURE 13.   In AOT-hydrocarbon-water systems hydrocarbon self-diffusion is one to two orders of magnitude more rapid than water and surfactant self-diffusion. (From Stilbs, P. and Lindman, B., *Prog. Colloid Polym. Sci.*, 69, 39, 1984. With permission.)

behavior[60] (studied by the FT PGSE NMR method). Hydrocarbon self-diffusion is rapid throughout while surfactant diffusion is 10 to 100 times slower. Water diffusion shows a different and unexpected behavior being quite rapid for many situations and in fact of the same order of magnitude as hydrocarbon diffusion. Water diffusion behavior is strikingly dependent on hydrocarbon. With hexane, octane, decane, and dodecane, D_{water} is 3 to 5 · 10^{-10} m²/sec at the lower water contents while it goes down by a factor of approximately 20 at the higher water contents. With tetradecane, water diffusion is rapid throughout (approximately 6 · 10^{-10} m²/sec), in fact more rapid than hydrocarbon diffusion. It is clear that in these systems one has in certain concentration regions the same pattern as in the AOT systems with a pronounced W/O structure but in others a bicontinuous situation similar to that found in the four-component systems with cosurfactant discussed later in this chapter; it appears that surfactant diffusion is more rapid in the bicontinuous concentration regions than in the W/O regions, again a behavior which is analogous to the cosurfactant systems. The observation that on increasing the water content one introduces a transition from connected to disconnected water domains is very striking and will be of great value in rationalizing the complex phase behavior of these surfactants.

D. Surfactant-Alcohol or Fatty Acid-Water Systems

Tracer diffusion studies were performed for the sodium octanoate-decanol-water, sodium octanoate-octanoic acid-water, and sodium cholate-decanol-water systems[61,62] while the FT PGSE technique was applied recently to a wide range of potassium oleate-alcohol-water systems.[63-65]

For the water-poor solution phase (L_2) of the sodium octanoate-decanol-water system, water diffusion is retarded by a factor of approximately 50 compared to pure water while decanol diffusion is only little retarded (see Figure 14). These microemulsions are of the W/O type except for the highest decanol (low water) contents where they become appreciably molecule-disperse.

For the L_2 phase of the potassium oleate-decanol-water system, D_{water} is low and decreases with increasing water content (Figure 15) corresponding to a W/O limiting case. With short-chain alcohols the behavior is very different with D_{water} being high and increasing with increasing water content, thus inconsistent with a W/O structure. Branching and shortening the alkyl chain length both favor a bicontinuous structure.

E. Tracer Self-Diffusion Studies of a System with Surfactant and Cosurfactant

The first multicomponent self-diffusion study of microemulsions was probably the tracer study of the system sodium *p*-octylbenzenesulfonate-pentanol-decane-sodium chloride-water performed in Montpellier.[62,66] Self-diffusion coefficients of surfactant, counterion, water, and hydrocarbon were obtained over wide concentration ranges in the extensive microemulsion phase (see Figure 16). The most striking observation was that for large concentration ranges both D_{water} and $D_{hydrocarbon}$ are rapid, in fact, of the same order of magnitude as for the neat liquids or simple solutions; surfactant diffusion is an order of magnitude slower. The slow surfactant diffusion points to an aggregated system rather than a molecule-disperse situation while the water and hydrocarbon D values rule out structures where either of these constituents are confined to closed domains. These observations instead favor a bicontinuous structure which, however, in view of other observations (viscosity, NMR relaxation, etc.), must be very flexible and dynamic.

These results also demonstrate that at high water contents $D_{hydrocarbon}$ approaches $D_{surfactant}$ while at high decane contents D_{water} approaches $D_{surfactant}$. Under these conditions, one encounters then O/W and W/O droplet structures, respectively.

Counterion self-diffusion is mainly determined by the water content so that at low water contents (at all surfactant-hydrocarbon mixing ratios) $D_{counterion}$ is low. This corresponds to

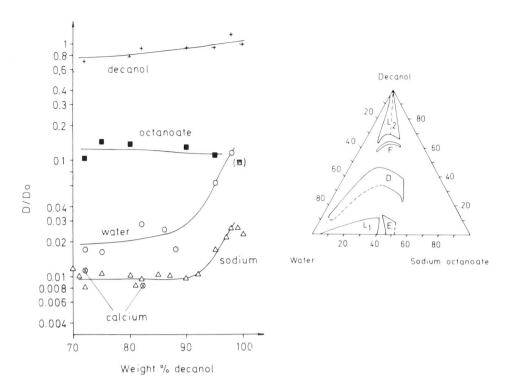

FIGURE 14. In the water-poor solution phase (L₂) of the system sodium octanoate-decanol-water, water self-diffusion is much retarded while that of decanol is rapid. (D values are given relative to the values of the pure liquids while those of the ions are related to dilute aqueous solutions.) A W/O structure applies over most of the concentration range. (From Fabre, H., Kamenka, N., and Lindman, B., *J. Phys. Chem.*, 85, 3493, 1981. With permission.)

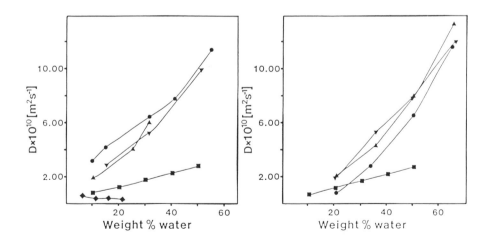

FIGURE 15. Water self-diffusion coefficients as a function of weight-% water in ternary microemulsions with different cosurfactants: (a) ■, 1-pentanol, weight ratio cosurfactant:potassium oleate 68:32; ▼, 1-butanol, weight ratio 63:37; ◆, 1-decanol, weight ratio 81:19; ▲, benzyl alcohol, weight ratio 68:32; ●, ethylene glycol monobutyl ether, weight ratio 80:20. (b) ■, 1-Pentanol, weight ratio cosurfactant:potassium oleate 68:32; ▼, 3-methyl-2-butanol, weight ratio 68:32; ▲, 2-methyl-1-butanol, weight ratio 67:33; ●, 2,2-dimethylpropanol, weight ratio 67:33. (From Wärnheim, T., Sjöblom, E., Henriksson, U., and Stilbs, P., *J. Phys. Chem.*, 88, 5420, 1984. With permission.)

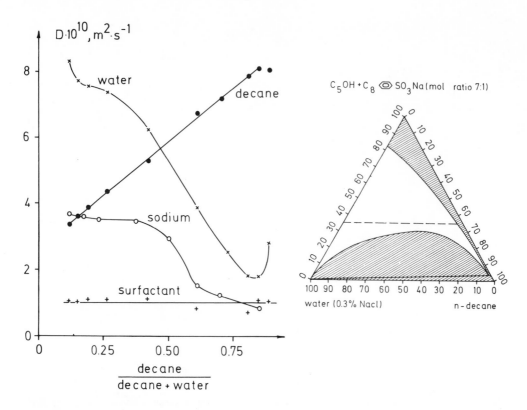

FIGURE 16. For the system sodium octylbenzenesulfonate-pentanol-decane-water-NaCl, both hydrophilic and hydrophobic components diffuse rapidly over wide concentration ranges giving evidence for a bicontinuous structure. (Sample compositions indicated by the dashed line in the schematic phase diagram.)[62,66]

a large fraction of counterions associated with the aggregates which is to be expected for a system with low effective dielectric constant.

F. FT PGSE NMR Studies of Complex Ionic Surfactant Systems

To characterize in any detail the self-diffusion behavior of all constituents of four- or five-component microemulsion systems with the tracer technique is a very laborious task, while here the FT NMR technique finds some of its best applications. Indeed a number of aspects of different systems have been investigated by this technique; studies[54,63-65,67-73] have confirmed and extended the conclusions from the tracer work.

For an overall visualization of the self-diffusion behavior, iso-diffusion lines are useful as exemplified for two quaternary systems with extensive microemulsion regions in Figures 17 and 18. We note again that over most of the regions of existence of the microemulsions, D_{water} and $D_{hydrocarbon}$ are of similar magnitude while $D_{surfactant}$ is much lower. At the highest water contents we encounter an oil-in-water structural type since $D_{water} \gg D_{hydrocarbon}$, while at the high decane contents the structure is of the water-in-oil type. Surfactant and cosurfactant diffusion are quite weakly concentration dependent, but at different levels. Surfactant molecules are mainly located in the aggregates while the cosurfactant molecules are partitioned rather evenly between the aggregates and the (continuous) medium.

Cosurfactant partitioning seems to be very important both for the nature of the phases formed (see for example, phase diagrams) and for the microemulsion structure. The latter, as illustrated in Figure 19, could be rather well illustrated from self-diffusion studies. This type of study, which involves measurements for different lengths of an *n*-alcohol cosurfactant at a fixed molar (or mass) ratio between the four components, was made for a number of

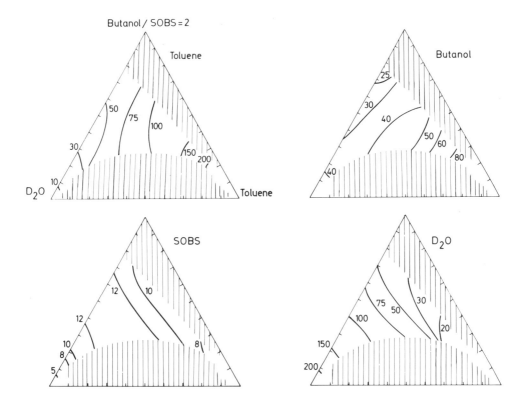

FIGURE 17.   Observed iso-diffusion lines within the isotropic solution region of the system sodium octylbenzenesulfonate-butanol-toluene-water (D_2O) (self-diffusion coefficients displayed in units of 10^{-11} m²/sec). (From Reference 56; based on data mainly from Reference 70.)

systems. With butanol and pentanol as cosurfactant we re-encounter the situation with D_{water} and $D_{hydrocarbon}$ only differing by a small factor. For a long chain alcohol, on the other hand, hydrocarbon diffusion is more rapid than water diffusion by a factor of more than 100. Consequently, a medium chain alcohol has a strong tendency to favor a bicontinuous structure while a long chain alcohol confers on the system a structure with discrete droplets (or a lamellar liquid crystal).

Dilution procedures are often used in studies of microemulsion systems, for example, to allow an elimination of interdroplet interaction effects by extrapolation. In such dilution procedures, one follows phase boundaries and the droplets and the composition of the interdroplet medium are argued to remain unchanged. The dilution procedure commonly used is based on successive alternate additions of oil and cosurfactant, the latter added to the point where a microemulsion is regenerated from the previously heterogeneous sample. Self-diffusion results[72,73] along such dilution lines are shown in Figures 20 and 21. With pentanol as cosurfactant we note first that $D_{hydrocarbon} \gg D_{water} \simeq D_{surfactant}$, demonstrating a W/O structure, and second that both D_{water} and $D_{surfactant}$ extrapolate to the droplet D_o mutual diffusion coefficient obtained in quasielastic light scattering studies. Comparing these results with the results given earlier, it should be noted that they are obtained at rather low water contents.

With butanol as cosurfactant, D_{water} is considerably larger than $D_{surfactant}$ and W/O structure with water confined to closed droplets is not applicable. Also we note that $D_{surfactant}$ decreases at a lower rate with increasing water content with butanol than with pentanol which can be taken to suggest less repulsive or attractive interdroplet interactions or admixture of a bicontinuous structure.

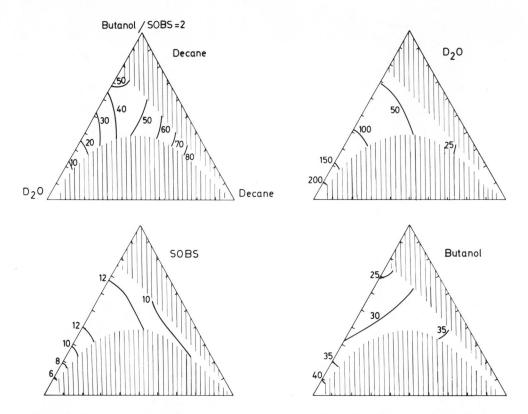

FIGURE 18.   The same information as in Figure 17, but with decane as hydrocarbon constituent. (From Stilbs, P. and Lindman, B., *J. Colloid Interface Sci.*, 99, 290, 1984. With permission.)

Very dramatic variations are observed in experiments where salinity is altered to induce phase changes. Presented in Figure 22 are self-diffusion coefficients for the microemulsion phase as one proceeds from a Winsor I (microemulsion in equilibrium with oil phase) via a Winsor III (microemulsion in equilibrium with both an oil and a water phase) to a Winsor II situation (microemulsion in equilibrium with an aqueous phase). At low salinities D_{water} » $D_{hydrocarbon}$ ≃ $D_{surfactant}$ and an O/W structure applies. At high salinities the opposite situation is encountered, i.e., $D_{hydrocarbon}$ » D_{water} ≃ $D_{surfactant}$, corresponding to a W/O structure. At intermediate salinities we infer a bicontinuous structure since D_{water} ≃ $D_{hydrocarbon}$, both being high.

G. Nonionic Surfactant-Hydrocarbon-Water Systems

For these systems, phase equilibrium studies by Shinoda and Friberg have demonstrated a rich and sensitive phase behavior (see, e.g., References 74 and 76). These interesting microemulsions show a complex self-diffusion behavior which is yet incompletely characterized.[17,54,77-79] Systems studied include tetraethylene glycol dodecyl ether ($C_{12}E_4$)-hexadecane-water, $C_{12}E_4$-decane-water, and tetraethylene glycol decyl ether ($C_{10}E_4$)-hexadecane-water. We recall the temperature-dependent phase behavior of these systems which is schematically shown in Figure 23; at low temperatures, with an extensive isotropic phase connected with the water corner (L_1); at high temperatures, with an extensive isotropic phase connected with the oil-surfactant base (L_2); and at intermediate temperatures, with a disconnected isotropic region (S). We review here the more significant self-diffusion observations which are

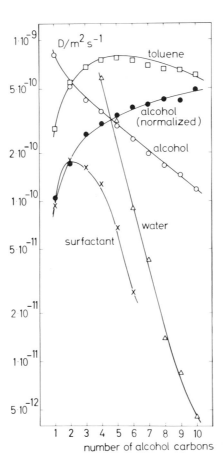

FIGURE 19.   The water self-diffusion coefficient decreases by more than two orders of magnitude as the cosurfactant is changed from butanol
to decanol. (SDS-alcohol-toluene-water system
with weight ratios 17.5:35:12.5:35.) (From Stilbs,
P. and Lindman, B., *J. Colloid Interface Sci.*,
99, 290, 1984. With permission.)

1. In a wide concentration range from the water corner (down to approximately 60%
    water), D_{water} is two orders of magnitude or more larger than $D_{hydrocarbon}$ and $D_{surfactant}$,
    these two being approximately equal. Here the structure is thus of the O/W type with
    water diffusion decreasing with increasing droplet volume fraction at a rate corre-
    sponding to obstruction and hydration of spherical particles. From $D_{surfactant}$ and
    $D_{hydrocarbon}$ it is possible from the extrapolated values to obtain the hydrodynamic radii
    which for the three systems mentioned are in the range 120 to 230 Å. By taking the
    area per polar group from low-angle X-ray measurements on the lamellar liquid crys-
    talline phases, it is possible from simple geometric considerations to estimate micellar
    radii which are generally in rough agreement with the hydrodynamic radii.
2. For the water-rich systems, $D_{surfactant}$ and $D_{hydrocarbon}$ in some cases decrease steadily
    up to high volume fractions of surfactant + hydrocarbon, which can be referred to
    droplet-droplet repulsive interaction, slowing down droplet motion. At higher con-
    centrations, $D_{hydrocarbon}$ and $D_{surfactant}$ have a minimum and then start to increase which
    is referred to an exchange of molecules between droplets. Very interestingly, for some
    systems this exchange starts to become dominant at very low volume fractions of

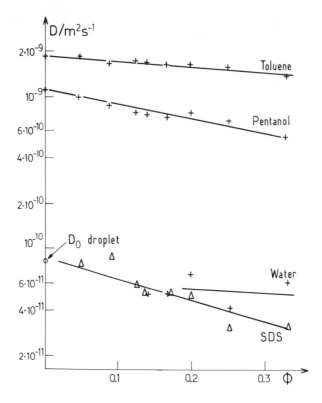

FIGURE 20.   Constituent self-diffusion coefficients for microemulsions composed of SDS, pentanol, toluene, and water. Microemulsion compositions follow a dilution line and data are given as a function of volume fraction of dispersed phase (water droplet model assumed). D_o is the dynamic light scattering diffusion coefficient extrapolated to infinite dilution.[73]

hydrocarbon + surfactant as is illustrated in Figure 24. We note especially the $C_{10}E_4$-hexadecane-water system where exchange of hydrocarbon molecules between micelles occurs more rapidly than surfactant exchange even at low concentrations. These types of observations can be correlated with the distance from the so-called phase inversion or hydrophilic-lipophilic balance temperature. The results demonstrate attractive droplet-droplet interactions and a rapid fusion of micelles since otherwise the surfactant self-diffusion coefficient should be equal to or higher than the hydrocarbon self-diffusion coefficient.

3.   Approaching the hydrocarbon corner (but few points measured) we obtain $D_{hydrocarbon}$ to be 10 to 100 times larger than D_{water}, thus demonstrating a W/O structure. Varying the water content but keeping the surfactant-to-hydrocarbon ratio constant at a high level gives a very regular variation in D_{water} especially for the $C_{10}E_4$-hexadecane-water system (Figure 25) showing no structuring associated with a confinement of water into droplets.

4.   Since these phase diagrams are so sensitive to temperature, it is possible to change the phase (L_1, S, or L_2) for a given sample composition by changing the temperature by only a few degrees. For the $C_{10}E_4$-hexadecane-water system, we observed only marginal increases in the different D values on increasing the temperature from 30 to 40°C which changes the phase region from L_1 to L_2. In particular, the water self-diffusion coefficients taken relative to those of pure water are closely constant. Thus,

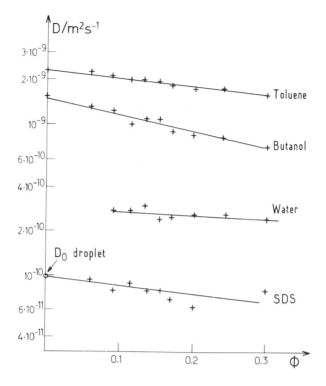

FIGURE 21.   Constituent self-diffusion coefficients for microemulsions composed of SDS, butanol, toluene, and water. Explanations are as for Figure 20.[73]

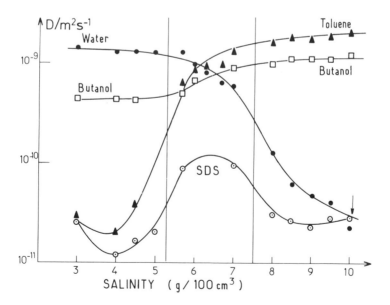

FIGURE 22.   Self-diffusion coefficients (at 20°C) of the constituents for microemulsions composed of SDS (1.99% by weight), butanol (3.96%), toluene (46.25%), and brine (46.8%). Brine salinity was varied between 3 and 10% whereby the system changed from Winsor I to Winsor II type. Between vertical lines there is a so-called "middle phase" microemulsion, i.e., the system is of type Winsor III.[73]

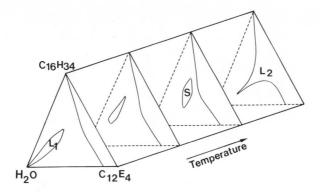

FIGURE 23.   Schematic composition-temperature diagram for a non-ionic surfactant-hydrocarbon-water system showing the approximate extension of the isotropic phases. The water-rich phase (L_1) and the hydrocarbon-rich phase (L_2) are connected by an isotropic channel, the surfactant phase (S). (From Nilsson, P. G. and Lindman, B., *J. Phys. Chem.*, 86, 27, 1982. With permission.)

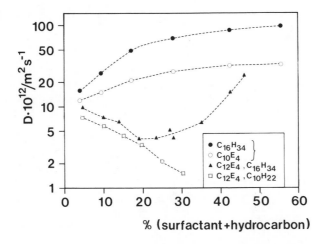

FIGURE 24.   Surfactant and hydrocarbon self-diffusion coefficients in the L_1 phase of various three-component nonionic surfactant systems. The experimental temperatures are 30°C ($C_{10}E_4$-$C_{16}H_{34}$), 19.6°C ($C_{12}E_4$-$C_{16}H_{34}$), and 4.7°C ($C_{12}E_4$-$C_{10}H_{22}$). (From Nilsson, P. G., Wennerström, H., Lindman, B., *Chem. Scr.*, 25, 67, 1985. With permission.)

there are no significant structural differences between the L_1 and L_2 phases at a fixed composition.

5.   In the surfactant phase (S), which is found in the center of the ternary phase diagram, the self-diffusion of all three components is rapid showing that no closed aggregates exist above the nanosecond range. Shinoda[74] has suggested the structure of the surfactant phase to be bicontinuous (both oil- and watercontinuous). Since the surfactant phase is a low-viscous, optically isotropic phase characterized by narrow proton NMR spectra, any large aggregates must be very flexible and dynamic in nature.

These studies of nonionic surfactant microemulsions are yet only sketchy and it is of great interest to perform more systematic studies which can elucidate in some detail the complex structural characteristics of these systems; particularly appropriate to investigate are the microemulsions formed at very low surfactant concentrations.[74]

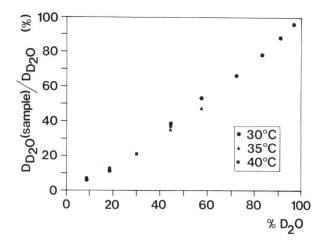

FIGURE 25.   Water self-diffusion coefficients relative to pure water for the system $C_{10}E_4$-$C_{16}H_{34}$-D_2O. The surfactant-to-hydrocarbon mass ratio was kept constant at 63:37. (From Nilsson, P. G. and Lindman, B., *J. Phys. Chem.*, 86, 27, 1982. With permission.)

V. CONCLUSIONS

It is demonstrated that the FT PGSE NMR technique permits a rapid and convenient characterization of the self-diffusion behavior of the different constituents of microemulsions. This is of value first since molecular motion over macroscopic distances is important in practical applications of microemulsions and second since it allows a direct characterization of microemulsion structure. From the discussion of the latter aspect, it is apparent that as a function of surfactant molecular structure, temperature, cosurfactant, composition, and added salt, both droplet-type structures of the O/W or W/O characters and bicontinuous structures are possible. Much remains to be done, however, in defining the conditions of appearance of different structures. A situation of much fundamental and practical interest is that of microemulsions formed at very low surfactant contents in both nonionic and ionic systems.[80]

REFERENCES

1. **Crank, J., Ed.,** *The Mathematics of Diffusion,* 2nd ed., Claredon Press, Oxford, 1975.
2. **Bird, R. B., Stewart, W. E., and Lightfoot, E. N., Eds.,** *Transport Phenomena,* Wiley-Interscience, New York, 1960.
3. **Cussler, E. L., Ed.,** *Diffusion; Mass Transfer in Fluid Systems,* Cambridge University Press, London, 1984.
4. **Hansen, J. P. and McDonald, I. R., Eds.,** *Theory of Simple Liquids,* Academic Press, London, 1976.
5. **Einstein, A., Ed.,** *Investigations in the Theory of Brownian Movement,* Dover, New York, 1956.
6. **Weingärtner, H.,** Self-diffusion in liquid water. A reassessment, *Z. Phys. Chem.,* 132, 129, 1982.
7. **Halle, B. and Karlström, G.,** Prototropic charge migration in water, *J. Chem. Soc. Faraday Trans. 2,* 79, 1031, 1047, 1983.
8. **Fitts, D. D., Ed.,** *Non-Equilibrium Thermodynamics,* McGraw-Hill, New York, 1962.
9. **Meares, P.,** Mass transfer and reactions at interfaces, *Faraday Discuss. Chem. Soc.,* 77, 7, 1984.
10. **Lindman, B., Kamenka, N., Puyal, M. C., Rymdén, R., and Stilbs, P.,** Micelle formation of anionic and cationic surfactants from Fourier transform hydrogen-1 and lithium-7 nuclear magnetic resonance and tracer self-diffusion studies, *J. Phys. Chem.,* 88, 5048, 1984.

11. **Stilbs, P.,** Solubilization equilibria determined through Fourier transform NMR self-diffusion measurements, *J. Colloid Interface Sci.,* 80, 608, 1981.

12. **Stilbs, P.,** Fourier transform NMR pulsed-gradient spin-echo (FT-PGSE) self-diffusion measurements of solubilization equilibria in SDS solutions, *J. Colloid Interface Sci.,* 87, 385, 1982; **Stilbs, P.,** A comparative study of micellar solubilization for combinations of surfactants and solubilizates using the Fourier transform pulsed-gradient spin-echo NMR multicomponent self-diffusion technique, *J. Colloid Interface Sci.,* 94, 463, 1983.

13. **Stilbs, P., Arvidson, G., and Lindblom, G.,** Vesicle membrane-water partition coefficients determined from Fourier transform pulsed-gradient spin-echo NMR based self-diffusion data. Application to anesthetic binding in tetracaine-phosphatidylcholine-water systems, *Chem. Phys. Lipids,* 35, 309, 1984.

14. **Stilbs, P.,** Micellar breakdown by short-chain alcohols, *J. Colloid Interface Sci.,* 89, 547, 1982.

15. **Bull, T. and Lindman, B.,** Amphiphile diffusion in cubic lyotropic mesophases, *Mol. Cryst. Liq. Cryst.,* 28, 155, 1974.

16. **Lindblom, G., Larsson, K., Johansson, L., Fontell, K., and Forsén, S.,** The cubic phase of mono-glyceride-water systems. Arguments for a structure based upon lamellar bilayer units, *J. Am. Chem. Soc.,* 101, 5465, 1979.

17. **Nilsson, P. G., Wennerström, H., and Lindman, B.,** Nonionic micelles. Size, shape, hydration and intermicellar interactions from self-diffusion studies, *Chem. Scr.,* 25, 67, 1985.

18. **Hahn, E. L.,** Spin echoes, *Phys. Rev.,* 80, 580, 1950.

19. **Freeman, R. and Hill, H. D. W.,** Determination of spin-spin relaxation times in high-resolution NMR, in *Dynamic Magnetic Resonance,* Jackman, L. M. and Cotton, F. A., Eds., Academic Press, New York, 1975.

20. **Stejskal, E. O. and Tanner, J. E.,** Spin diffusion measurements: spin echoes in the presence of a time-dependent field-gradient, *J. Chem. Phys.,* 42, 288, 1965.

21. **Fukushima, E. and Roeder, S. B. W., Eds.,** *Experimental Pulse NMR. A Nuts and Bolts Approach,* Addison-Wesley, Reading, Pa., 1981.

22. **Vold, R. L., Waugh, J. S., Klein, M. P., and Phelps, D. E.,** Measurement of spin relaxation in complex systems, *J. Chem. Phys.,* 48, 3831, 1968.

23. **James, T. L. and McDonald, G. G.,** Measurement of the self-diffusion coefficient of each component in a complex system using pulsed-gradient Fourier transform NMR, *J. Mag. Reson.,* 11, 58, 1973.

24. **Stilbs, P. and Moseley, M. E.,** Multicomponent self-diffusion measurements by the pulsed-gradient spin-echo method on a standard Fourier transform NMR spectrometer, *Chem. Scr.,* 15, 176, 1980.

25. **Stilbs, P. and Moseley, M. E.,** Carbon-13 pulsed-gradient spin-echo studies. A method for the elimination of J-modulation and proton exchange effects in self-diffusion measurements, *Chem. Scr.,* 15, 215, 1980.

26. **Callaghan, P. T., Trotter, C. M., and Jolley, K. W.,** A pulsed field gradient system for a Fourier transform spectrometer, *J. Mag. Reson.,* 37, 247, 1980.

27. **Mazei, F.,** Neutron spin-echo: ten years after, *Physica (Amsterdam),* 120, 51, 1983.

28. **de Lara, E. C., Kahn, R., and Mazei, F.,** Determination of the intercrystalline diffusion coefficient of methane in A zeolites by means of neutron spin-echo experiments, *J. Chem. Soc. Faraday Trans. 1,* 79, 1911, 1983.

29. **Anderson, J. S. and Saddington, K.,** The use of radioactive isotopes in the study of the diffusion of ions in solution, *J. Chem. Soc.,* S381, 1949.

30. **Stokes, R. H.,** An improved diaphragm-cell for diffusion studies, and some tests of the method, *J. Am. Chem. Soc.,* 72, 763, 1950.

31. **Poo, M. M. and Cone, R. A.,** Lateral diffusion of rhodopsin in the photoreceptor membrane, *Nature (London),* 247, 438, 1974.

32. **Ware, B. R., Cyr, D., Gorti, S., and Lanni, F.,** Electrophoretic and frictional properties in complex media measured by laser light scattering and fluorescence photobleaching recovery, in *Measurement of Suspended Particles by Quasi-Elastic Light Scattering,* Dahneke, B. E., Ed., Wiley-Interscience, New York, 1983, 255.

33. **Lanni, F. and Ware, B. R.,** Modulation detection of fluorescence photobleaching recovery, *Rev. Sci. Instrum.,* 53, 905, 1982.

34. **Rondelez, F.,** Forced Rayleigh scattering in fluids, in *Light Scattering in Liquids and Macromolecular Solutions,* Degiorgio, V., Corti, M., and Giglio, M., Eds., Plenum Press, New York, 1980, 243.

35. **Lalanne, J. R., Pouligny, B., and Sein, E.,** Transport properties of diluted inverted micelles and microemulsions, *J. Phys. Chem.,* 87, 696, 1983.

36. **Svensson, H. and Thompson, T. E.,** Translational diffusion methods in protein chemistry, in *A Laboratory Manual of Analytical Methods in Protein Chemistry,* Vol. 3, Alexander, P. and Block, R. J., Eds., Pergamon Press, New York, 1961.

37. **Sundelöf, L.-O. and Bonner, F. J.,** New arrangement of reference marks for interferometrically recorded sedimentation, diffusion and thermal diffusion experiments, *Chem. Scr.,* 10, 187, 1976.

38. **Nicholson, J. D. and Clarke, J. H. R.,** Photon correlation techniques in investigation of water-in-oil microemulsions, in *Surfactants in Solution,* Vol. 3, Mittal, K. L. and Lindman, B., Eds., Plenum Press, New York, 1984.

39. **Pecora, R.,** Quasi-elastic light scattering of macromolecules and particles in solution and suspension, in *Measurement of Suspended Particles by Quasi-Elastic Light Scattering,* Dahneke, B. E., Ed., Wiley-Interscience, New York, 1983, 3.

40. **Cazabat, A. M., Langevin, D., Meunier, J., and Pouchelon, A.,** Critical behaviour in microemulsions, *Adv. Colloid Interface Sci.,* 16, 175, 1982.

41. **Jönsson, B., Wennerström, H., Nilsson, P. G., and Linse, P.,** Self-diffusion of small molecules in colloidal systems, *Colloid Polym. Sci.,* 264, 77, 1986.

42. **Nilsson, P. G. and Lindman, B.,** Nuclear magnetic resonance self-diffusion and proton relaxation studies of nonionic surfactant solutions. Aggregate shape in isotropic solutions above the clouding temperature, *J. Phys. Chem.,* 88, 4764, 1984.

43. **Evans, G. T. and James, C. P.,** A calculation of the self-diffusion coefficient for a dilute solution of Brownian particles, *J. Chem. Phys.,* 79, 5553, 1983.

44. **Lekkerkerker, H. N. W. and Dhont, J. K. G.,** On the calculation of the self-diffusion coefficient of interacting Brownian particles, *J. Chem. Phys.,* 80, 5790, 1984.

45. **Van Megen, W. and Snook, I.,** Brownian-dynamics simulation of concentrated charge-stabilized dispersions. Self-diffusion, *J. Chem. Soc. Faraday Trans. 2,* 80, 383, 1984.

46. **Snook, I., van Megen, W., and Tough, R. J. A.,** Diffusion in concentrated hard sphere dispersions: effective two particle mobililty tensors, *J. Chem. Phys.,* 78, 5825, 1983.

47. **Nilsson, P. G., Wennerström, H., and Lindman, B.,** Structure of micellar solutions of nonionic surfactants. NMR self-diffusion and proton relaxation studies of poly(ethyleleoxide) alkylethers, *J. Phys. Chem.,* 87, 1377, 1983.

48. **Mackay, R. A.,** Properties of high emulsifier content O/W microemulsions, in *Microemulsions,* Robb, I. D., Ed., Plenum Press, New York, 1982, 207.

49. **Mackay, R. A., Dixit, N. S., Agarwal, R., and Seiders, R. P.,** Diffusion measurements in microemulsions, *J. Dispersion Sci. Technol.,* 4, 397, 1983.

50. **Eicke, H. F., Hilfiker, R., and Holz, M.,** Percolative phenomena in microemulsions of the ''one-component macrofluid'' type, *Helv. Chim. Acta,* 67, 361, 1984.

51. **Eicke, H. F. and Kubik, R.,** Concentrated dispersions of aqueous polyelectrolyte-like microphases in nonpolar hydrocarbons, *Faraday Discuss. Chem. Soc.,* 76, 305, 1983.

52. **Rouvière, J., Couret, J. M., Lindheimer, M., Dejardin, J. L., and Marrony, R.,** Structure des agrégats inversés d'AOT. I. Forme et taille des micelles, *J. Chim. Phys.,* 76, 289, 1979.

53. **Rouvière, J., Couret, J. M., Marrony, R., and Dejardin, J. L.,** Etude par viscosité et autodiffusion des agregats inversés dans les systémes di(2 éthylhexyl)sulfosuccinate de sodium-eau-hydrocarbures, *C.R. Acad. Sci. Paris,* 286, 5, 1978.

54. **Lindman, B., Stilbs, P., and Moseley, M. E.,** Fourier transform NMR self-diffusion and microemulsion structure, *J. Colloid Interface Sci.,* 83, 569, 1981.

55. **Stilbs, P. and Lindman, B.,** Aerosol OT aggregation in water and hydrocarbon solutions from NMR self-diffusion measurements, *J. Colloid Interface Sci.,* 99, 290, 1984.

56. **Stilbs, P. and Lindman, B.,** NMR measurements on microemulsions, *Prog. Colloid Polym. Sci.,* 69, 39, 1984.

57. **Angel, L. R., Evans, D. F., and Ninham, B. W.,** Three-component ionic microemulsions, *J. Phys. Chem.,* 87, 538, 1983.

58. **Ninham, B. W., Chen, S. J., and Evans, D. F.,** Role of oils and other factors in microemulsion design, *J. Phys. Chem.,* 88, 5855, 1984.

59. **Chen, S. J., Evans, D. F., and Ninham, B. W.,** Properties and structure of three-component ionic microemulsions, *J. Phys. Chem.,* 88, 1631, 1984.

60. **Blum, F. D., Pickup, S., Ninham, B., Chen, S. J., and Evans, D. F.,** Structure and dynamics of three-component microemulsions, *J. Phys. Chem.,* 89, 711, 1985.

61. **Fabre, H., Kamenka, N., and Lindman, B.,** Aggregation in three-component surfactant systems from self-diffusion studies. Reversed micelles, microemulsions and transitions to normal micelles, *J. Phys. Chem.,* 85, 3493, 1981.

62. **Lindman, B., Kamenka, N., Brun, B., and Nilsson, P. G.,** On the structure and dynamics of microemulsions. Self-diffusion studies, in *Microemulsions,* Robb, I. D., Ed., Plenum Press, New York, 1982, 115.

63. **Wärnheim, T., Sjöblom, E., Henriksson, U., and Stilbs, P.,** Phase diagrams and self-diffusion behaviour in ionic microemulsion systems containing different cosurfactants, *J. Phys. Chem.,* 88, 5420, 1984.

64. **Sjöblom, E. and Henriksson, U.,** The importance of the alcohol chain length and the nature of the hydrocarbon for the properties of ionic microemulsion systems, in *Surfactants in Solution,* Vol. 3, Mittal, K. L. and Lindman, B., Eds., Plenum Press, New York, 1984, 1867.

65. **Sjöblom, E., Henriksson, U., and Stilbs, P.,** Microemulsions stabilized by oleate/pentanol, in *Reverse Micelles: Biological and Technological Relevance of Amphiphilic Structures in Apolar Media,* Luisi, P. L. and Straub, B. E., Eds., Plenum Press, New York, 1984, 131.

66. **Lindman, B., Kamenka, N., Kathopoulis, T.-M., Brun, B., and Nilsson, P. G.,** Translational diffusion and solution structure of microemulsions, *J. Phys. Chem.,* 84, 2485, 1980.

67. **Stilbs, P., Moseley, M. E., and Lindman, B.,** Fourier transform NMR self-diffusion measurements on microemulsions, *J. Mag. Reson.,* 40, 401, 1980.

68. **Lindman, B. and Stilbs, P.,** Characterization of microemulsion structure using multi-component self-diffusion data, in *Surfactants in Solution,* Mittal, K. L. and Lindman, B., Eds., Plenum Press, New York, 1984, 1651.

69. **Stilbs, P., Rapacki, K., and Lindman, B.,** Effect of alcohol cosurfactant length on microemulsion structure, *J. Colloid Interface Sci.,* 95, 583, 1983.

70. **Lindman, B., Ahlnäs, T., Söderman, O., Walderhaug, H., Rapacki, K., and Stilbs, P.,** Fourier transform carbon-13 relaxation and self-diffusion studies of microemulsions, *Faraday Discuss. Chem. Soc.,* 76, 317, 1983.

71. **Lindman, B. and Stilbs, P.,** Diffusion in and the structure of microemulsions, in Proc. World Surfactant Congress, Munich, 1984, 159.

72. **Chatenay, D., Guéring, P., Urbach, W., Cazabat, A. M., Langevin, D., Meunier, J., Léger, L., and Lindman, B.,** Diffusion coefficients in microemulsions, in *Surfactants in Solution,* Bothorel, P. and Mittal, K. L., Eds., Plenum Press, New York, 1986.

73. **Guéring, P. and Lindman, B.,** Droplet and bicontinuous structures in microemulsions from multi-component self-diffusion measurements, *Langmuir,* 1, 464, 1985.

74. **Shinoda, K.,** Solution behaviour of surfactants: the importance of surfactant phase and the continuous change in HLB of surfactant, *Prog. Colloid Polym. Sci.,* 68, 1, 1983.

75. **Kunieda, H. and Shinoda, K.,** Phase behaviour in systems of nonionic surfactant/water/oil around the hydrophile-lipophile-balance-temperature (HLB-Temperature), *J. Dispersion Sci. Technol.,* 3, 233, 1982.

76. **Friberg, S., Lapczynska, I., and Gillberg, G.,** Microemulsions containing nonionic surfactants — the importance of the PIT value, *J. Colloid Interface Sci.,* 56, 19, 1976.

77. **Bostock, T. A., McDonald, M. P., Tiddy, G. J. T., and Waring, L.,** Phase diagrams and transport properties in nonionic surfactant/paraffin/water systems, in Surf. Act. Agents, Symp., Soc. Chem. Ind., London, 1979, 181.

78. **Nilsson, P. G. and Lindman, B.,** Solution structure of nonionic surfactant microemulsions from NMR self-diffusion studies, *J. Phys. Chem.,* 86, 271, 1982.

79. **Klose, G., Bayerl, T., Stilbs, P., Brückner, S., Zirwer, D., and Gast, K.,** Diffusion in three- and four-component microemulsions containing nonionic surfactant studied by pulsed field gradient NMR and QELS, *Colloid Polym. Sci.,* 263, 81, 1985.

80. **Shinoda, K., Kunieda, H., Arai, T., and Saijō, H.,** Principles of attaining very large solubilization (microemulsion): inclusive understanding of solubilization of oil and water in aqueous and hydrocarbon media, *J. Phys. Chem.,* 88, 5126, 1984.

81. **Nilsson, P. G. and Lindman, B.,** Water self-diffusion in non-ionic surfactant solutions — hydration and obstruction effects, *J. Phys. Chem.,* 87, 4756, 1983.

Chapter 6

DYNAMICS OF MICROEMULSIONS

Raoul Zana and Jacques Lang

TABLE OF CONTENTS

I. INTRODUCTION

Microemulsions can be defined as transparent or translucent, thermodynamically stable, liquid monophasic, and optically isotropic systems made of oil, water, and "surfactant"[1a] (the "surfactant" is often a mixture of surfactant and cosurfactant, generally a medium chain length alcohol). Microemulsions are currently the subject of many investigations because of their wide range of potential and actual utilizations.[1a,1b,2-5] For the most part, these utilizations rest on (1) the unique capability of microemulsions to unite in a single phase of low viscosity up to 45% water and 45% oil, the remaining 10% being "surfactant", and (2) their very low interfacial tension with excess oil or water.

Most of the investigations performed on microemulsions deal with such aspects as stability, phase diagrams, structure, and interactions.[1a,1b] There has been also a number of studies dealing with the dynamics of microemulsions and the interest in such studies has been increasing in recent years. Indeed, it is hoped that in addition to providing knowledge on the dynamics of these systems per se, such studies will also help in the understanding of widely differing topics concerning microemulsions such as:

1. Equilibrium properties of microemulsions (structure, interactions, critical-like behavior)
2. Solubilization of various compounds by microemulsions
3. Use of microemulsions as media for chemical reactions (clearly, the dynamics of microemulsions should have some influence on the rate of chemical reactions performed in the systems)

This review is restricted to those studies relating to the dynamics of microemulsions. Studies of the kinetics of chemical reactions performed in microemulsions will be reviewed only when they were performed in order to obtain information on the dynamics of microemulsions. The examination of the literature suggests that the studies of the dynamics of microemulsions can be broadly divided into three groups.

1. Dynamics of the internal motions in the molecules of the various components of the microemulsion, with particular emphasis on the surfactant and cosurfactant
2. Dynamics of the exchanges of some of the microemulsion components between "free" and "bound" states
3. Dynamics of process involving the whole microemulsion droplets of O/W or W/O encountered in some regions of the water/surfactant/cosurfactant/oil phase diagram

A variety of methods, both direct (NMR, ESR, chemical relaxation, fluorescence decay, pulse radiolysis) and indirect (rate of dissolution of oil and water in microemulsions, rate of chemical reactions in microemulsions, transport of probes by microemulsions), have been used for these studies. These techniques, however, will not be reviewed here. They will be simply mentioned, with the appropriate references.

As will be seen later in the chapter, the picture of microemulsion which emerges from the various results is one of very labile structures. This highly dynamic character of microemulsions may very well be at the origin of their thermodynamic stability. Also, some of these studies appear capable of yielding information on the structure of microemulsions (bicontinuous structure[8] vs. droplet structure[7]).

However, before starting to review the various aspects of the dynamics of microemulsions, it is appropriate in this introduction to briefly recall the main processes involved in the dynamics of simpler but closely related systems, namely pure micellar solutions and mixed alcohol + surfactant micellar solutions. The first two processes mentioned before are also involved in the description of the dynamics of these systems in addition to the process of micelle formation formation-dissolution.[9-14] Indeed, micelles (and mixed micelles) are in dynamic equilibrium with the surrounding solution which contain free surfactant (and free alcohol) and will thus constantly form and disappear (dissolution). Depending on the experimental conditions, the micelle formation-dissolution can occur stepwise with association/dissociation of *one* surfactant at a time to/from a micelle (see Process I),[9-12] and/or through progressive coagulation of submicellar aggregates into micelles and its reverse process, i.e., fragmentation of micelles into submicellar aggregates (see Process II).[13,14]

Process I: Stepwise micelle formation-dissolution

$$A + A \rightleftarrows A_2$$

$$A_2 + A \rightleftarrows A_3$$

$$A_{i-1} + A \rightleftarrows A_i$$

$$A_{N-1} + A \rightleftarrows A_N$$

A_i stands for an aggregate containing i surfactants. A_N stands for a full micelle with an aggregation number equal to that obtained using classical methods.

Process II: Coagulation-fragmentation

$$A_i + A_j \rightleftarrows A_{i+j}$$

These reactions may involve full micelles. Then $i + j = N$ and $j = N - i$.

II. INTERNAL MOTIONS IN THE MOLECULES OF THE MICROEMULSION COMPONENTS

Most studies have been performed by means of NMR techniques and involved measurements of line widths and relaxation times. Recall that changes of line widths yield information on the rate of exchange of nuclei between various states, whereas relaxation times are related to correlation times which in turn yield information on the relative mobilities of the various parts of the molecules under investigation.

A. Water-in-Oil (W/O) Microemulsions

These systems are considered first as they have been the topic of most of the investigations. Studies of systems at constant composition (the four-component water/potassium oleate/1-hexanol/benzene system,[15] and the three-component water/sodium di-2-ethylhexylsulfosuccinate (AOT)/benzene system[16]) indicate that the mobility in the surfactant chain increases progressively in going from the ionic head group to the terminal methyl group. However, the mobility of the 2-ethyl side chain in the AOT molecule was found to be very restricted. A restricted mobility of the head group of the nonionic surfactant Triton® X-100 with respect to that of its alkyl chain was also reported in a study of the four-component system: water/Triton® X-100/1-hexanol/cyclohexane.[17] Thus the increase of mobility in going from the surfactant head group to the carbon atoms of its alkyl chain, farther and farther away from the head group, appears to be independent of the nature of the surfactant and the presence or absence of the cosurfactant. Note that a similar observation was made for AOT aggregates in chloroform or benzene, in the absence of water.[18]

As the water content of four component W/O microemulsions was increased, line width measurements suggested a progressive decrease of the surfactant head group mobility. Hansen[15] attributed this effect to an increase of order in the interfacial film, with less freedom for internal motion for most of the surfactant molecule, including the head group, as the water content increased. A similar restriction of motion of the surfactant head group was reported[17] for microemulsions containing the nonionic surfactant Triton® X-100 on increasing the water content of the microemulsions. However, the local motion of the surfactant chain did not appear to be much affected by the build-up of the water core. This difference in behavior for an ionic and a nonionic surfactant may reflect the much stronger anchoring of the carboxylate head group of the oleate at the water core surface, as compared to that of the larger and more flexible polyethyleneoxide head group of Triton® X-100. In the oleate system the restriction of motion progressively disappears as one moves away from the head group. The terminal methyl group can reorient freely in the organic phase.[15]

Contrary to the previously mentioned two systems, the addition of water to water/AOT/ organic solvent microemulsions which do not contain alcohol was found to result in an increased mobility of the carbon atoms of the surfactant, particularly those close to the head group, in ^{13}C-NMR[16,19] and relaxation time T_1 of the protons of the AOT molecule[20] investigations. The largest increase of mobility took place at low water content, i.e., where the water molecules were used to hydrate the sodium counterions and the sulfonate head groups and where the water core was not yet formed.[16,19,20] (For AOT in organic solvents the transition from reverse micelles to true microemulsions was reported to occur at a molar ratio [H$_2$O]/[AOT] of about 10.[19,21] A similar value was found for this transition in the water/ potassium oleate/benzene system.[22])

This difference of behavior between AOT-containing microemulsions (no alcohol) and classical four-component microemulsions which contain alcohol is not easy to explain. Indeed, there is a complex relationship between internal motions in the surfactant and alcohol molecules, exchange of these molecules between the interfacial film separating the oil and water phases and these phases, the fluidity and dynamics of the interfacial film, and finally, the oil penetration in this film. Some aspects of these relationships are discussed in Sections III and IV.

The studies reviewed so far concerned the internal motions in the surfactant molecules. There is not much literature on the motions in the cosurfactant (alcohol) molecules. No broadening of the resonance lines of the protons adjacent to the hydroxyl group of 1-hexanol was observed,[15] contrary to the surfactant proton resonance lines, in the water/potassium oleate/1-hexanol/benzene system. This was attributed to the very fast exchange of the alcohol between various environments. ^{13}C-NMR studies may be able to provide information on the internal motions in the cosurfactant molecule but remain to be performed.

B. Oil-in-Water (O/W) Microemulsions

The investigations on these systems are not numerous. In a study of water/sodium dodecylsulfate (SDS)/1-butanol/toluene W/O and also of O/W microemulsions, Lalanne et al.[23] noted that the T_1 values for the methylene protons of SDS depend little on composition and concluded that the surfactant mobility is restricted, owing to the formation of the interfacial layer between oil and water and the anchoring of the ionic head group at this interface, whichever the nature of the microemulsions (O/W or W/O). A continuous evolution was observed when going from W/O to O/W microemulsions and possibly to lamellar structures. No conclusion was inferred about the mobility in the 1-butanol chain owing to its fast exchange between various environments.[23,24]

III. PROCESSES OF EXCHANGE OF THE MICROEMULSION
COMPONENTS BETWEEN VARIOUS ENVIRONMENTS

This section reviews the studies relevant to the dynamics of exchange of the microemulsion components between various environments. These exchanges all can be considered to occur between a "bound" state (the bound surfactant and cosurfactant molecules are those constituting the interfacial film between oil and water, the bound water molecules are those hydrating the counterions and the surfactant head groups, and the bound counterions being those trapped in the Stern layer of the interfacial film) and a "free" state (alcohol and surfactant present in the oil and/or the water phases, water or oil molecules making up the water or oil core, and counterions in the dispersed or continuous water phase).

The results relative to water, counterion, surfactant, and alcohol will be examined successively. In each case the relationship with results pertaining to simpler systems (aqueous solutions of electrolytes, micellar solutions of pure surfactant or surfactant + alcohol) will be pointed out.

A. Water

The state of water in W/O microemulsions has been the subject of many investigations involving both three-component alcohol-free microemulsions and four-component microemulsions. All these studies conclusively showed that at low water content (for a molar concentration ratio W = [water]/[surfactant] below, say, 6 to 10) all the water is bound to the counterions and to the surfactant head groups.[20,25-29] At W > 6 to 10, free water is detected.[19,21,22] This corresponds to the onset of formation of the water core.

The dynamic of the exchange of water between the bound state (hydration water) and the free state was investigated for various microemulsions by measuring the relaxation times T_1 and T_2 of the hydroxylic protons. Of course when alcohol is present in the system, these times involve a contribution of the alcohol OH group. However, at low water content, when the contribution of the alcohol is significant, as well as at high water content, when the contribution of the water is largely predominant and determines the relaxation time, a single resonance line was found for the OH group.[15,17,24,30,31] This result indicates a rapid exchange on the NMR time scale between the OH protons of water and alcohol. On the other hand, as single exponential relaxations were obtained in T_1 measurements[15] and since the OH line is always narrow,[15,17,20] it was concluded that the exchange between "bound" and "free" water is rapid on the NMR time scale (i.e., exchange time shorter than 10^{-4} sec[17,20]). In agreement with this conclusion, Lindman and Wennerstrom[32] cite unpublished NMR results indicating that water molecules have on the micelle surface a residence time in the nanosecond range. Since counterions and surfactant head groups are similar in micellar systems and microemulsions, no drastic difference should be expected between these two types of systems as far as water exchange is concerned. Also, going one step lower in the complexity of the system, it is worth recalling that in aqueous solution the average time spent by a water molecule in the hydration shell of such counterions as Na^+ or K^+, which are the most frequently found in microemulsions, is about 10^{-9} sec,[33] that is, the value found in micellar systems.

A dynamic study which did not deal with the exchange of water between bound and free states but which dealt with the state of water was recently published. In this study,[34] the effect of water on the ultrasonic absorption of AOT solutions in decane revealed the existence of a relaxation process associated with an equilibrium between a water-separated ion pair, and a contact ion pair at low water content (W < 4) when the Na^+ hydration shell is not complete. The characteristic time of this process was about 10^{-7} to 10^{-8} sec, that is, larger than in bulk water, as to be expected because the Na^+-H_2O interaction is much stronger when the Na^+ hydration shell is not complete.

B. Counterions

The width of the ^{23}Na NMR absorption line was found to decrease dramatically upon addition of water to AOT solutions in heptane,[20] or to sodium octanoate solution in 1-hexanol.[35] This decrease was attributed to the hydration of Na$^+$ by the added water, until completion of the counterion hydration shell at around $[H_2O]/[Na^+] = 6$. At higher water content, a water core builds up and some counterions dissociate from the interfacial layer and move into the water core, where they are in dynamic equilibrium with the counterions still in the interfacial layer.[20] A study of the temperature dependence of the ^{23}Na line width[20] indicated a fast exchange between free and bound Na$^+$ counterions on the NMR time scale (characteristic time $<10^{-4}$ sec). At this point, it is noteworthy recalling two results reported for micellar solutions. First, unpublished results quoted in a recent review[32] indicated an exchange time for alkali metal counterions in the nanosecond range. On the other hand, in an electrical field jump study of aqueous micellar solutions of cetylpyridinium iodide, Grünhagen[36] concluded that the dissociation rate constant of I$^-$ from the micelle is about 10^7/sec and, thus, the residence time of I$^-$ in the micelle Stern layer is of about 10^{-7} sec.

C. Cosurfactants

All NMR studies of alcohol-containing O/W and W/O microemulsions concluded with a fast exchange of the alcohol between the interfacial film, the continuous phase, and/or the dispersed phase (depending on whether the alcohol is soluble in these two phases, or just in one of them) on the NMR time scale, thereby indicating a characteristic time well below 10^{-4} sec for this exchange.[15,17,20,23,24,31,37] In the case of a W/O microemulsion made of water/SDS/butanol/toluene, the addition of NaCl to the dispersed water was found to result in a slowdown of the alcohol exchange,[31] perhaps because the interfacial film was then more compact.

Recall that the exchange rate of alcohol between mixed alcohol + surfactant aqueous micelles and the intermicellar solution was measured in pulse-radiolysis[38] and ultrasonic relaxation[39,40] studies, and was found to be extremely large (10^9/sec for 1-pentanol). The effect of addition of oil to these systems, in order to generate O/W microemulsions and, at higher oil content, W/O microemulsions, was found to affect only slightly exchange of the alcohol, which remained extremely rapid.[41,42] The relaxation time for the alcohol exchange in a variety of microemulsions, both O/W and W/O, was found to be of the order of 10^{-8} sec and below, in agreement with the exchange rate constants quoted previously for mixed micellar systems. On the basis of the results obtained for these systems,[39,40,43] one would expect this relaxation time to decrease upon increasing alcohol chain length. However, there has been no report of the dependence of the alcohol exchange rate on its chain length in four-component microemulsions probably because the experiments are difficult both to perform and to interpret.

In the case of microemulsions where the alcohol/surfactant weight ratio was kept constant, the relaxation time for the cosurfactant exchange showed a continuous change upon increasing oil content including the range where the O/W microemulsion turned into a W/O microemulsion.[42] On the contrary, for microemulsions where the water/surfactant weight ratio was kept constant, the relaxation time went through a broad maximum at the water volume fraction corresponding to the W/O-O/W phase inversion.[42] This difference of behavior has not been explained yet. It may be related to the fact that in the later system, the droplet size changes only slightly with the water content in the W/O microemulsion range, whereas large changes of droplet size are expected to occur in the former system. These changes may modify the variation of the alcohol exchange relaxation time with composition.

Here, we review some studies dealing with the dynamics of exchange of neutral spin probes solubilized in microemulsions. Such studies are indeed very similar to those dealing with the alcohol exchange. Thus, the exchange frequency of a neutral spin probe between benzene and the water core in the W/O microemulsion water/dodecylammonium propionate/benzene was reported[44] to be greater than 6×10^6/sec. In another study,[45,46] the exchange rate of a neutral probe between the interfacial film and the water core in water/AOT/heptane microemulsion was found to be 10^7/sec at W = $[H_2O]/[AOT]$ = 31, with an activation energy of 7.5 kcal/mol. The exchange of charged spin probes has also been investigated (Reference 44 and references therein), but will not be considered here as it is very different from that of the alcohol.

D. Surfactants

The ultrasonic relaxation method was used to detect the exchange of the surfactant between the interfacial film and the water phase in O/W microemulsion (for instance, the water/SDS/1-butanol/toluene system with a weight ratio butanol/surfactant = 2).[41,42] The relaxation time for the surfactant exchange was found to be around 3×10^{-8} sec and to change only slightly when the amount of toluene in the system was increased.[42] Note that this value is smaller than that found for pure SDS in aqueous solutions[47] simply because the surfactant concentration was much larger in the investigated microemulsions, and also because the presence of the alcohol somewhat accelerates the surfactant exchange.[40] However, the rate constants for the association-dissociation of the surfactant to/from the interfacial film are probably not very different from their values in pure or mixed micellar solutions. Recall that extensive studies, mostly by means of chemical relaxation[12,48] but also by pulse radiolysis,[38] have been performed on these systems. They showed that the rate constant k^- for the dissociation of one surfactant from its micelle depends very heavily on its chain length[12,48] (k^- decreases by a factor of 3 per additional CH_2 group) and little on the surfactant head group and the presence of alcohol.[40] The rate constant k^+ for the association of a surfactant to its micelle is in the range 5×10^8 to 5×10^9 M sec, that is, only a little bit below the value for a diffusion-controlled process. k^+ decreases slightly upon increasing chain length,[12,48] and is only marginally affected by the presence of alcohol.[40]

Unambiguous evidence for the surfactant exchange process in W/O microemulsions has not been reported thus far. Ultrasonic relaxation studies[41,42] of these systems did show the existence of a relaxation process with a relaxation time longer than 3×10^{-8} sec in addition to that associated with the alcohol exchange. The comparison to the results relative to O/W systems as well as the continuity of the changes in going from O/W to W/O systems suggested that this process may indeed correspond to the surfactant exchange.[41,42] Another possibility, however, was that the observed process was due to the self-association of alcohol molecules in the oil phase through H-bonding. Indeed such processes[49] were shown to give rise to an ultrasonic relaxation in approximately the same time scale (10^{-8} to 10^{-9} sec).

In some NMR investigations, it was concluded that the surfactant is restricted to the interface.[15,23] At first sight this conclusion appears to contradict the ultrasonic studies which reveal a fast exchange of the surfactant, at least in O/W microemulsions. It must be recalled, however, that NMR allows the detection of an exchange process between two states only when both states are sufficiently populated. The ultrasonic relaxation method is much more sensitive. The exchange process is detected even when the population in the least populated state is well below that required by NMR.

IV. DYNAMICS OF THE INTERFACIAL FILM SEPARATING OIL AND WATER

This section has been placed after one dealing with exchange processes between the film and the bulk and/or dispersed phases, and before that dealing with whole droplets. Indeed, the dynamic character of the interface results partly from the exchanges between this interface and the adjacent phases on the one hand, and determines the dynamics of the processes involving whole droplets on the other.

The studies reviewed in this section do not concern processes such as those reviewed in Sections II, III, and V. They deal with the fluidity and flexibility of the interfacial film which are more difficult to quantify but nevertheless very important in understanding the behavior of microemulsions. Indeed, these properties reflect the ease with which the film rearranges upon a constraint, and thus its dynamics. They also reflect the spontaneous fluctuations of the film, which are just starting to be studied quantitatively. It must be noted that the lateral diffusion of the molecules making up the interfacial film has been investigated (see, for instance, Reference 52). However, such studies are not directly related to those dealing with the flexibility or fluidity of the film.

The fluidity of the interfacial film in microemulsions was inferred from measurements of the spin-lattice relaxation time T_1 of the alcohol in the water/potassium oleate/1-pentanol or 1-hexanol/cyclohexane[50] and water/SDS/1-butanol/toluene[23,30] systems. The measurement of the order parameter of the interfacial film using spin labels also led to the conclusion that under some conditions the interfacial film may be very fluid.[51,52] More important were the results which showed that, for a given system, the fluidity was largely increased when the alcohol chain length was decreased.[50,51] This increase of fluidity was accompanied by drastic changes in many of the properties of the system such as electrical conductivity, dielectric constant, and intensity of the scattered light. In some cases the behavior of the system was totally altered by a change in the alcohol chain length, going from that of a system of interacting hard-spheres for a long chain alcohol, to that of a system close to a critical point for a shorter chain alcohol. Such a transition in behavior was obtained by decreasing the alcohol chain length by only one methylene group for a given surfactant and oil. The effect of the alcohol chain length was interpreted[51,54] in terms of the disorder introduced by the alcohol in the interfacial film, owing to the difference of chain length between the alcohol and the surfactant. It can also be understood in terms of the exchange of the alcohol between the film and the oil and water phase.[24,55] These exchanges make the film porous[24] and thus increase its fluidity.

No distinctive times were given to characterize the fluidity of the interface. More recently, however, the fluctuations of the interface were evidenced in the so-called "birefringent microemulsions" by means of spin labels. Even though such systems are not true microemulsions if one strictly adheres to the definition of such systems given in Section I,[1] it is interesting to note that the characteristic time for such fluctuations, which correspond to local reorganizations of the film, appeared to be longer than 0.1 μsec.[56] In a transient electrical birefringence study of various W/O microemulsions, a negative birefringence component observed at volume fraction of the dispersed phase larger than 0.1 was attributed to such film fluctuations and found to have a characteristic time of about 1 μsec,[57] in agreement with the conclusion of the spin label study.

Finally, in recent works, the flexibility of the interfacial film was related to the structure adopted by the microemulsion system (droplet or bicontinuous structure).[58,59] It was also shown that the role of the alcohol is to increase the flexibility of the film.[60]

V. DYNAMICS OF PROCESSES INVOLVING WHOLE MICROEMULSION DROPLETS

It is generally agreed that in the water-rich region as well as in the oil-rich region of the water/surfactant/alcohol/oil phase diagrams, the O/W and W/O microemulsions contain small droplets of oil and water, respectively, coated by a mixed film of surfactant and alcohol. The hydrophobic side of this film is penetrated by the oil up to about the terminal methyl groups of the alcohol.

As seen before, the surfactant and alcohol are rapidly exchanging between the interfacial film and the oil and/or water phase. This means that both surfactant and alcohol can be exchanged between droplets via the solution surrounding the droplets, i.e., essentially water in the case of O/W microemulsions and essentially oil in the case of W/O microemulsions. However, if these processes were the only ones by which matter is exchanged between droplets, one would except no interdroplet transfer or only an extremely slow transfer in W/O microemulsion. Indeed, usual surfactant ions (such as dodecylsulfate ion), counterions (such as Na^+), and water are all insoluble in oil. They are therefore expected to dissociate very slowly from the interfacial film or the core (the case of AOT is peculiar as it is soluble in oil and is thus expected to exchange very rapidly between droplets, perhaps carrying some water molecules in this process). Likewise, oil should be very slowly transferred from droplet to droplet in O/W microemulsions as it is usually nearly insoluble in water. Contrary to these expectations, evidence has recently accumulated which shows that the components of some microemulsions can be redistributed between droplets fairly rapidly and even extremely rapidly by two types of processes:

1. Droplet collisions accompanied by temporary merging of the droplets into a larger droplet (fusion) followed by breakdown of this larger droplet (fission)
2. Partial breakdown (or fragmentation) of droplets with loss of droplet fragments, which can later associate with other droplets (coagulation)

These two processes, which are schematically represented on Figure 1, are probably the most important ones for understanding the properties of microemulsions and will be considered in detail in this review.

A. W/O Microemulsions

For these systems, only the process of exchange through droplet collisions and merging was discussed in the literature. Note, however, that in a stopped flow study of W/O microemulsion it was concluded[61] that the droplets are very stable and do not form or dissolve as micelles in aqueous solution.

Most of the earlier studies were performed on water/AOT/organic solvent systems. The main two reasons for the use of AOT were that AOT can form microemulsions in the absence of alcohol thus leading to simpler systems from the physicochemical point of view, and that the existence of water droplets in these systems had been well demonstrated and the droplets well characterized.

The first evidence for the rapid exchange of the droplet content was reported[62] in 1973. It was observed that upon mixing two identical water/AOT/octane W/O microemulsions, one containing imidazole and the other containing p-nitrophenyl, p-guanidinobenzoate-HCl, both solubilized in the droplets, the ester was hydrolyzed. The authors concluded that either the droplets merged partially or totally and exchanged their contents during the lifetime of the "transient droplet dimer", which later broke down into two droplets, or, more unlikely, that the reagents had been exchanged through the octane phase. This ambiguity was eliminated in a similar study of the complexation of Tb^{3+} by hydrophenylacetic acid (HPA) which is accompanied by a large fluorescence enhancement.[63] This enhancement occurred as fast as the mixing of two identical water/AOT/isooctane microemulsions, one containing water-solubilized Tb^{3+} and the other water-solubilized HPA, could be performed. However, the Tb^{3+} fluorescence was not modified when the two microemulsions were separated by a dialysis membrane, with a mean pore diameter smaller than the droplet diameter. This observation clearly showed that the complexation reaction involved droplet collision and temporary merging.[63] It also suggested that the droplets in the investigated system have a very long lifetime, in contradistinction to micelles which have been shown to form and dissolve rapidly in aqueous solutions, and probably also in organic solutions (for both types of systems the word micelle refers to aggregates of the pure surfactant).[12-14]

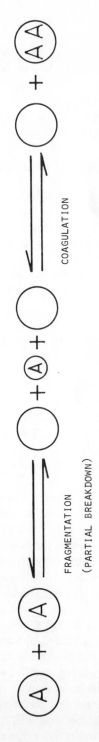

FIGURE 1.   Processes of exchange of matter (other than via the continuous phase) in microemulsions. A refers to a molecule or ion solubilized in the droplet core which monitors the exchanges of matter between droplets. A may be a chemical reagent or a luminescence probe. Notice that, whereas process I involves collisions between full micelles or droplets (that is, aggregates having a size equal to that determined by means of methods sensitive to particle size), process II involves collisions between a full aggregate and a subaggregate.

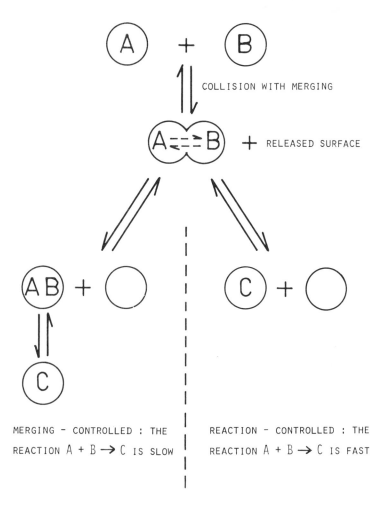

FIGURE 2.   Limiting situations for the kinetics of chemical reactions in microemulsion droplets, according to Robinson et al.[64]

The rapid exchange of the droplet contents was demonstrated in three additional studies. In the first study, fluorescence probes were used to study molecular processes in AOT microemulsions.[16] This investigation confirmed the dynamic character of these microemulsions and the occurrence of rapid merging and breakdown of the droplets.[16] In the second investigation, Robinson et al.[64] studied the rate of the complex formation between the water soluble reagents A ≡ Ni^{2+}, B ≡ murexide to give C ≡ Ni-murexide complex in water/AOT/heptane microemulsions by means of stopped flow. Recall that this technique permits a mixing of reagents in a time of 1 to 2 msec. The authors developed the rate equations for the limiting cases where the rate of reaction is limited either by the transfer of reagents between collided droplets, or by the rate of reaction in the droplets (see Figure 2). In the case of the slow complexation of Ni^{2+} by murexide, all the results were found to be consistent with the second situation. This study did not permit the rate constant for the droplet exchange to be obtained, but led to the conclusion that the rate of droplet collision with transfer of reagent must be large, at least of the order of $10^6/M$sec.[64] Finally, the third study was concerned with the rate of dissolution of water in water/SDS/1-pentanol/dodecane W/O microemulsions.[65] In these experiments, a small volume of water was rapidly dispersed in a microemulsion and the rate of dissolution measured by monitoring the change of intensity of the light transmitted through the water-microemulsion mixture. It was observed that the turbidity vanished as fast as the mixing was performed (~2 msec). The implication of this observation as regards the rate of splitting of the injected water droplets is interesting. Indeed, these droplets were initially quite large (several microns in diameter) and were split into smaller droplets as they became progressively coated with surfactant and cosurfactant through collisions with microemulsion droplets. A large number of fissions, probably above 10^3, therefore had to occur in a time equivalent to the mixing time. This set the time for a single fission to about 10^{-6} sec, and the rate constant for collision with merging to about 10^8 to $10^9/M$sec in view of the droplet concentration in these experiments.[65]

The aforementioned studies did not permit a quantitative determination of the rate constant for collision with merging. Nevertheless, Eicke et al.[63] developed a model which assumed the opening of a water channel at the point of collision of two droplets (transient dimer). Theoretical calculations showed that an exchange of water-solubilized additive could indeed take place via diffusion through the channel during the collision time, taken as the period of the slowest vibration (5×10^{-11} to 5×10^{-10} sec, depending on the droplet radius) of a droplet upon collision with another droplet.

Quantitative determination of the rate constant for collision with merging, k_e, was reported only recently. In all these studies, the droplet merging was detected through the reaction of some probes dissolved in the water core: true chemical reactions and fluorescence or phosphorescence quenching. Thus, Robinson et al.[66,67] investigated the fast complexation of Zn^{2+} by murexide, the protonation of 4-nitrophenol-2-sulfonate, and the electron transfer between $IrCl_6^{2+}$ and Fe^{2+} or $Fe(CN)_6^{4-}$ in the droplets present in the water/AOT/heptane microemulsion. In all instances, the compartmentalization of the reagents in the droplets was found to slow down the reaction indicating a coupling between the chemical reaction and the rate of droplet merging. The kinetic results were analyzed to yield the overall bimolecular exchange rate constant k_e for the reaction

$$Ⓐ + Ⓑ \xrightarrow{k_e} Ⓒ + ◯$$

This rate constant is a complex one as an exchange reaction involves a collision between two droplets, the opening of a water channel (merging), the diffusion of and the reaction between reagents, and the dimer breakdown. The second step is probably rate limiting. Nevertheless, k_e was found to be of the order of 10^7/sec at 25°.[66,67] Thus, about one collision per thousand results in a transient merging of the collided droplets, with an exchange of matter between them. This process was found to have a large entropy of activation, attributed to the fact that upon merging (activated state), some AOT is released from the interfacial film. The authors also concluded that the lifetime of the transient dimer could not be longer than microseconds. Longer lifetimes would lead to phase separation because other droplets would merge with the transient dimer during its lifetime. They also pointed out that the uneven splitting of the transient dimer is likely to show as a polydispersity of droplet size.[67]

Atik and Thomas[68,69] investigated the same system by measuring the decay of the luminescence of water-soluble $Ru(bipy)_3^{2+}$ in the presence of various water-soluble quenchers. The lifetime of $Ru(bipy)_3^{2+}$ was found to be decreased upon addition of the quencher. This effect indicated unambiguously a redistribution of the quencher on the time scale of the fluorescence lifetime of $Ru(bipy)_3^{2+}$ that is about 0.5 μsec. The analysis of the data in terms of droplet collisions with temporary merging yielded nearly the same value of k_e as in the stopped-flow study.[66,67] Furthermore, Atik and Thomas showed the extreme sensitivity of k_e to the presence of various additives. In particular, benzylalcohol was found to increase k_e by a factor of nearly 20, thereby emphasizing the role of the cosurfactant on the dynamic behavior of microemulsions. This was also demonstrated in another fluorescence probing study of the four-component W/O microemulsion water/potassium oleate/alcohol/hexadecane.[70] Atik and Thomas[70] showed that k_e is extremely sensitive to the alcohol chain length, being much larger for 1-pentanol than 1-hexanol containing microemulsions. The strongly differing conductivities of the two systems were interpreted in terms of differences in k_e.

More recently, time-resolved luminescence quenching was extensively used to study a number of W/O microemulsions.[71-74] The results are listed in Table 1. In the next paragraph, these results are discussed in relation to the properties of microemulsions. At this stage, only one comment is made on the very large values of k_e, up to $1.5 \times 10^9/M$sec, found for some systems.[71,74,75] Such large values are only slightly lower than the rate of collisions (about $5 \times 10^9/M$sec at 25°C). This indicates that the exchange between droplets is, in these instances, no longer limited by the rate of opening of the interfacial film, but rather by the diffusion of the droplets.

The last evidence for fast changes of droplet size, as involved in fusion-fission processes, was recently obtained in a study of water/AOT/isooctane systems by means of time-resolved microwave conductivity.[76] Ionic species formed during radiolysis of the system brought about an adjustment of the droplet size on a time scale of micro- to milliseconds.[76]

B. O/W Microemulsions

It has been possible to follow the changes of the relaxation time for the aggregate formation-dissolution when generating an O/W microemulsion starting from a concentrated micellar solution by successive additions of alcohol and then of oil. The addition of the medium chain length alcohols used for microemulsion formulation (propanol to hexanol) was found to result in an extreme labilization of the micelles.[40,77-79] This effect was attributed to the contribution of coagulation/fragmentation processes (see Figure 1) to the mixed micelle formation-dissolution. The fusion-fission process was not invoked to explain these results because the strong electrostatic repulsions were thought to prevent droplet collisions. Contrary to alcohols, the addition of oil resulted in exactly the opposite effect, i.e., a stabilization of the oil droplets, with a considerable lengthening of the relaxation time associated with the droplet formation/dissolution (see Figure 3) as if the coagulation/fragmentation no longer contributed to the droplet formation-dissolution.[78]

The absence of fusion-fission processes was also inferred in a stopped-flow study of the rate of dissolution of oil by O/W microemulsions. Indeed, it was observed[65] that the dissolution of oil by fairly dilute O/W microemulsions is very slow. This was attributed to the rapid adsorption of some of the surfactant ions present in the aqueous phase by the large injected oil droplets which thus take an electrical charge of the same sign as that of the microemulsion droplets initially present in the system. Electrostatic repulsions then greatly reduce the rate of collisions between injected oil droplets and microemulsion droplets, and thus the rate of redistribution of surfactant and alcohol between microemulsion droplets and injected oil droplets, thereby resulting in a slow rate of disappearance of the latter.[65]

Time-resolved fluorescence of pyrene was also used to probe coagulation-fragmentation processes in O/W microemulsion.[80] Thus a rapid intermicellar exchange of micelle-solubilized pyrene was observed in mixed alcohol + surfactant micellar solutions. Owing to the strong electrostatic repulsions between droplets, this exchange was assumed to occur through coagulation-fragmentation (see Figure 1) with a fragmentation rate constant k^- in the 10^5 to 10^6/sec range. Upon increasing oil additions, the interdroplet exchange of pyrene was progressively slowed down, and vanished at high oil content (about 10%). This result suggested, in agreement with chemical relaxation-studies,[78] that coagulation-fragmentation processes contributed less and less to the dynamics of the O/W microemulsion, as the oil core progressively built up.[80]

C. Miscellaneous Studies

In this paragraph are examined results which concern microemulsions showing a critical-

Table 1
W/O MICROEMULSIONS WITH EXCHANGE PROCESS

Microemulsions	Methods	$10^{-7} \times k_e$ (/Msec)	Ref.
Water/AOT/n-heptane w = 15; 25°	Rate of chemical reactions (proton transfer and metal ligand substitution)	1.6	66
w = 11; 25°	Quenching of triplet emission of pyrenetetrasulfonate ion by Fremy's salt	1.7	69
w = 33; 30°	Quenching of Ru(bipy)$_3^{2+}$ fluorescence by Fe(CN)$_6^{3-}$	140	71
Water/AOT/isooctane w = 15	Quenching of triplet emission of Mg-tetraphenylporphyrin by methyl viologen (MV^{2+})	19	72
	Quenching of hydrated electron by NaNO$_3$	No exchange on the time scale of the experiments ($k_e <$ 10^8/Msec)	73
Water/SDS/1-pentanol/toluene (volume fraction 0.04—0.12)	Quenching of Ru(bipy)$_3^{2+}$ fluorescence by Fe(CN)$_6^{3-}$	25	71
Water/CTAB/1-hexanol/dodecane w = 20	Quenching of Ru(bipy)$_3^{2+}$ fluorescence by Fe(CN)$_6^{3-}$	14	71
	Quenching of Ru(bipy)$_3^{2+}$ fluorescence by MV^{2+}	12	71
Water/potassium/oleate/1-hexanol/hexadecane w = 16	Quenching of Ru(bipy)$_3^{2+}$ fluorescence by Fe(CN)$_6^{3-}$	22	70
Water/CTAB/1-hexanol/hexane w = 20	Quenching of Ru(bipy)$_3^{2+}$ fluorescence by Fe(CN)$_6^{3-}$	1.1	71
Water/potassium oleate/1-pentanol/hexadecane	Quenching of Ru(bipy)$_3^{2+}$ by Fe(CN)$_6^{3-}$	10	70
Water/dodecylammonium propionate/c-hexane w = 2.75	Quenching of 1-methylnaphthalene fluorescence by KI	130	74

Abbreviations: AOT, sodium diethylhexylsulfosuccinate; MV^{2+}, methylviologen; Ru(bipy)$_3^{2+}$, ruthenium (II) tribipyridine; w = water/surfactant in mol/mol.

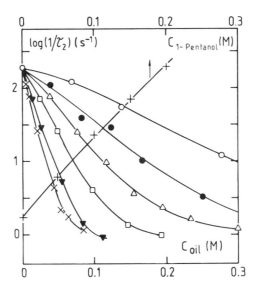

FIGURE 3.   Effect of oil addition on the relaxation time τ_2 of the micelle formation-breakdown equilibrium in the mixed micellar solution of 0.3 M hexadecylpyridinium chloride + 0.2 M 1-pentanol:(○) cyclohexane; (●) n-hexane; (△) n-heptane; (□) n-octane; (▼) n-decane; and (X) n-dodecane at 25°. The effect of 1-pentanol additions on the values of τ_2 for the 0.3 M solution of hexadecylpyridinium chloride is also represented (symbol (+); upper scale).

like behavior of several of their properties, and systems where the structure may not be of the droplet type (microemulsions with high and nearly equal volume fractions of both water and oil, middle phase microemulsions).

1. Microemulsions Showing a Critical-Like Behavior

Ultrasonic absorption has been used to investigate a number of microemulsion systems, both O/W and W/O.[41,42] These studies failed to reveal the existence of fast composition fluctuations similar to those existing in binary systems close to the critical conditions.[41,42]

Two W/O microemulsions, the ATB and ATP systems, made of water/SDS/toluene/1-butanol (ATB) or 1-pentanol (ATP), respectively, were studied in detail by means of various techniques[53] and their dynamics examined by means of ultrasonic relaxation.[81] The ATB system which showed a critical-like behavior of several of its properties[53] and a percolation of conductivity upon increasing water volume fraction also showed, contrary to the ATP system, a peculiar change of its ultrasonic absorption.[81] A similar peculiar behavior of the ultrasonic absorption was observed[82] when the critical-like behavior and the percolation of conductivity were induced by increasing the salinity of the system. The changes of ultrasonic absorption were assigned to the modification of the exchange processes of the surfactant and alcohol between the interfacial film and the oil and/or water phases (see Section III) upon collision and merging of the droplets in the ATB systems, due to the release of some surfactant and alcohol from the interfacial film. The same interpretation had been previously given[66,67] to the large entropy of activation of the droplet collision-merging process.

No peculiarity in the ultrasonic absorption was observed with (H_2O + NaBr)/cetyltrimethylammonium bromide (CTAB)/1-butanol/octane O/W microemulsions, which showed a critical-like behavior in quasielastic light scattering (QELS) studies.[81] Again, this is in agreement with the fact that in O/W microemulsions, the electrostatic repulsions between droplets prevent collisions and thus merging of droplets. The droplet concentration can fluctuate, i.e., the droplets can cluster, but they retain their individuality in these clusters, with little if any modification of the exchange processes and thus of the ultrasonic absorption.[81]

2. Middle Phase Microemulsions and Microemulsions with Comparable Volume Fractions of Oil and Water

The ultrasonic relaxation spectrum of a middle phase microemulsion was found to contain the features of the spectra of both the lower and upper phase microemulsions of the same system (brine/SDS/1-butanol/toluene).[82] This result suggested a bicontinuous structure of the middle phase microemulsion.

The results of dynamic studies of some microemulsions containing comparable volumes of oil and water led to the same conclusion. Thus both oil and water were found to dissolve as fast as their mixing with such microemulsions could be performed in the stopped-flow experiments.[65] Finally, the rates of exchange of surfactant and alcohol as well as their fairly small change with the oil or water volume fraction are consistent with this assumption.[42]

D. Relationship Between the Properties and the Dynamic Behavior of Microemulsions

In this section, an attempt is made to interpret/predict some peculiar properties of microemulsions in terms/on the basis of their dynamic behavior.

1. Polydispersity of Microemulsion Droplets

It is clear that both droplet merging/breakdown as well as coagulation/fragmentation processes should result in polydispersity in droplet size as they involve systems of droplets of different sizes in dynamic equilibrium.[67] A contrario, the disappearance of these processes by a modification of the system should result in a decrease of the polydispersity. In agreement with this last prediction we refer to a QELS investigation of the effect of alkane addition on the polydispersity of the aggregates present in the O/W H_2O-0.1 M KBr/CTAB/1-pentanol/ alkane systems.[83] QELS showed that the initial mixed CTAB/pentanol micelles were characterized by a large variance, and thus polydispersity. The addition of alkane resulted in a decrease of variance and thus of polydispersity. At a given alkane volume fraction, this decrease was the largest for the longer chain alkane.[83] Thus there is a very good correlation between the QELS results and the chemical relaxation results of Figure 3 which show a progressive decrease of the contribution of the coagulation/fragmentation processes.

2. Self-Diffusion of the Microemulsion Components

Extensive measurements of self-diffusion coefficients of the microemulsion components (oil, water, counterion, surfactant ion, and alcohol) were performed in the whole range of composition of the phase diagram of various microemulsions.[84-88] It was observed that in a large range of composition, the self-diffusion coefficients of the various species, except the surfactant, were very large. Likewise, Stilbs et al.[89] investigated the effect of alcohol chain length for a given microemulsion at a given composition on the self-diffusion of the water and oil. Below a certain alcohol chain length, the self-diffusion coefficients of both the oil and water were large and comparable. Both observations were interpreted as indicating that the systems were then either bicontinuous or cosolubilized.[84-89] The dynamic results reviewed in Section V.C.2 do not disagree with this interpretation. However, one cannot discard the possibility that very fast interdroplet exchanges by merging/breakdown or coagulation/fragmentation processes are responsible for a part of the changes of the self-diffusion coefficients observed in some of these studies. In agreement with such an interpretation, we refer to the effect of the alcohol chain length on the changes of k_e in luminescence quenching studies[70] and on the ultrasonic absorption behavior in the case of the ATP and ATB systems.[81]

3. Critical Behavior of Microemulsions and Interdroplet Exchanges

The results in Section V.A showed the sensitivity of the k_e value to the chain length of the alcohol. Likewise, the results of Table 1 show that k_e is very dependent on the chain length of the oil, with a decrease of over a factor of 10 going from dodecane to hexane.[71] These results on the dynamics of microemulsions correlate extremely well with those of recent studies of interdroplet interactions and critical-like behavior of microemulsions. More precisely, it appears that the larger the k_e values, the larger the interaction coefficients determined from elastic and QELS. At this point, recall that as the interaction coefficient increases, the microemulsion behaves more and more like a critical system.[90,91] It is therefore tempting to state to that there must exist a correlation between the increases of k_e, of the size polydispersity of the droplets, of their mutual interactions, and of the critical-like character of the microemulsions.

More work along these lines, both theoretical and experimental, should be done in order to check this statement.

REFERENCES

1. **Danielsson, I. and Lindmann, B.,** The definition of microemulsion, *Colloids Surf.,* 3, 391, 1981.
1a. **Robb, I. D., Ed.,** *Microemulsions,* Plenum Press, New York, 1982.
1b. **Mittal, K. L., Ed.,** *Micellization, Solubilization, and Microemulsions,* Vol. 2, Plenum Press, New York, 1977; **Mittal, K. L., Ed.,** *Solution Chemistry* of Surfactants, Vol. 2, Plenum Press, New York, 1979; **Mittal, K. L. and Fendler, E. J., Eds.,** *Solution Behavior of Surfactants,* Plenum Press, New York, 1982; **Mittal, K. L. and Lindman, B., Eds.,** *Surfactants in Solution,* Vols. 1 and 3, Plenum Press, New York, 1984.
2. **Bavière, M.,** Les microémulsions, *Rev. Inst. Fr. Pet.,* 29, 41, 1976.
3. **Friberg., S.,** Microemulsions and their potentials, *Chem. Technol.,* 6, 124, 1976.
4. **Prince., L. M., Ed.,** *Microemulsions: Theory and Practice,* Academic Press, New York, 1977.
5. **Bothorel, P.,** Microemulsions, *Cour. CNRS,* 48, 39, 1982.
6. **Langevin, D.,** Technological relevance of microemulsions and reverse micelles in apolar media, in *Reverse Micelles: Biological and Technological Relevance of Amphiphilic Structures in Apolar Media,* Luisi, P. L. and Straub, B. E., Eds., Plenum Press, New York, 1984, 287.
7. **Ruckenstein, E.,** Stability, phase equilibria and interfacial free energy in microemulsions, in *Micellization, Solubilization, and Microemulsions,* Vol. 2, Mittal, K. L., Ed., Plenum Press, New York, 1977, 755.
8. **Scriven, L. E.,** Equilibrium bicontinuous structures, in *Micellization, Solubilization, and Microemulsions,* Vol. 2, Mittal, K. L., Ed., Plenum Press, New York, 1977, 877.
9. **Aniansson, E. A. G. and Wall, S.,** On the kinetics of stepwise micelle association, *J. Phys. Chem.,* 78, 1024, 1974.
10. **Aniansson, E. A. G. and Wall, S.,** Kinetics of stepwise micelle association. Correction and improvement, *J. Phys. Chem.,* 79, 857, 1975.
11. **Aniansson, E. A. G.,** A treatment of the kinetics of mixed micelles, in *Techniques and Applications of Fast Reactions in Solution,* Gettins, W. J. and Wyn-Jones, E., Eds., D. Reidel, Hingham, Mass., 1979, 249.
12. **Aniansson, E. A. G., Wall, S., Almgren, M., Hoffmann, H., Kiehlman, I., Ulbricht, W., Zana, R., Lang, J., and Tondre, C.,** Theory of the kinetics of micellar equilibria and quantitative interpretation of chemical relaxation studies of micellar solutions of ionic surfactants, *J. Phys. Chem.,* 80, 905, 1976.
13. **Kahlweit, M.,** Kinetics of formation of association colloids, *J. Colloid Interface Sci.,* 90, 92, 1982; What do we know about micelles and which questions are still open?, *Pure Appl. Chem.,* 53, 2069, 1981.
14. **Lessner, E., Teubner, M., and Kahlweit, M.,** Relaxation experiments in aqueous solutions of ionic micelles. II. Experiments on the system H_2O-NaDS-$NaClO_4$ and their theoretical interpretation, *J. Phys. Chem.,* 85, 3167, 1981.
15. **Hansen, J. R.,** High resolution and pulsed NMR studies of microemulsions, *J. Phys. Chem.,* 78, 256, 1974.
16. **Eicke, H.-F. and Zinsli, P. E.,** Nanosecond spectroscopic investigations of molecular processes in W/O microemulsions, *J. Colloid Interface Sci.,* 65, 131, 1978.
17. **Kumar, C. and Balasubramanian, D.,** Spectroscopic studies on the microemulsions and lamellar phases of the system Triton X-100:hexanol:water in cyclohexane, *J. Colloid Interface Sci.,* 74, 64, 1980.
18. **Ueno, M., Kishimoto, H., and Kyogoku, Y.,** ^{13}C-NMR study of Aerosol-AOT in aqueous and organic solutions, *J. Colloid Interface Sci.,* 63, 113, 1978.

19. **Martin, C. and Magid, L.,** ^{13}C NMR Investigations of AOT water-in-oil microemulsions, *J. Phys. Chem.,* 85, 3938, 1981.
20. **Wong, M., Thomas, J. K., and Nowak, T.,** Structure and state of H_2O in reversed micelles. III., *J. Am. Chem. Soc.,* 99, 4730, 1977.
21. **Zulauf, M. and Eicke, H.-F.,** Inverted micelles and microemulsions in the ternary system H_2O/Aerosol-OT/Isooctane as studied by photon correlation spectroscopy, *J. Phys. Chem.,* 83, 480, 1979.
22. **Sjöblom, E. and Friberg, S.,** Light-scattering and electron microscopy determinations of association structures in W/O microemulsions, *J. Colloid Interface Sci.,* 67, 16, 1978.
23. **Lalanne, P., Biais, J., Clin, B., and Bellocq, A.-M.,** Etude de microémulsions par mesure des temps de relaxation RMN des protons à 270 MHz, *J. Chim. Phys.,* 75, 236, 1978.
24. **Bellocq, A.-M., Biais, J., Clin, B., Lalanne, P., and Lemanceau, B.,** Study of dynamical and structural properties of microemulsions by chemical physics methods, *J. Colloid Interface Sci.,* 70, 524, 1979.
25. **Sunamoto, J., Hamada, T., Seto, T., and Yamamoto, S.,** Microscopic evaluation of surfactant-water interaction in apolar media, *Bull. Chem. Soc. Jpn.,* 53, 583, 1983.
26. **Mathews, M. and Hirschhorn, E.,** Solubilization and micelle formation in a hydrocarbon medium, *J. Colloid Interface Sci.,* 8, 86, 1953.
27. **Rouvière, J., Couret, J.-M., Lindheimer, M., Dejardin, J.-L., and Marrony, R.,** Structure des agrégats inverses d'AOT. I., *J. Chim. Phys.,* 76, 289, 1979.
28. **Wong, M., Thomas, J. K., and Grätzel, M.,** Fluorescence probing of inverted micelles. The state of solubilized water in alkane/Aerosol OT solutions, *J. Am. Chem. Soc.,* 98, 2391, 1976.
29. **Correll, G., Cheser, R. N., III, Nome, F., and Fendler, J. H.,** Fluorescence probe in reversed micelles. Luminescence intensities, lifetimes, quenching, energy transfer and depolarization of pyrene derivatives in cyclohexane in the presence of dodecylammonium propionate aggregates, *J. Am. Chem. Soc.,* 100, 1254, 1978.
30. **Gillbert, G., Lehtinen, H., and Friberg, S.,** NMR and IR investigation of the conditions determining the stability of microemulsions, *J. Colloid Interface Sci.,* 33, 40, 1970.
31. **Biais, J., Clin, B., Lalanne, P., and Lemanceau, B.,** Etude de microemulsions en RMN à haut champ, *J. Chim. Phys.,* 74, 1197, 1977.
32. **Lindman, B. and Wennerström, H.,** Structure and dynamics of micelles and microemulsions, in *Solution Behavior of Surfactants,* Vol. 1, Mittal, K. L., and Fendler, E. J., Eds., Plenum Press, New York, 1982, 3.
33. **Eigen, M.,** Fast elementary steps in chemical reaction mechanisms, *Pure Appl. Chem.,* 6, 97, 1963.
34. **Zana, R.,** Ultrasonic absorption studies of solutions of ionic amphiphiles in organic solvents, in *Solution Chemistry of Surfactants,* Vol. 1, Mittal, K. L., Ed., Plenum Press, New York, 1979, 473.
35. **Fujii, H., Kawai, T., Nishikawa, H., and Ebert, G.,** ^{13}C and Na-NMR of sodium octanoate in reversed micelles, *Colloid Polym. Sci.,* 261, 340, 1983.
36. **Grünhagen, H.,** Chemical relaxation of cetylpyridinium iodide micelles in high electric fields, *J. Colloid Interface Sci.,* 53, 282, 1975.
37. **Nguyen, T. and Hadj Ghaffarie, H.,** Etude de la distribution du cosurfactant dans une microémulsion par RMN ^{19}F, *J. Chim. Phys.,* 76, 513, 1979.
38. **Almgren, M., Grieser, F., and Thomas, J. K.,** Rates of exchange of surfactant monomer radicals and long chain alcohols between micelles and aqueous solutions, *J. Chem. Soc. Faraday Trans. 1,* 75, 1674, 1979.
39. **Yiv, S. and Zana, R.,** Fast exchange of alcohol molecules between micelles and surrounding solution in aqueous micellar solutions of ionic surfactants, *J. Colloid Interface Sci.,* 65, 286, 1978.
40. **Yiv, S., Zana, R., Ulbricht, W., and Hoffmann, H.,** Effect of alcohol on the properties of micellar systems. II. Chemical relaxation studies of the dynamics of mixed alcohol + surfactant micelles, *J. Colloid Interface Sci.,* 80, 224, 1981.
41. **Lang, J., Djavanbakht, A., and Zana, R.,** Ultrasonic absorption studies of microemulsions, *J. Phys. Chem.,* 84, 145, 1980.
42. **Lang, J., Djavanbakht, A., and Zana, R.,** Ultrasonic relaxation studies of microemulsions, in *Microemulsions,* Robb, I. D., Ed., Plenum Press, New York, 1982, 233.
43. **Gettins, J., Denver, H., Jobling, P., Rassing, J., and Wyn-Jones, E.,** Thermodynamics and kinetic parameters associated with the exchange process involving alcohols and micelles, *J. Chem. Soc. Faraday Trans.,* 74, 1957, 1978.
44. **Lim, Y. Y. and Fendler, J. H.,** Spin probes in reversed micelles. EPR spectra of 2,2,5,5-tetramethyl-pyrrolidone-1-oxyl derivatives in benzene in the presence of dodecylammonium propionate aggregates, *J. Am. Chem. Soc.,* 100, 7490, 1978.
45. **Yoshioka, H.,** Exchange of the position of a spin probe in AOT reversed micelles, *J. Colloid Interface Sci.,* 95, 81, 1983.
46. **Yoshioka, H. and Kazama, S.,** Spectral simulation of the positional exchange of a spin probe in AOT reversed micelles, *J. Colloid Interface Sci.,* 95, 240, 1983.

47. **Folger, R., Hoffmann, H., and Ulbricht, W.,** To the mechanism of the formation of micelles in SDS solutions, *Ber. Bunsenges. Phys. Chem.,* 78, 986, 1974.
48. **Hoffmann, H.,** The dynamics of micelle formation, *Ber. Bunsenges. Phys. Chem.,* 82, 988, 1978.
49. **Djavanbakht, A., Lang, J., and Zana, R.,** Ultrasonic absorption studies in relation to H-bonding in solutions of alcohols. II. Ultrasonic relaxation spectra of solutions of alcohols in cyclohexane, *J. Phys. Chem.,* 81, 2620, 1977.
50. **Shah, D., Walker, R., Hshieh, W., Shah, N., Dwivedi, S., Nelander, J., Pepinski, R., and Deamer, D.,** Some structural aspects of microemulsions and co-solubilized systems, *Proc. Symp. Improved Oil Recovery,* paper SPE 5815, Society of Petroleum Engineers of AIME, Tulsa, Okla., 1976.
51. **Bansal, V., Chinnaswamy, K., Ramachandram, C., and Shah, D.,** Structural aspects of microemulsions using dielectric relaxation and spin label techniques, *J. Colloid Interface Sci.,* 72, 524, 1979.
52. **Tabony, J., Llor, A., and Drifford, M.,** Quasielastic neutron scattering measurements of monomer molecular motions in micellar aggregates, *Colloid Polym. Sci.,* 261, 938, 1983.
53. **Cazabat, A.-M. and Langevin, D.,** Diffusion of interacting particles: light scattering study of microemulsions, *J. Chem. Phys.,* 74, 3148, 1981.
54. **Maelstaf, P. and Bothorel, P.,** Correlations d'orientation moléculaires dans des groupes hydrophobes de composés tensio-actifs usuels, *C.R. Acad. Sci. Paris,* 288, 13, 1979.
55. **Zana, R.,** Effect of alcohols on the equilibrium properties and dynamics of micellar solutions, in *Surface Phenomena in Enhanced Oil Recovery,* Shah, D. O., Ed., Plenum Press, New York, 1981, 521.
56. **Di Meglio, J.-M., Dvolaitzky, M., Ober, R., and Taupin, C.,** Defects and curvatures of the interfaces in the "birefringent microemulsions", *J. Phys. Lett.,* 44, L229, 1983.
57. **Guering, P. and Cazabat, A.-M.,** Water in oil microemulsions, transient electric birefringence response, *J. Phys. Lett.,* 44, L601, 1984.
58. **De Gennes, P.-G. and Taupin, C.,** Microemulsions and the flexibility of oil/water interfaces, *J. Phys. Chem.,* 86, 2294, 1982.
59. **Taupin, C., Dvolaitzky, M., and Ober, R.,** Structure of microemulsions; role of the interfacial flexibility, *Nuovo Cimento,* 3D, 62, 1984.
60. **Di Meglio, J.-M., Dvolaitzky, M., and Taupin, C.,** Determination of the rigidity constant of the amphiphilic film in "birefringent microemulsion"; role of the cosurfactant, *J. Phys. Chem.,* 89, 871, 1985.
61. **Tamura, K. and Schelly, Z. A.,** Reversed micelles of AOT in benzene. III. Dynamics of the solubilization of picric acid, *J. Am. Chem. Soc.,* 103, 1018, 1981.
62. **Menger, F., Donahue, J., and Williams, R.,** Catalysis in water pools, *J. Am. Chem. Soc.,* 95, 286, 1973.
63. **Eicke, H.-F., Shepherd, J.-C., and Steinemann, A.,** Exchange of solubilized water and aqueous electrolyte solutions between micelles in apolar media, *J. Colloid Interface Sci.,* 56, 168, 1976.
64. **Robinson, B. H., Steyler, D., and Tack, R.,** Ion-reactivity in reversed micellar systems, *J. Chem. Soc. Faraday Trans. 1,* 75, 481, 1979.
65. **Tondre, C. and Zana, R.,** Rate of dissolution of n-alkanes and water in microemulsions of water/SDS/1-pentanol/dodecane in rapid mixing experiments, *J. Dispersion Sci. Technol.,* 1, 179, 1980.
66. **Robinson, B. H. and Fletcher, P. D.,** Dynamic processes in W/O microemulsions, *Ber. Bunsenges. Phys. Chem.,* 85, 863, 1981.
67. **Fletcher, P. D., Howe, A., Perrins, N., Robinson, B. H., Toprakcioglu, C., and Dore, J.,** Structural and dynamic aspects of microemulsions, in *Surfactants in Solution,* Vol. 3, Mittal, K. L. and Lindman, B., Eds., Plenum Press, New York, 1984, 1745.
68. **Atik, S. and Thomas, J. K.,** Transport of ions between water pools in alkanes, *Chem. Phys. Lett.,* 79, 351, 1981.
69. **Atik, S. and Thomas, J. K.,** Transport of photoproduced ions in water in oil microemulsions: movement of ions from one water pool to another, *J. Am. Chem. Soc.,* 103, 3543, 1981.
70. **Atik, S. and Thomas, J. K.,** Abnormally high ion exchange in pentanol microemulsions compared to hexanol microemulsions, *J. Phys. Chem.,* 85, 3921, 1981.
71. **Lianos, P., Lang, J., Cazabat, A.-M., and Zana, R.,** Luminescent probe study of W/O microemulsions, *Surfactants in Solution,* Mittal, K. L. and Bothorel, P., Eds., Plenum Press, New York, 1986.
72. **Furois, J. M., Brochette, P., and Pileni, M.-P.,** Photoelectron transfer in reverse micelle: 2-photooxidation of magnesium porphyrin, *J. Colloid Interface Sci.,* 97, 552, 1984.
73. **Pileni, M.-P., Brochette, P., Hickel, B., and Lerebours, B.,** Hydrated electrons in reverse micelles. II. Quenching of hydrated electron by sodium nitrate, *J. Colloid Interface Sci.,* 98, 549, 1984.
74. **Geladé, E. and De Schryver, F. C.,** Fluorescence quenching in dodecylammonium propionate micelles, *J. Photochem.,* 18, 223, 1982.
75. **Geladé, E. and De Schryver, F. C.,** Energy transfer in inverse micelle, *J. Am. Chem. Soc.,* 106, 5871, 1984.
76. **Bakale, G. and Warman, J.,** Rearrangement of inverse micelles following charge scavenging observed in time-resolved microwave conductivity, *J. Phys. Chem.,* 88, 2928, 1984.

77. **Uehara, H.,** The effect of additives on the relaxation of the micellization of sodium dodecylsulfate, *J. Sci. Hiroshima Univ. Ser. A,* 40, 305, 1976.
78. **Lang, J.,** Effect of alcohol and oil on the dynamics of micelles, in *Surfactants in Solution,* Mittal, K. and Bothorel, P., Eds., Plenum Press, New York, 1986.
79. **Bayer, O., Hoffmann, H., and Ulbricht, W.,** Influence of solubilized additives on surfactant systems containing rod-like micelles, in *Surfactants in Solution,* Mittal, K. and Bothorel, P., Eds., Plenum Press, New York, 1985.
80. **Lang, J. and Zana, R.,** Unpublished data, 1984, submitted.
81. **Zana, R., Lang, J., Sorba, O., Cazabat, A.-M., and Langevin, D.,** Ultrasonic investigation of critical behavior and percolation phenomena in microemulsions, *J. Phys. Lett.,* 43, L829, 1982.
82. **Hirsch, E., Debeauvais, F., Candau, F., Lang, J., and Zana, R.,** Effect of salinity on the ultrasonic absorption and flow birefringence of microemulsions, *J. Phys.,* 45, 257, 1984.
83. **Hirsch, E., Candau, S., and Zana, R.,** Effect of added oil in aqueous alkyl-trimethylammonium bromide micelles in the presence of alcohol, in *Surfactants in Solution,* Mittal, K. L. and Bothorel, P., Eds., Plenum Press, New York, 1986.
84. **Lindman, B., Stilbs, P., and Moseley, M.,** Fourier transform NMR self-diffusion and microemulsion structure, *J. Colloid Interface Sci.,* 83, 569, 1981.
85. **Larché, F., Rouvière, J., Delord, P., Brun, B., and Dussossoy, J.,** Existence of a bicontinuous zone in microemulsion systems, *J. Phys. Lett.,* 41, L437, 1980.
86. **Nilsson, P.-G. and Lindman, B.,** Solution structure of nonionic surfactant microemulsions from NMR self-diffusion studies, *J. Phys. Chem.,* 86, 271, 1982.
87. **Lindman, B., Kamenka, N., Kathopoulis, T.-M., Brun, B., and Nilsson, P.-G.,** Translational diffusion and solution structure of microemulsion, *J. Phys. Chem.,* 84, 2485, 1980.
88. **Fabre, H., Kamenka, N., and Lindman, B.,** Aggregation in three-component surfactant systems from self-diffusion studies. Reversed micelles, microemulsions and transitions to normal micelles, *J. Phys. Chem.,* 85, 3493, 1981.
89. **Stilbs, P., Rapacki, K., and Lindman, B.,** Effect of alcohol cosurfactant length on microemulsion structure, *J. Colloid Interface Sci.,* 95, 583, 1983.
90. **Roux, D.,** Influence des interactions intermicellaires sur le comportement critique des microemulsions et leur diagramme de phases and references therein, D. Sc. thesis, University of Bordeaux, 1984.
91. **Skoulios, A. and Guillon, D.,** Les microemulsions relèvent-elles des phénomènes critiques?, *J. Phys. Lett.,* 38, L137, 1977.

Chapter 7

LOW INTERFACIAL TENSIONS IN MICROEMULSION SYSTEMS

D. Langevin

TABLE OF CONTENTS

I. INTRODUCTION

Microemulsions are dispersions of oil and water made with surfactant molecules.[1] They spontaneously form when the surfactant has reduced the oil-water interfacial tension to very low values: from about 50 dyn/cm without surfactant to about 10^{-2} dyn/cm. This was recognized early by Schulman and co-workers, who initiated the fundamental research on these systems. In order to explain the spontaneous emulsification of oil and water, they first proposed that the interfacial tensions became negative, thus allowing the free energy to decrease when the oil-water interfacial area is increasing.[2] It was shown later on by Ruckenstein and co-workers that negative surface tensions were not necessary to form microemulsions. They introduced the contribution of the dispersion entropy in the free energy and were able to show that small but positive tensions gave reasonable values for the spatial scale of the dispersion.[3] This scale is noticeably smaller than in ordinary emulsions: typically 100 Å instead of 1 μm. They demonstrated in this way that microemulsions were thermodynamically stable systems.

The concept of ultralow tensions found an interesting application in enhanced oil recovery processes.[4] In the primary oil recovery process, oil is recovered due to the pressure of natural gases which force the oil out through production wells. In secondary processes, water is injected to force further oil out. The average recovery at this stage is about 30 to 50% of the total oil. Enhanced or tertiary oil recovery deals with processes which allow improvement of the total recovery. Processes using surfactants were patented around 1960. They were based on the lowering of the oil-water interfacial tension, allowing release of oil ganglia trapped at small pore throats of the reservoir rock. In the former processes, small amounts of surfactants were used and the recovery efficiency was greatly reduced by surfactant adsorption on the surface of the rocks. Later on, patents using microemulsions were proposed. In these processes the microemulsions had to coexist with excess oil and excess water because they could not solubilize all the reservoir oil and the injected water. The process was efficient when the interfacial tensions between the microemulsions and both the oil and the water were very low, below 10^{-2} dyn/cm. This was possible in well-chosen systems called "optimal". Let us mention that although low tensions are a requirement, other microemulsions properties can favor recovery enhancements: improvement of rock wetting, small interfacial viscosities encouraging the oil ganglia to coalesce and to form an oil bank,[5,6] partial solubilization of heavy oils, which could be effective even in tar sand extraction,[7] etc.

The mechanisms leading to low tensions were early investigated. The usual methods for measuring interfacial tensions — Wilhelmy plate, drop shapes — were not suitable for tensions below 10^{-2} dyn/cm, and the first studies by Schulman and others were not accurate enough. The subject progressed rapidly when Princen and co-workers[8] proposed to use the spinning drop tensiometer with which tensions as low as 10^{-4} to 10^{-5} dyn/cm can be measured. In the first studies, commercial surfactants were used and it was reported that the lowest tensions were reached at large enough concentrations leading to microemulsion formation[9] (two- or three-phase equilibria). This suggested that the mechanisms responsible for low tensions were different for oil-water interfaces and microemulsions-oil or water interfaces. In the first case, the low tensions were explained by the presence of a surfactant monolayer of high surface pressure.[2] In the second case, different mechanisms were proposed including vicinity of critical points,[10-13] and presence of liquid crystalline layers at the interfaces.[14] The work carried out in our laboratory on pure surfactant systems contributed to clarifying the situation. We showed that most of the time, even when microemulsions phases were involved, the low tensions were associated with surfactant monolayers.[15-17]

Although the mechanisms responsible for low tensions are now qualitatively well understood, quantitative predictions are difficult to formulate. Many theories are presently available,[18-26] but they all rely on several hypotheses relative to microemulsion structure, which is still poorly known from the experimental point of view. A great deal of work remains to be done to achieve a complete understanding of the low tensions in microemulsions systems.

In the following we will present these problems in more detail. First we will recall the existing experimental methods to measure low tensions (Section II). Second, we will present some experimental data for different model systems (Section III). Finally, we will discuss these data in relation to the existing theories (Section IV).

II. EXPERIMENTAL TECHNIQUES

Interfacial tensions below 10^{-2} dyn/cm cannot be measured with the aid of conventional ring or Wilhelmy plate methods. Sessile and pendent drop methods require a contact between the droplet and a solid surface. If this solid surface is not perfectly homogeneous and smooth, the drop can lose its cylindrical symmetry and thus introduce an uncertainty about the relationship droplet shape-tension.

One of the most reliable methods is the so-called spinning drop method where a droplet of the lightest phase is introduced into a capillary tube containing the denser phase, and made to spin around the length axis of the tube. Under the action of centrifugal forces, the drop elongates, but remains a finite length due to the opposite surface tension forces. In the limited case of very elongated or cylindrical-like drops, the relationship tension-shape is very simple:

$$\gamma = \frac{\omega^2 \, \Delta\rho \, r_0^3}{4} \qquad (1)$$

where γ is the tension, ω the angular velocity, $\Delta\rho$ the density difference between the two phases, and r_0 the radius of the cylindrical part of the drop.

It must be pointed out that precise and reliable data often cannot be obtained if the two phases have not been previously carefully equilibrated. Indeed the drop in this case can be dissolved in the denser phase and disappear during the experiment. Before this, the measured tensions appear to vary with time.[9,14]

Such problems are less crucial in a more sophisticated technique: the quasielastic surface light scattering,[27] where the volumes of the two phases are similar. In this method, the light of a laser beam is scattered by interfacial roughness created by Brownian motion. The roughness is very small, about 100 Å for tensions of 10^{-2} dyn/cm. Photomultipliers can detect the scattered light and distinguish it from bulk light scattering, provided the scattering angle is not too large: below a few degrees. As a first method, one can measure the intensity of the scattered light.[28] However, the stray light scattered by the cell windows can be the source of considerable error. The spectrum analysis of the scattered light avoids this difficulty. Indeed the surface roughness changes with time. If the interface profile is Fourier transformed into sinusoidal components of wavelength $2\pi/q$ (Figure 1), these components behave like surface waves. If the surface tension, which is again the restoring force, is small compared to damping effects due to bulk viscosities of the two liquids, the surface waves are over-damped. The scattered light on this moving grating is frequency shifted due to the Doppler effect. If the incident light is monochromatic, the spectrum of the scattered light is broadened (Figure 1), its width $\Delta\omega$ being related to the damping time of the waves τ: $\Delta\omega = \tau^{-1}$ and:

$$\Delta\omega = \frac{\gamma \, q}{2(\eta_1 + \eta_2)} \qquad (2)$$

where η_1 and η_2 are the bulk viscosities of the coexisting phases. Observation of a given

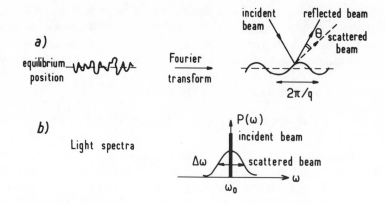

FIGURE 1.   (a) Corrugated interface between two liquids, and scattering of light by a Fourier component. (b) Spectra of the incident and scattered light; case of over-damped surface waves.

Fourier component of wave vector q is very simple; scattering theory relates the scattering angle θ to q by:

$$q = \frac{2\pi}{\lambda_0} \theta$$

where λ_0 is the laser wavelength, if θ is small and the scattered beam perpendicular to the plane of incidence.

This method is very suitable to measurements at high temperatures and pressures; the only requirement for the cell is that it have two optical windows.*

Let us mention that a comparison between this technique and the spinning drop technique has been performed and that the measured tensions on several model systems were the same within experimental errors: about 5% for the two methods.[29]

Although these two methods are the more widely used for the measurements of low tensions in microemulsion systems, several other methods have been used to measure very low tensions including capillary rise method close to a critical point[30] and drop shapes in aqueous solutions with density gradients.[31]

III. RESULTS ON MODEL SYSTEMS

A. Earliest Monolayer Studies

In the earliest studies by Schulman and co-workers, the accuracy of the surface tension measurements was limited. They observed that before microemulsion formation the oil-water interfacial tension was below measurable values.[32] Their evidence for negative tensions was indirect.[2] They studied systems where the amount of oil was reduced in order to obtain a very thin oil phase, about 100-Å thick. They called these systems "duplex films" (Figure 2) and measured the air-liquid surface tension $\gamma_{A/l}$. They deduced the oil-water $\gamma_{O/w}$ interfacial tension from:

$$\gamma_{O/w} = \gamma_{a/l} - \gamma_{A/O}$$

where $\gamma_{A/O}$ is the oil surface tension. In this picture, although negative $\gamma_{O/w}$ were found, the duplex films were stabilized by the opposite tension of the oil/air interface $\gamma_{A/O}$.

* A commercial light scattering apparatus operating at high pressures is now available (Langley-Ford).

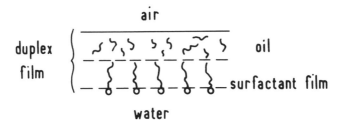

FIGURE 2.   Duplex film, after Schulman and co-workers.

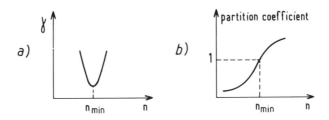

FIGURE 3.   (a) Schematic surface tension variation vs. alkane carbon number of the oil n. (b) Partition coefficient (surfactant aqueous concentration divided by surfactant in alkane concentration) vs. n: schematic variation.

Although negative tensions are not required to form microemulsions, this picture had the advantage of showing that surfactant films had to develop large surface pressures $\Pi = \gamma^0_{O/W} - \gamma_{O/W}$ (where $\gamma^0_{O/W}$ is the oil-water tension without surfactant) in order to decrease $\gamma_{O/W}$ to small enough values. These studies were performed with fatty acids or their salts as surfactants and alcohols or amines as cosurfactants.

B. Petroleum Sulfonate Model Systems for Oil Recovery

Since the development of the spinning drop technique, extensive studies have been done on model systems for oil recovery: alkane-water-petroleum sulfonate systems. Commercial petroleum sulfonate surfactants were widely studied.

1. Influence of the Oil Chain Length, EACN Concept

A deep minimum of interfacial tension was found for a well-defined number of carbon atoms of the oil[9] (Figure 3).

The minimum was specific, not only for normal alkanes, but for branched, cyclic, and aromatic hydrocarbons as well. From correlations between oil composition and measured tensions it was possible to assign an "equivalent alkane carbon number" (EACN) to any hydrocarbon or mixture of hydrocarbons including crude oils[33] (most crude oils behave as if they were mixtures of alkanes in the range hexane-nonane). This empirical finding stimulated the experimental work on pure alkane systems.

The existence of the minimum in the interfacial tension can be qualitatively understood: it has been shown to be associated with a partition coefficient of the surfactant between oil and water equal to one[9,34] (Figure 3). This means that the surfactant is as soluble in oil as in water. It will then concentrate at the interface, thus increasing the surface pressure of the interfacial film (Figure 4). Moreover, according to the Bancroft rule,[35] the spontaneous curvature of this film is close to zero and then the positive curvature contribution to the interfacial tension is minimum (see Section IV).

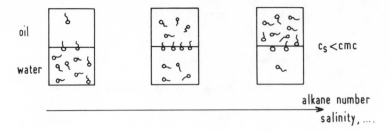

FIGURE 4.   Evolution of the surfactant solubility in the oil and in the water
and of the adsorption at the interface vs. alkane number or salinity. Case of
small aqueous surfactant concentrations c_s < cmc.

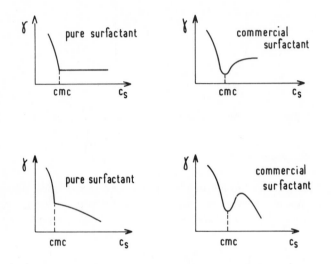

FIGURE 5.   Schematic surface tension variation with surfactant aqueous
concentration c_s for pure and commercial surfactants. Aqueous surfactant solutions above and oil-water-surfactants solutions below.

2. Influence of the Surfactant Concentration

Let us now turn to the problem of surfactant concentration. It has been reported that the
lowest tensions were reached at concentrations much higher than the critical micellar concentration (cmc) of the surfactant in water, for commercial as well as for pure surfactants[9]
(Figure 5). This was apparently very different from the case of aqueous surfactant solutions,
where after the cmc the surface tension remains constant because the extra amounts of
surfactant are incorporated into micelles. The answer to the paradox is in fact simple; swollen
micelles are indeed formed after the cmc in oil-water-surfactant mixtures, but the tension
will remain constant only if the droplet concentration is varied, their size remaining constant.
This cannot be merely obtained by increasing the surfactant concentration because the
cosurfactant partitions in a complex manner between the aqueous and oil phases and the
droplets. If a proper variation of droplets concentration is performed, a constant tension is
indeed obtained.[16] Let us point out that these measurements have been done on a model
system water-sodium chloride-toluene-sodium dodecyl sulfate (SDS)-butanol, i.e., with pure
chemicals in order to suppress the tension minimum around the cmc observed with commercial surfactants.

The apparent decrease of the tension after the cmc led many authors to the conclusion
that the interfacial film was not responsible for the low tension as in simpler micellar
solutions. In other words, for these authors the Schulman concept of high film pressures π

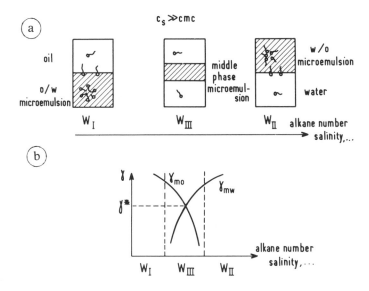

FIGURE 6. (a) Evolution of the phase behavior with alkane number or salinity. Case of large aqueous surfactant concentrations $c_s \gg$ cmc. (b) Corresponding schematic interfacial tension variations.

balancing almost exactly the initial oil-water tension could not be appropriate at high surfactant concentration. They proposed other mechanisms:

1. Formation of liquid crystalline layers at the interface[14]
2. Vicinity of a critical consolute point in the phase diagram[10-13]

We will now present, in more detail, the type of results obtained at high surfactant concentration (Figure 6). If the surfactant is more soluble into the water, an O/W microemulsion is formed and coexists with excess oil. The corresponding phase equilibria has been named Winsor I (WI). If the surfactant is more soluble into the oil, a W/O microemulsion is formed and coexists with excess water: Winsor II (WII) equilibrium. In the intermediate case a middle phase microemulsion coexists with both excess oil and water: Winsor III (WIII) equilibria. The lowest interfacial tensions are found in the WIII region where the surfactant partition coefficient is close to unity. This is the region of interest for enhanced oil recovery. The "optimal" conditions correspond to the point where the two interfacial tensions involving the microemulsion phase are equal (Figure 6). The optimal alkane number is the same as the one corresponding to the minimum tension at low surfactant concentration. The same remark is valid for the optimal salinity.

In order to investigate the origins of the low tensions it was necessary to study systems with pure compounds. The petroleum sulfonates were difficult to purify and several workers began to study other surfactants: shorter sulfonates, like the octyl benzene sulfonate (OBS),[36] sulfates like SDS,[15-17] etc.

C. SDS Systems

We have studied in our laboratory a model system showing the successive phase equilibria WI → WIII → WII with increasing salinity. The composition of the mixture is

Water + sodium chloride	46.8 wt%
Toluene	47.2 wt%
SDS	2 wt%
n-Butanol	4 wt%

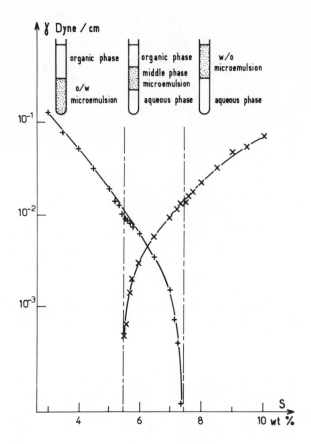

FIGURE 7. Interfacial tensions variation with salinity in the SDS system; + = tensions between microemulsions and top phases; x = tensions between microemulsions and bottom phases. Dotted lines show the limit of the three-phase domain.

The water salinity S was varied between 3 wt% and 10 wt%, and the transitions WI → WIII were observed for S_1 = 5.4 wt% and WIII → WII for S_2 = 7.4 wt%.

The interfacial tension measurements are represented on Figure 7. The optimal salinity is S☆ = 6.3 wt% for which γ☆ = 4.5 10^{-3} dyn/cm.

1. Surfactant Layer Properties

A dilution procedure was found for the microemulsion phases in the WI and WII regions. The interfacial tensions between the excess phases and the continuous phases of the microemulsions are represented on Figure 8. In the WIII region, the points correspond to the tension between the two excess phases. An important remark can be made. These tensions are equal to the largest ones involving the microemulsion phases in Figure 7. The amount of surfactant present in the microemulsion continuous phases or in the excess phases is very small close to the cmc (about 10^{-5} by weight). As in simple micellar systems, the tension in the WI and in the WII regions is independent of the droplets concentration. By analogy it can be concluded that the low tension is only due to the high surface pressure of the surfactant layer at the interface.

Although the middle phase microemulsion in the WIII region is not dilutable, the low tensions in Figure 8 are also likely to be due to a surfactant layer. The presence of a thin middle phase layer responsible for the low tension can be excluded for the following reasons:

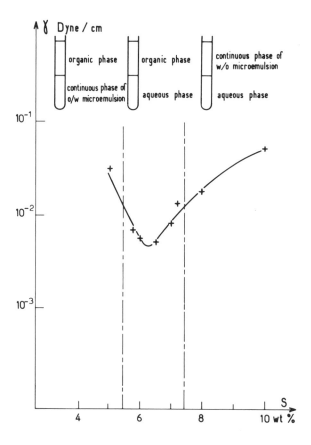

FIGURE 8.   Interfacial tension between organic and aqueous phases
containing no micelles: top and bottom phases in the three-phase domain; top phase and continuous phase of the microemulsion on the
left; bottom phase and continuous phase of the microemulsion on the
right.

- γ_{OW} is smaller than the sum $\gamma_{mw} + \gamma_{mo}$; if a thin middle phase layer were present,
  the measured γ_{OW} would be equal to the sum.
- The middle phase does not wet the oil-water interface (Figure 9); this would not be
  the case if a thin layer of this phase were already present.

2. Critical Behavior

Tensions smaller than $\gamma^\star$ were attributed to a different origin. Indeed when the salinity
S approaches S_1 the excess water phase becomes turbid before it disappears; the microemulsion turbidity also increases. When S approaches S_2, the excess oil phase behaves in
a similar way. This resembles critical behavior, although if S_1 and S_2 where the exact critical
points, the interface between the microemulsions and the excess phases would disappear
without moving toward the bottom or the top of the sample tube.

Detailed investigation of this critical behavior has been performed through bulk light
scattering experiments, both static and dynamic. Three independent values of the correlation
length of the concentration fluctuations ξ have been determined close to S_2: from the angular
variations of the scattered intensity and of the diffusion coefficient, and from the diffusion
coefficient extrapolated at zero scattering angle. The three determinations were in satisfactory
agreement. Values of ξ up to 800 Å were measured (Figure 10). It must be pointed out,
however, that the microemulsion droplet size in the WII region as determined also by bulk

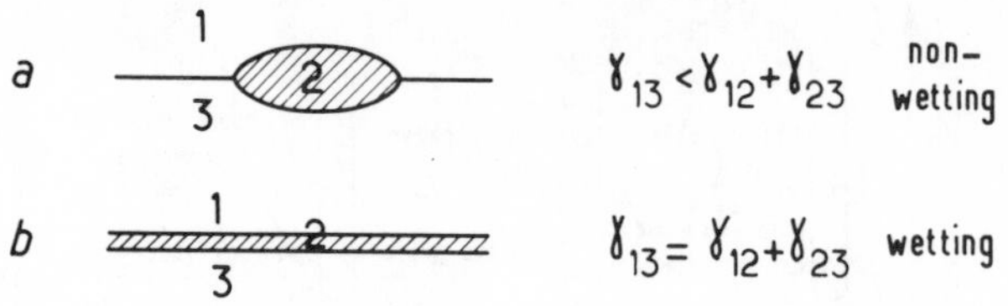

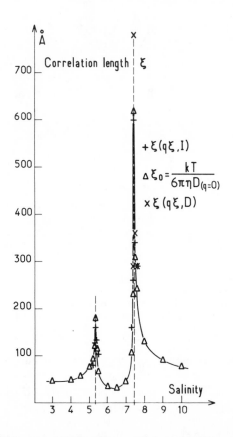

FIGURE 9.   Illustration of the Antonov relation. (a) Nonwetting case: phase 2 forms a lens at the interface between phases 1 and 3; (b) wetting case: phase 2 spreads at the 1-3 interface.

FIGURE 10.   Correlation lengths vs. salinity in the SDS system: + = from the diffusion coefficient limit at zero scattering angle; O = from the angular anisotropy of the scattered intensity; ☆ = from the angular anisotropy of the diffusion coefficient.

light scattering experiments approaches 200 Å close to S_2 (Table 1). The largest measured ξ is only four times the droplet size; this means that S_2 is still relatively far from the true critical point. The other boundary S_1 is still farther from the other critical point. Bulk light scattering experiments cannot be interpreted with the critical phenomena theories even very close to S_1, where the largest ξ value is only about the droplet size (~200 Å). The data also indicate that the droplets of the WI region become elongated and polydisperse close to S_1[16] (Table 1).

These two points are critical end points (CEP) (Figure 11). They have been precisely located in the phase diagram of another model system: water + NaCl + dodecane + OBS

Table 1

**MICROEMULSION PHASES DATA IN
THE SDS SYSTEM**

S	R (Å)	R_H (Å)	B	α	φ
3.5	85	95	5	− 1.5	0.16
4	—	120	$\gtrsim 0$	− 1	0.16
4.5	—	180	$\gtrsim 0$	− 1.5	0.19
5	130	190	3	− 5	0.21
5.2	155	240	$\gtrsim 0$	− 6	0.24
8	180	200	2	− 7	0.18
10	125	150	6	− 2	0.14

Abbreviations: R, droplets radius; R_H, hydrodynamic B, osmotic virial coefficient; α, diffusion coefficient virial coefficient; φ, droplets volume fraction. N.B.: hard sphere droplets correspond to $R = R_H$, B = 8, and α = 1.5, interpenetrating or elongated droplets to $R < R_H$, and attractive droplets to B < 8 and α < 1.5.

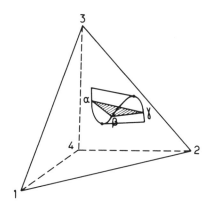

FIGURE 11.   Schematic phase diagram of a four-component system. The three-phase domain consists of a stack of triangles α, β, and γ. The points where α = β and β = γ are two CEP.

+ pentanol.[36] For quaternary mixtures, two critical end points may exist at the boundary surface of a three-phase domain. For five-component mixtures, CEP lines may exist.

The critical behavior will also affect the interfacial tensions. Theory predicts that if ε is the distance to the critical point:

$$\epsilon = \left| \frac{X - X_c}{X_c} \right| \tag{3}$$

where X is a field variable (same value in the coexisting phases) and X_c its value at the critical point; then:

$$\gamma = \gamma_0 \, \epsilon^\mu \tag{4}$$

where γ_0 is a scale factor, and μ is a critical exponent.

The variable X can be the temperature, the chemical potential of a given chemical species, etc. Concentrations of the different species, the salt concentration in particular, are not good field variables because they are different in the different phases. As the true field variables are unknown, several methods to overcome this difficulty have been proposed.

Fleming and co-workers[11] related theoretically the field variables to salt concentration. They obtained expressions of the form:

$$\gamma = \gamma_0' \, | \, S - S_c |^{\mu'} \qquad \mu' = \mu/(1 - \alpha) \tag{5}$$

α being the critical exponent for the specific heat and μ' a renormalized exponent.

Equation 5 contains three adjustable parameters. But $\mu/(1 - \alpha)$ is expected to be universal. Generally the fits with Equation 5 lead to different μ' values. In our system:

$$\mu' = 1.25 \qquad \text{for the WIII} \rightarrow \text{WII transition}$$
$$\mu' = 1.56 \qquad \text{for the WI} \rightarrow \text{WIII transition}$$

The theoretical values would be

$$\mu' = 1.41 \qquad \text{3d Ising models}$$
$$\mu' = 1.5 \qquad \text{mean field models}$$

In our opinion, these fits are not accurate enough because they involve too many adjustable parameters. Moreover, they are usually done in the whole range of interfacial tension data, although we have seen that values above γ☆ are not associated with critical phenomena. The number of adjustable parameters is sufficient to account for the γ variation with salinity, and it is not surprising that the method introduces artifacts in the determination of μ'.

We decided to reduce the number of adjustable parameters by using another physical property of the system, experimentally measurable, and showing critical behavior. The density difference between the two coexisting phases is one of such variables:

$$\Delta\rho = \Delta\rho_0 \, \epsilon^\beta \tag{6}$$

By eliminating ϵ between Equations 4 and 6, one obtains:

$$\gamma = \gamma_0 \left(\frac{\Delta\rho}{\Delta\rho_0} \right)^{\mu/\beta} \tag{7}$$

A log-log plot of γ vs. $\Delta\rho$ is shown in Figure 12. The slope μ/β is now the same for the two interfaces, γ_{om} and γ_{wm}:

$$\frac{\mu}{\beta} = 4$$

in agreement with the 3d Ising exponent. The mean field value of μ/β is 3.

From Figure 12 one sees clearly the change in slope above γ☆, indicating that critical behavior is no longer followed.

A last remark can be made about critical behavior. Theory predicts that some combinations of scale factors are also universal. For instance, in 3d Ising models:[37]

$$\frac{\gamma_0 \, \xi_0^2}{kT} = 0.20 \tag{8}$$

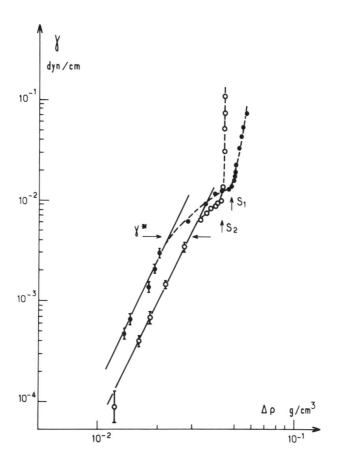

FIGURE 12.   Interfacial tensions vs. density difference between the phases:
● = microemulsion, top phase; ○ = microemulsion, bottom phase. The
lines are theoretical; their slope is 4.

where ξ_0 is the scale factor for the correlation length ξ:

$$\xi = \xi_0 \, \epsilon^{-\nu} \tag{9}$$

where ϵ is unknown, and γ_0 and ξ_0 cannot be determined. Theory also predicts $\mu = 2\,\nu$, so that $\gamma_0 \xi_0^2 = \gamma \xi^2$ whatever ϵ is.

The experimental numbers that we get are

$$\frac{\gamma \, \xi^2}{kT} \sim 0.012 \text{ to } 0.025$$

which is much smaller than the theoretical value. It must be pointed out that experiments on pure fluids and binary mixtures close to critical points lead to a very reasonably universal value:

$$\frac{\gamma_0 \, \xi_0^2}{kT} \sim 0.5 \text{ to } 0.7$$

although slightly larger than predicted by the 3d Ising theory.

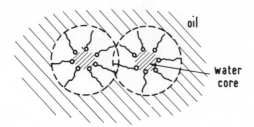

FIGURE 13.   Interpenetration of the surfactant layers of W/O microemulsion droplets.

The origin of the discrepancy found in microemulsion systems is not understood. It might be due to the fact that the critical points are not approached close enough. It might also indicate that microemulsion systems cannot be described by the same Ising models than those of simple fluids and binary mixtures of small molecules. The reason for that could be the presence of oil-water interfaces covered by surfactant layers, or the transient character of the microstructures.

In order to discuss models to describe critical behavior, it is necessary to understand the mechanisms responsible for phase separation. Experiments indicate that the interactions between droplets in the WI and WII regions become more attractive as the phase boundaries are approached.[17] Therefore, it seems reasonable to attribute the phase separation and the critical behavior to increasingly attractive interactions between droplets.

In a mean field model proposed by Miller et al.,[18] this problem was discussed. In the WI region (O/W microemulsions) the interactions between droplets were attractive van der Waals' forces supplemented by repulsive screened electrostatic forces. In the WII region, attractive forces were still van der Waals forces, and the repulsive ones, steric forces.

In the second case, the attractive van der Waals' forces are never strong enough to produce phase separation unless the droplets are allowed to interpenetrate (Figure 13). The interpenetration facilitates the exchanges of the water cores of the droplets. Above a certain droplet concentration of about 20% in volume fraction, a percolation phenomenon is observed indicating that the aqueous phase becomes continuous as one approaches and enters into the WIII region (Figures 14 and 15). The middle phase microemulsion is therefore probably bicontinuous rather than a dispersion of isolated droplets. The model which predicts a phase separation between a droplets-rich (the middle phase) and a droplets-poor microemulsion (the lower phase, turbid close to S_2) is therefore approximate. It has also the disadvantage of predicting mean field exponents. We will discuss these points later in the chapter.

The transition WI → WIII is still more complex at first sight. The microemulsion also becomes bicontinuous as one enters into the WIII region (Figure 16). The reason for the prior elongation of the droplets is not well understood.

The critical point is a lower critical consolute point, i.e., phase separation is obtained also upon heating a microemulsion in the WI region. Such phase separation cannot be driven with simple van der Waals' and electrostatic forces. Other attractive forces like hydration forces have to be introduced, or the closest distance of approach of two droplets have to be assumed to be temperature dependent. The rough character of the initial theory is illustrated in Figure 17. The theory better describes a region where a single phase microemulsion transforms into a two-phase microemulsion rather than the Winsor equilibria. The Hamaker constants are even much too large for these systems (for oil-water systems H $\sim 5\ 10^{-14}$ erg).

In order to better understand this phase separation, a simpler model system has been studied.

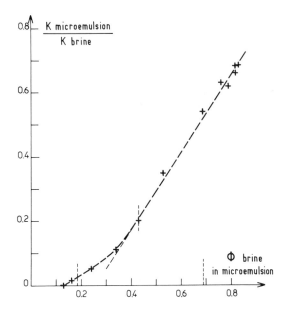

FIGURE 14.   Electrical conductivity vs. salinity in the SDS system.

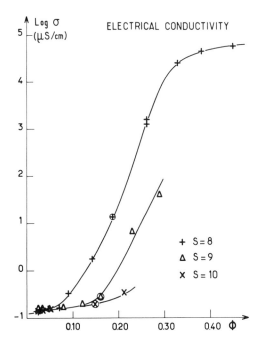

FIGURE 15.   Electrical conductivity vs. droplets volume fraction for three series of W/O microemulsions of different salinites. Circles correspond to the microemulsions obtained from phase separation (points of Figure 14).

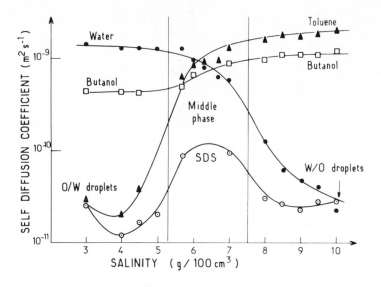

FIGURE 16.   Self diffusion coefficients of the different chemical species in the SDS system as measured by NMR spin echo techniques (after Reference 51). Oil and water mobilities are high in the WIII region, as expected in bicontinuous media.

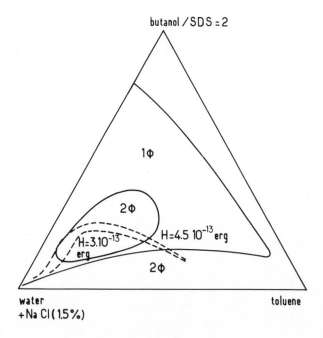

FIGURE 17A.

FIGURE 17.   Phase diagrams for alcohol/SDS ratio = 2 and two different salinities: (a) S = 1.5%; (b) S = 6%. The dotted lines are theoretical and H is the corresponding Hamaker constant; we have taken a droplets charge of 1.33 10^{-3} Cb/m². The full lines are experimental.

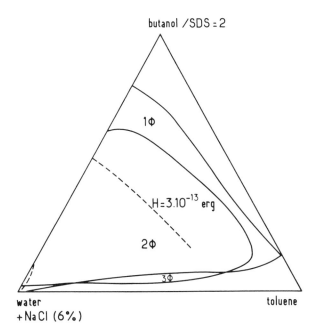

FIGURE 17B.

Table 2
SAMPLES COMPOSITION

Sample	TS	B$_1$	B$_2$	DT
Oil	47.2	—	—	40.6
Butanol	3.96	7.77	8.13	3.6
Surfactant	1.99	3.92	4.46	1.8
Brine	46.8	88.31	87.41	54
S	3—10	5.5—8	6.6	0.4—0.9
T	20°	20°	20°—30°	20°
S$_1$	5.4	6.6	6.6	0.52
S$_2$	7.4	—	—	0.61
T$_C$	—	—	24.85	—

Note: Surfactant is SDS in samples TS, B$_1$, and B$_2$, Texas #1
in samples DT. Temperature of the measurements is T.

D. SDS Systems Without Oil

In the initial microemulsion system, there were many critical points close to the critical end point of interest.[38] It was hopeless, therefore, to approach a single point by varying the temperature. We then decided to simplify the problem and to suppress the oil cores of the droplets since we suspected that the interaction forces involved mainly the outer part of the droplets containing the polar heads of the surfactant molecules. Indeed the mixture of the same components without oil phase separated at a salinity S = 6.6 wt% very close to S$_1$. It was also close to critical.

We have studied both interfacial and bulk properties of this system together with those of slightly different compositions but closer to critical[39] (Table 2). A temperature variation was performed, thus allowing a proper determination of the scale factors. The results are the following:

$$\mu = 1.24 \qquad\qquad \gamma_0 = 0.48 \text{ dyn/cm}$$
$$\nu = 0.62 \qquad\qquad \xi_0 = 7 \text{ Å}$$

The exponents are again close to Ising exponents, but the product $\gamma_0\xi_0^2$ is smaller by a factor 2 than the value for simple fluids and binary mixtures of small molecules.

Further work is planned on this system to better understand the nature of the attractive forces responsible for phase separation.

E. Systems With No Surfactant

Three-phase equilibria are commonly encountered in mixtures without surfactant or with short chain surfactants.

The phase diagram of one of such mixtures has been carefully investigated by Lang et al.[40] Two CEP were found. As temperature increases, the three-phase region shrinks and the two points merge into a tricritical point where the three-phase region disappears. Interfacial tension measurements were in good agreement with critical phenomena theory in the whole three-phase region. Moreover, the Antonov rule was also obeyed:

$$\gamma_{13} = \gamma_{12} + \gamma_{23} \tag{10}$$

where 1, 2, and 3 represent the upper, middle, and lower phases, respectively.

This is very different from the microemulsion system behavior where the Antonov rule was not obeyed:

$$\gamma_{13} < \gamma_{12} + \gamma_{23} \qquad \text{Antonov inequality} \tag{11}$$

and where the middle phase did not wet the oil-water interface.

Other systems were studied later by Davis, Scriven, and co-workers.[41,42] They were mixtures of brine-hydrocarbon-alcohol and their phase behaviors were quite similar to those of the model systems for oil recovery. A salinity scan could induce, for instance, the sequence 2-phase $\rightarrow$ 3-phase $\rightarrow$ 2-phase equilibria. Interfacial tensions were typically two orders of magnitude larger than corresponding mixtures with surfactants.[43] Again, the Antonov rule was satisfied in these systems.[42]

It can be postulated, therefore, that in microemulsions systems, the surfactant layers play an important role and distinguish these systems from mixtures of small molecules. The surfactant layers build up large surface pressures that are responsible most of the time for the low tensions. The unusual wetting properties and critical behavior also seem related to the existence of these layers.

Let us finally mention the intermediate cases of low molecular weight surfactants. Kahlweit and co-workers[12] explained the three-phase formation in oil-water-nonionic short chain surfactants by considering the position of the critical points in the binary mixtures water-nonionic and oil-nonionic. They attribute the origin of low tensions to critical phenomena exclusively. It would be interesting to confirm this with tension measurements and to investigate what happens when the surfactant chain length increases enough to allow for surfactant layer formation.

IV. THEORETICAL PREDICTIONS

A. Surfactant Monolayers

When a surfactant monolayer is present at the oil-water interface, we have seen that the interfacial tension is lowered by a quantity equal to the surface pressure developed by the monolayer:

$$\gamma = \gamma_0 - \Pi \tag{12}$$

π is the derivative of the surfactant free energy $h(\Sigma)$, Σ being the area covered by a surfactant molecule. The total free energy is[44]

$$F = F_{bulk} + \gamma_0 A + n_s h(\Sigma) \tag{13}$$

where A is the total interfacial area and n_s the number of surfactant molecules: $n_s = A/\Sigma$. From $\gamma = \partial F/\partial A$ and Equations 12 and 13, one has indeed:

$$\Pi(\Sigma) = -\frac{\partial h}{\partial \Sigma} \tag{14}$$

The chemical potential of the surfactant is

$$\mu_s = \left.\frac{\partial F}{\partial n_s}\right|_A = h(\Sigma) - \Pi(\Sigma)\Sigma \tag{15}$$

This value is equal to the chemical potential of the surfactant in the bulk which, of course, depends on the shape of the aggregates, on their concentration, and on the interactions between the aggregates.

1. Spherical Droplets Without Interactions

Let us first consider the case where one bulk phase contains spherical droplets without interactions. γ has been derived in this case by Israelachvili[25] in the following way: if N is the aggregation number,

$$\mu_s = \mu_N^0 + \frac{kT}{N} \ln \frac{X_N}{N} \tag{16}$$

where μ_N^0 contains the interaction terms between polar and nonpolar parts of the surfactant in the droplet; X_N is the surfactant concentration present in the droplet (volume fraction).

By using Equations 13 to 15 one can show that:

$$\mu_s = \mu_\infty^0 - \gamma \Sigma \tag{17}$$

μ_∞^0 is the limit of μ_N^0 for an infinite aggregation number: flat monolayer. The interfacial tension is then:

$$\gamma = \frac{1}{\Sigma} \frac{kT}{N} \ln \frac{X_N}{N} + \frac{1}{\Sigma} (\mu_\infty^0 - \mu_N^0) \tag{18}$$

The first term is an entropy contribution γ_E arising from mixing. If the droplets have a radius R, then:

$$\gamma_E = -\frac{kT}{4\pi R^2} \ln \frac{X_N}{N} \tag{19}$$

This expression is analogous to the one derived by Ruckenstein on different thermodynamical bases.[21]

It could be noted that the order of magnitude of γ_E is kT/R^2, to be compared to the tension between simple liquids which are of the order of kT/a^2, "a" being a molecular length. This means that if R is 100 times larger than a molecular length, γ will be 10^4 times smaller than usual surface tensions.

The second term of Equation 18 is a curvature contribution γ_C since it includes the difference between surfactant interactions in spherical and flat aggregates. This contribution has been explicitly calculated by several authors.[19,23,24] As shown recently,[44] these calculations are equivalent to:

$$\gamma_C = \frac{K}{2R^2} \tag{20}$$

where K is a curvature modulus. As K is expected to be of the order of kT, γ_C is expected to be of the same order of magnitude as γ_E. As both contributions are inversely proportional to R^2, it will be difficult to distinguish experimentally between the two contributions.

The experimental values of γ in the SDS system are of the same order of magnitude as those calculated from the preceding theories by using the R experimental values in the WI and WII regions.

2. Bicontinuous Models

The entropic contribution has been calculated by using a Voronoi tesselation model for the structure:[22]

$$\gamma_E \sim \frac{kT}{\xi^2} \tag{21}$$

where ξ is the size of the elementary units (the persistence length in the De Gennes model[42]). Again one sees that γ_E is of the order of kT divided by the square of the characteristic size.

Curvature contributions have not been calculated in this case, but they can reasonably be expected to be negligibly small.

ξ measurements were recently performed with X-ray techniques in a dodecane-water-NaCl-butanol-hexadecyl benzene sulfonate (Texas #1) system.[45,46] At the optimal salinity a value of $\xi \sim 386$ Å has been measured, close to the theoretical value of ξ which is[44]

$$\xi = \frac{6\phi_o\phi_w}{n_s\Sigma}$$

where ϕ_o and ϕ_w are the oil and the water volume fractions and n_s the number of surfactant molecules per unit volume. At the optimal salinity $\phi_o = \phi_w \sim 0.5$ and:

$$\xi_{th}^* = 363 \text{ Å}$$

with $\Sigma = 110$ Å^2, as measured from light and X-ray scattering.[47]

We have recently performed surface tension measurements on this system[47] (see Figure 18). At the optimal salinity $\gamma^* \sim 8.10^{-4}$ dyn/cm.

ξ measurements are not available for the SDS systems where the experiments deviate seriously from the predictions of the Talmon-Prager model.[48] The calculated value at optimal salinity is $\xi^* \sim 220$ Å, smaller than in the Texas #1 system. The ratio of the two calculated values for the two systems is about 1.6, allowing one to predict a ratio of γ^* values of about 2.7. This is only in qualitative agreement with the data. The experimental ratio is noticeably larger: 5.6.

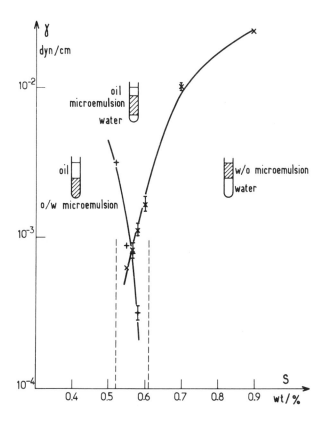

FIGURE 18.   Interfacial tensions vs. salinity in the Texas # 1 system, the composition of which is indicated in Table 2.

It must be noted, however, that theory applies to an equilibrium between two microemulsions of symmetrical compositions ϕ_w and $1 - \phi_w$. Moreover, the theory predicts an oil-rich region close to the oil-rich phase and a water-rich region close to the water-rich phase. Recent experiments with ellipsometric techniques on various model systems including the SDS system[49,50] indicate rather the opposite situation.

B. Critical Behavior

Interactions have been incorporated in some of the droplet models.[18-23] They led to tensions that would obviously depend on the volume fraction of the droplets. This is also true for the entropic term (Equation 18) but as the concentration appears in a logarithm, the variation of tension with the concentration is very small. Let us recall that this was indeed observed in the WI and WII regions far from the WIII region.[16] On the contrary, when the phase boundaries are approached, the interactions between the droplets were observed to increase. The tensions follow the scaling law sufficiently close:

$$\gamma = \gamma_0 \, \epsilon^\mu$$

and the product $\gamma \xi^2$ is a universal constant of the order kT. Thus, again $\gamma \sim kT/\xi^2$ forms similar to what was found for surfactant layers. This probably explains the smoothness of the surface tension variation when crossing the frontier between the two regimes: surfactant monolayer and critical behavior.

Many problems remain to be solved in relation to critical behavior:

- What is the origin of the attractive forces responsible for phase separation?
- Are microemulsions relevant to the Ising model for simpler mixtures?
- If not, what is the nature of the order parameter?

Again in relation to critical exponents, Widom[26] predicted recently that the CEP associated with Winsor equilibria are analogous to He^3-He^4 mixtures and that μ/β should be equal to 2 or 2.5 instead of 4 as observed. The studied systems could be too far from the true CEP, and closer to simpler critical points, thus explaining that μ/β could be equal to 4. All these important questions deserve further studies.

V. CONCLUSION

The quantitative predictions for surface tension in microemulsions are still incomplete, especially close to critical points. The critical behavior might be very different from that of mixtures of simple molecules because of the presence of surfactant films. Up to now, the theories are unable to explain the observed features. The interaction forces leading to phase separation in O/W systems remain mostly unknown. The fact that microemulsions could pertain to a new universality class of Ising models might explain their unusual wetting properties.

On the other hand, even far from critical points, the relationship between surface tension and microemulsion structure still deserves more work, especially in the middle phase domain. Some systematic trends, however, have been tested: γ is roughly inversely proportional to the square of the characteristic length of the structure (droplets radius or persistence length).

ACKNOWLEDGMENTS

A great deal of the work described in this paper has been performed in the Physics Laboratory of the Ecole Normale Supérieure in collaboration with A. M. Cazabat, J. Meunier, D. Chatenay, A. Pouchelon, O. Abillon, and D. Guest.

Helpful collaboration with the other groups of the French cooperative action in microemulsions, especially with P. Lalanne, A.-M. Bellocq, D. Roux, and R. Zana, is gratefully acknowledged.

We have also benefited from very fruitful discussions for the interpretation of the low interfacial tensions with B. W. Ninham, J. Israelachvili, and D. J. Mitchell. We gratefully thank B. Widom for his suggestions relative to the wetting experiments.

REFERENCES

1. **Prince, L. M., Ed.,** Microemulsions: Theory and Practice, Academic Press, New York, 1977.
2. **Schulman, J. M. and Montagne, J. B.,** Formation of microemulsions by amino alkyl alcohols, *Ann. N.Y. Acad. Sci.,* 92, 366, 1961.
3. **Ruckenstein, E. and Chi, J.,** Stability of microemulsions, *J. Chem. Soc. Faraday Trans. 2,* 71, 1690, 1975.
4. **Shah, D. O., Ed.,** *Surface Phenomena in Enhanced Oil Recovery,* Plenum Press, New York, 1981.
5. **Wasan, D. T., McNamara, J. J., Shah, S. M., Samptah, K., and Aderangi, N.,** The role of coalescence phenomena and interfacial rheological properties in enhanced oil recovery: an overview, *J. Rheol.,* 23, 181, 1979.
6. **Clint, J. H., Neustadter, E. L., and Jones, T. J.,** Dynamic interfacial phenomena related to EOR, *Chem. Flood.,* p. 135, 1982.
7. **Desnoyers, J. E., Sarbar, M., and Lemieux, A.,** The tar sand extractions with microemulsions. II. The dispersion of bitumen in microemulsions, *Can. J. Chem. Eng.,* 61, 680, 1983.

8. **Princen, M. M., Zia, J. Y. Z., and Mason, S. G.,** Measurement of interfacial tension from the shape of a rotating drop, *J. Colloid Interface Sci.,* 23, 99, 1967.

9. **Wade, W. H., Morgan, J. C., Schechter, R. S., Jacobson, J. K., and Salager, J. L.,** Interfacial tension and phase bahavior of surfactant systems, paper SPE 6844, *Soc. Pet. Eng. J.,* 18, 242, 1978.

10. **Davis, H. T. and Scriven, L. E.,** The origins of low interfacial tensions for enhanced oil recovery, paper SPE 9278, presented at 55th Annu. Fall and Exhibition of the Society of Petroleum Engineers of AIME, Dallas, Sept. 21 to 24, 1980.

11. **Fleming, P. D., Vinatieri, J. E., and Glinsman, G. R.,** Theory of interfacial tensions in multicomponent systems, *J. Phys. Chem.,* 84, 1526, 1980.

12. **Kahlweit, M., Lessner, E., and Strey, R.,** Influence of the properties of the oil and the surfactant on the phase behavior of systems of the type H_2O-oil-non ionic surfactant, *J. Phys. Chem.,* 87, 5032, 1983.

13. **Kim, M. W., Bock, J., and Huang, J. S.,** Interfacial and critical phenomena in microemulsions, in *Waves on Fluid Interfaces,* Academic Press, New York, 1983, 151.

14. **Puig, J. E., Franses, E. I., Davis, H. T., Miller, W. G., and Scriven, L. E.,** On interfacial tensions measured with alkyl aryl sulfonate surfactants, *Soc. Pet. Eng. J.,* p. 71, 1979.

15. **Pouchelon, A., Meunier, J., Langevin, D., Chatenay, D., and Cazabat, A. M.,** Low interfacial tensions in three-phase systems obtained with oil-water surfactant mixtures, *Chem. Phys. Lett.,* 76, 277, 1980.

16. **Pouchelon, A., Chatenay, D., Meunier, J., and Langevin, D.,** Origin of low interfacial tensions in systems involving microemulsion phases, *J. Colloid Interface Sci.,* 82, 418, 1981.

17. **Cazabat, A. M., Langevin, D., Meunier, J., and Pouchelon, A.,** Critical behavior in microemulsions, *Adv. Colloid Interface Sci.,* 16, 175, 1982.

18. **Miller, C. A., Jwan, R., Benton, W. H., and Fort, T.,** Ultralow interfacial tensions and their relation to phase separation in micellar solutions, *J. Colloid Interface Sci.,* 61, 554, 1977.

19. **Robbins, M. L.,** Theory for the phase behavior of microemulsion, in *Micellization, Solubilization, and Microemulsions,* Vol. 2, Mittal, K. L., Ed., Plenum Press, New York, 1977.

20. **Antoniewicz, P. R. and Rodriguez, R.,** A model for the low interfacial tension of the hydrocarbon-water-surfactant system, *J. Colloid Interface Sci.,* 64, 320, 1978.

21. **Ruckenstein, E.,** Evaluation of the interfacial tension between a microemulsion and the excess dispersed phase, *Soc. Pet. Eng. J.,* 593, 1981; The origin of the middle phase of its chaotic structure and of the low interfacial tension, *Surfactants in Solution,* Vol. 3, Mittal, K. L. and Lindman, B., Eds., Plenum Press, New York, 1984.

22. **Talmon, Y. and Prager, S.,** The statistical thermodynamics of microemulsions. II. The interfacial region, *J. Chem. Phys.,* 76, 1535, 1982.

23. **Huh, C.,** Equilibrium of a microemulsion that coexists with oil and brine, SPE paper 10728, presented at the 3rd SPE/DOE Symp. on Enhanced Oil Recovery, Tulsa, 1982; Formation of a middle phase from a lower or upper phase microemulsion, *J. Colloid Interface Sci.,* 97, 201, 1984.

24. **Mitchell, D. H. and Ninham, B. W.,** Electrostatic curvature contributions to interfacial tension of micellar and microemulsion phases, *J. Phys. Chem.,* 87, 2996, 1983.

25. **Israelachvili, J.,** Physical principles of surfactant self association into micelles, bilayers, vesicles and microemulsion droplets, in *Surfactants in Solution,* Mittal, K. L. and Bothorel, P., Eds, Plenum Press, New York, 1986.

26. **Widom, B.,** Theories of surface tension, in *Surfactants in Solution,* Mittal, K. L. and Bothorel, P., Eds., Plenum Press, New York, 1986.

27. **Langevin, D., Meunier, J., and Chatenay, D.,** Light scattering by liquid surfaces, in *Surfactants in Solution,* Vol. 3, Mittal, K. L. and Lindman, B., Eds., Plenum Press, New York, 1984.

28. **Lachaise, J., Graciaa, A., Martinez, A., and Rousset, A.,** Measurement of low interfacial tensions from the intensity of the light scattered by liquid interfaces, *J. Phys. Lett.,* 40, L599, 1979.

29. **Chatenay, D., Langevin, D., Meunier, J., Bourbon, D., Lalanne, P., and Bellocq, A.-M.,** Measurement of low interfacial tension. Comparison between a light scattering technique and the spinning drop technique, *J. Dispersion Sci. Technol.,* 3, 245, 1982.

30. **Warren, C. and Webb, W. W.,** Interfacial tension of near critical cyclohexane-methanol mixtures, *J. Chem. Phys.,* 50, 3694, 1969.

31. **Lucassen, J.,** The shape of an oil droplet suspended in anaqueous solution with density gradient, *J. Colloid Interface Sci.,* 70, 355, 1979.

32. **Cooke, C. E. and Schulman, J. H., Eds.,** *Surface Chemistry,* Munksgaard, Copenhagen, 1965, 231.

33. **Cash, R. L., Cayias, J. L., Fournier, G., MacAllister, D. J., Schares, T., Schechter, R. S., and Wade, W. L. H.,** The application of low interfacial tension scaling rules to binary hydrocarbon mixtures, *J. Colloid Interface Sci.,* 59, 39, 1977.

34. **Chan, K. S. and Shah, D. O.,** The molecular mechanism for achieving ultralow interfacial tension minimum in a petroleum sulfonate oil brine system, *J. Dispersion Sci. Technol.,* 1, 55, 1980.

35. **Bywilder, D., Bancroft, W. D., and Tucker, C. W.,** Gibbs on emulsification, *J. Phys. Chem.,* 31, 1680, 1927.

36. **Bellocq, A. M., Bourbon, D., Lemanceau, B., and Fourche, G.,** Thermodynamic, interfacial and structural properties of polyphasic microemulsion systems, *J. Colloid Interface Sci.,* 89, 427, 1982.
37. **Brezin, E. and Feng, S.,** Amplitude of the surface tension near the critical point, *Phys. Rev.,* B29, 472, 1984.
38. **Abillon, O.,** Etude d'un Systeme Micellaire pres d'un Point Critique de Demixion, Thesis, University of Paris, 1984.
39. **Abillon, O., Chatenay, D., Langevin, D., and Meunier, J.,** Light scattering study of a lower critical consolute point in a micellar system, *J. Phys. Lett.,* 45, L223, 1984.
40. **Lang, J. C., Lim, P. K., and Widom, B.,** Equilibrium of three liquid phases and approach to the tricritical point in benzene-ethanol-water-ammonium sulfate mixtures, *J. Phys. Chem.,* 80, 1719, 1976.
41. **Knickerbocker, B. M., Pesheck, C. V., Davis, H. T., and Scriven, L. E.,** Patterns of three liquid-phase behaviors illustrated by alcohol-hydrocarbon-water-salt mixtures, *J. Phys. Chem.,* 86, 393, 1982.
42. **Seeto, Y., Puig, J. E., Scriven, L. E., and Davis, H. T.,** Interfacial tensions in systems of three liquid phases, *J. Colloid Interface Sci.,* 96, 360, 1983.
43. **Bellocq, A.-M., Bourbon, D., and Lemanceau, B.,** Thermodynamic, interfacial and structural properties of alcohol-brine-hydrocarbon systems, *J. Dispersion Sci. Technol.,* 2, 27, 1981.
44. **De Gennes, P. G. and Taupin, C.,** Microemulsions and the flexibility of oil/water interfaces, *J. Phys. Chem.,* 86, 2294, 1982.
45. **Kaler, E. W., Davis, H. T., and Scriven, L. E.,** Towards understanding microemulsion microstructure. II., *J. Chem. Phys.,* 79, 5685, 1983.
46. **Guest, D., Auvray, L., and Langevin, D.,** *J. Phys. Lett.,* 46, L1055, 1985.
47. **Abillon, O., Chatenay, D., Guest, D., Langevin, D., and Meunier, J.,** Low interfacial tensions in microemulsion systems, in *Surfactants in Solution,* Mittal, K. L. and Bothorel, P., Eds., Plenum Press, New York, 1986.
48. **Auvray, L., Cotton, J. P., Ober, R., and Taupin, C.,** Concentrated Winsor microemulsions: a small angle X-ray scattering study, *J. Phys.,* 45, 943, 1984.
49. **Beaglehole, E., Clarkson, M. T., and Upton, A.,** Structure of the microemulsion/oil/water interfaces, *J. Colloid Interface Sci.,* 330, 101, 1984.
50. **Tenebre, L., Haouche, G., and Brun, B.,** Ellipsometry in microemulsions, in *Surfactants in Solution,* Mittal, K. L. and Bothorel, P., Eds., Plenum Press, New York, 1986.
51. **Chatenay, E., Guering, P., Urbach, W., Cazabat, A. M., Langevin, D., Meunier, J., Leger, L., and Lindman, B.,** Diffusion coefficients in microemulsions, in *Surfactants in Solution,* Mittal, K. L. and Bothorel, P., Eds., Plenum Press, New York, 1986.
52. **Bellocq, A. M., Biais, J., Clin, B., Gelot, A., Lalanne, P., and Lemanceau, B.,** Three dimensional phase diagram of the brine-toluene-butanol-sodium dodecyl sulfate system, *J. Colloid Interface Sci.,* 74, 311, 1980.

Chapter 8

OIL RECOVERY AND MICROEMULSIONS

P. Neogi

TABLE OF CONTENTS

I. INTRODUCTION

The process for using surfactants to effect the recovery of crude petroleum oil from underground porous oil fields is far from being realized. However, the concepts and the aims are important and have given rise to a great interest in the surfactants capable of producing ultralow surface tensions in surfactant-brine-oil systems. The purpose here is to illustrate the difficulties in recovering the oil and discuss how, with systems such as microemulsions, some of these difficulties may be overcome. Since the understanding of the behavior in such systems is only recent, the research on their applications to oil recovery is very new. Still some observations have been made which, although lacking a coherent form at present, are of considerable interest.

Petroleum is known to be of animal origin, particularly of marine form, which is important because brine is inevitably found along with petroleum. The oil migrates until it is trapped in porous rock formations, such as limestone or sandstone, surrounded by impermeable rock. (It is noteworthy that limestone is also of marine origin). The pores here are less than 1 μm in diameter. The pore sizes provide an estimate for the permeabilities (k) as $k^{1/2}$ permeabilities range typically from 1 to 100 mD,[1] where $1 D = 10^{-8} cm^2$. The notation D stands for Darcy, the common unit for permeability in such systems. The void volume fractions in oil fields are of the order of 10 to 20%, and 10 to 25% of the pore volume is occupied by brine which is called the connate water. The rest is oil, referred to as the original oil in place or OOIP.

There are many books, monographs, and manuals on how oil is extracted from these reservoirs.[1-3,5-8,10-12] Oil is usually found at very high pressures and somewhat elevated temperatures. Consequently, when the first wells are dug, the oil rushes out under its own pressure. This spontaneous production is later supplemented by pumping action; these two stages constitute the primary recovery. The primary recovery manages to extract about 10 to 25% of the OOIP. For the management of the pressures, terrain, and the oil distribution in the reservoirs, the wells are drilled in what are called spot patterns. The area is divided into squares or triangles and the wells are driven at the corners. Thus, one has 4-, 5-, or 7-spot patterns, with many variations.[1] There are additional means for enhancing recovery at this stage. When the oil field is sloped, the oil is extracted from the bottom wells. The drop in pressure is compensated by the volatiles which vaporize from the oil and collect at the top and put pressure on the oil to leave through the bottom. This is known as the solution gas drive.

In the next stage, the secondary recovery, a different phase is introduced to sweep out the remaining oil. Water is most often used in this phase, as it is the least expensive, although carbonated water has also been tried.[13] The water is forced in through the injection well and moves outward in a piston-like fashion, displacing the oil. The latter is collected at the production well. The process has several problems because of which only about 10% of OOIP can be recovered. If the water-oil front is sharp and stable, the sweep efficiencies can be good. However, it has long been known that in a porous medium, the displacement front is unstable if the displacing liquid has a lower viscosity.[14] The low viscosity fluid channels through the high viscosity oil leaving much of the oil undisturbed. This is known as viscous fingering. Fingering also occurs because the reservoirs do not have uniform permeabilities; as different strata have differing permeabilities, the water moves far ahead through the strata that have high permeabilities.

Further, when the displacing fluid is lighter than oil as in carbon dioxide drives, it tends to rise above the oil and finger out through the top instead of displacing the oil in a piston-like fashion. This is known as the gravity override. Instabilities of this kind are also anticipated in water drives where water is heavier and the reverse situation takes place. Because of the instability of the moving front and the inhomogeneity of the reservoir, practical considerations require knowledge of both areal and vertical sweep efficiencies.[1] These range from 50 to 80% for the areal, and usually much less for the vertical sweep efficiencies.

Coupled with the problem of poor sweep efficiencies is that of capillarity. The latter prevents 100% displacement of the oil even when the displacement front is perfectly stable; i.e., it moves in a piston-like fashion. Since this feature is important here, it will be saved for detailed discussion in a later section. Simple considerations suggest that if the capillary forces prevent the displacement of oil then perhaps the water at higher pressures should be used. In water flooding, there is a limit to the amount of pressure that can be used, because high pressures fracture the reservoir rocks. Such fractures decrease the drive efficiencies. It is also seen that most of the pressure drop occurs very near the well and not in the interior of the reservoir. To rectify this, the well walls are treated with acids which corrode the rocks and form bigger pores. In this way, the permeability is increased near the well area, and the large pressure drops there are lowered. This process is known as acidization.[15]

Besides water flooding, flooding with carbonated water has also been used.[13] Carbon dioxide swells the oil and helps to move it out of small capillaries. Steam drives are often used and provide better results. First, the heat of condensing steam decreases the viscosity of the oil. This reduces the viscous fingering. The latter is further reduced because of the decrease in volume on condensation.[16] When the oil is very heavy and viscous, the above processes do not work well; the oil is set on fire at the injection well and the flame is fed by oxygen or air. The heat makes the oil less viscous and flow more easily, and is then recovered at the production well. Sometimes, when the situation permits, solution gas drive is used in conjunction with water flood;[5] the former pushes the oil down from the top and the latter pushes it up from the bottom. The oil is taken out through the middle.

The methods of oil recovery have been summarized briefly in the previous sections. A great deal of ingenuity goes into oil production since oil is such a valuable resource today. The resources are dwindling, and the publications by many government agencies in the U.S. can be used to find the latest information on the past history and future prospects. The resource is a strategic one as well, brought to focus in a report by Rand prepared for the U.S. Central Intelligence Agency.[9] Roughly 450 billion barrels of oil have been discovered. Of these, 118 billion barrels have been extracted and the conventional methods will be used to extract 28 billion barrels more. This leaves 304 billion barrels or 68% of the OOIP to be recovered by the tertiary oil recovery methods, of which the surfactant flooding is a strong candidate.

As has been discussed earlier, the two problems facing this last stage of recovery are the stability of the displacement and capillarity. In the next sections, the fluid flow and the problems of stability are described and the problem of capillarity is introduced.

II. MODELS

One very interesting aspect of oil recovery is that the completion of a drive or a phase takes months or years. Whereas many forms of monitoring the drive exist, it becomes very important to have a foreknowledge of the results. Thus, theoretical calculations have been stressed in this area where only the salient features have been discussed below.

When a liquid flows through a porous medium, the flow is governed by Darcy's law:[2,3]

$$\underline{V} = -\frac{k}{\mu} \nabla(p + \rho gz) \tag{1}$$

where $\underline{V}$ is the superficial velocity (the velocity calculated on the basis of an empty cross-section), k is the permeability, μ is the viscosity, p is the pressure and ρgz is the hydrostatic head where the positive direction of z is the negative of the direction of gravity g, and ρ is the density. The operator ∇ is the gradient operator; i.e., the vector differentiation in three dimensions. Whereas Equation 1 represents the force balance equation, the conversation of mass requires that

$$\nabla \cdot \underset{\sim}{V} = 0 \tag{2}$$

Substituting Equation 1 into Equation 2, one has for constant k and μ

$$\nabla^2(p + \rho gz) = 0 \tag{3}$$

Equation 3 is the well-known Laplace equation and its solutions have been explored in detail for applications in oil recovery.[6]

The case where there are two immiscible fluids in a porous medium is more complicated. Consider a single pore filled with air at atmospheric pressure. At one end, a wetting liquid is introduced, also at atmospheric pressure. The pressure at the meniscus in the liquid is, however, less than atmospheric by an amount $2\gamma/R$ where γ is the interfacial tension and R is the tube radius. Thus, the liquid is imbibed spontaneously into the tube with a velocity given by Hagen-Poiseuille's equation

$$V = \frac{\gamma}{4\mu Rl} a^2 \tag{4}$$

where l is the length of liquid in the tube. As $V = dl/dt$, one has on subsequent manipulations

$$V = \left(\frac{\gamma a^2}{8\mu R}\right)^{1/2} t^{-1/2} \tag{5}$$

Equation 5 is known as the Washburn-Green-Ampt equation.[4] If a porous medium is imagined to be made up of a bundle of capillaries, Equation 4 should describe the rate of penetration. The validity of the Washburn equation has been demonstrated in segregated porous media,[4,17] and finds application in the study of ground-water hydrology where the porous medium-soil is of the segregated or the unconsolidated kind. One reason for its success has been pointed out to be the fact that in such porous media, the two fluid interfaces are sharp.[18]

A different approach is needed to study the case of the consolidated porous media such as the oil bearing rocks. If p_o is the pressure in the oil phase and p_w is the pressure in the water phase, then an equation similar to Laplace equation is used to relate the two:

$$p_c = p_o - p_w \tag{6}$$

where p_c, the capillary pressure, is a function of the volume fraction of the water, S_w. Some success[1] has been reported in writing $(k/\phi)^{1/2} p_c/\gamma_{ow}\cos\lambda$ as a universal function of S_w. Here γ_{ow} is the oil-water surface tension, λ is the equilibrium contact angle, and ϕ is the porosity. In general, p_c depends on the lithologic type of the rock as well as its wettability. Further, $p_c (S_w)$ curves differ in their imbibition from their drainage characteristics. In general, they show that even if the water is forced in at high pressures into the oil bearing rock, it cannot displace all oil at equilibrium. The residual oil is left at $S_o \sim 0.15$ to 0.2, depending on the wettability.[1] Modern literature stresses the random nature of the pore structure and is able to unify some of these concepts and some others that appear below.[19]

Equation 1 is adapted to

$$\underline{V}_o = - \frac{k_{Ro}k}{\mu_o} \nabla P_o \tag{7a}$$

$$\underline{V}_w = - \frac{k_{Rw}k}{\mu_w} \nabla P_w \tag{7b}$$

where k_R are the relative permeabilities which are functions of S_w. In Equation 7, $P_i = p_i + \rho_i gz$. Since neither oil nor water can be driven off entirely, k_{Ro} and k_{Rw} reach values of zero at $S_o \sim 0.10$ to 0.2 and $S_w \sim 0.10$ to 0.2, respectively. When $k_{Ro} = 0$, $k_{Rw} = 1.0$ and vice versa.[1] When $k_{Ro} = 0$, the oil content is so low that droplets of oil are trapped in the form of discontinuous ganglia by capillary action.

For a one-dimensional flow in the horizontal direction, the effect of gravity is zero, and if the capillary pressure is neglected then the fractional flow

$$f_w = \frac{V_w}{V_w + V_o} \tag{8}$$

becomes on using Equation 7, and with $p_o = p_w = p$,

$$f_w = \frac{k_{Rw}/\mu_w}{[k_{Rw}/\mu_w + k_{Ro}/\mu_o]} \tag{9}$$

where k_{Ri}/μ_i is the mobility of the ith phase. Like p_c, f_w can also be measured experimentally.[8] In general, f_w is assumed to be a function of S_w and the ratio μ_o/μ_w.

The equation for conservation of mass for water is

$$\frac{\partial}{\partial t}(\phi S_w) = - \nabla \cdot \underline{V}_w \tag{10}$$

If the displacement is one-dimensional, as shown in Figure 1, then S_w is a function of x and t. That is

$$dS_w = \left(\frac{\partial S_w}{\partial x}\right)_t dx + \left(\frac{\partial S_w}{\partial t}\right)_x dt \tag{11}$$

If we wish to follow a point $X(t)$ in space where S_w is a constant, then from Equation 11 at constant S_w,

$$\left(\frac{dX}{dt}\right)_{Sw} = - \frac{(\partial S_w/\partial t)_x}{(\partial S_w/\partial x)_t} \tag{12}$$

At constant ϕ, Equation 10 is $(\partial S_w/\partial t)x = - 1/\phi \; \partial V_w/\partial x$. Further, $(\partial V_w/\partial x) = (\partial V_w/\partial S_w)(\partial S_w/\partial x)$ by chain rule and with these modifications, Equation 12 becomes

$$\frac{dX}{dt} = \frac{1}{\phi}\left(\frac{\partial V_w}{\partial S_w}\right)_t \tag{13}$$

Now $v_w = Vf_w$, where if the fluids are incompressible V, the net velocity is a constant. Substituting for V_w in Equation 13 and integrating, one has

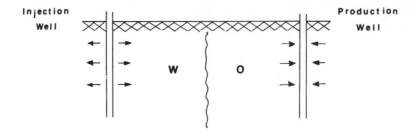

(a) Stable Displacement

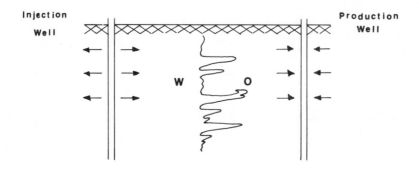

(b) Viscous Fingering

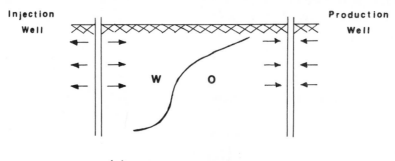

(c) Gravity Override

FIGURE 1.   One-dimensional displacement process. Water (W) is displacing oil (O) in (a) stable displacements, and unstable displacement with (b) viscous fingering and (c) gravity override (in case c, a lighter displacing fluid is used instead of water).

$$X = \frac{1}{A\phi} \left(\frac{df_w}{dS_w} \right)_{S_w} \int VAdt \qquad (14)$$

where the integral is the total volume of water injected, W, and A is the area of flow. If W and A are known, then one picks S_w, finds $(df_w/dS_w)S_w$, and calculates X. In this way, the entire $S_w(x,t)$ can be found. As suggested earlier, the solution shows that $S_o \sim 0.1$ to 0.2 is left behind since k_{Ro} becomes zero there. This is referred to as the residual oil. One problem arises in that $f_w - S_w$ curves are such that for two different values of S_w, the slopes, and hence the calculated values of X, are the same. The phenomenon represents a discontinuity. Here the mathematics show that the water concentration jumps down suddenly from a higher value of S_{w1} to a lower value S_{w2}. Thus, the water displacing the oil is in the

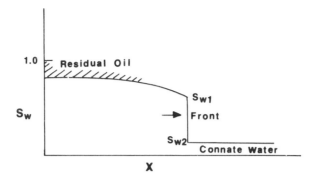

FIGURE 2. A solution under Buckley-Leverett scheme is shown; particularly noteworthy are the retention of the residual oil, the piston-like front, and the reservoir connate water.

form of a front, even though the displacement is not 100%. The nature of the front is sketched in Figure 2.

This solution is called the Buckley-Leverett solution,[2] and it satisfies the equations of the balance of forces and the conversation of mass. In real cases, the reservoir heterogeneities, conservation of species if there are more than one in a phase, interphase mass and heat transfer, capillary pressure, etc. have to be contended with.[1,4] Besides the basic flow, the stability considerations also come into the picture.

The stability analysis that is conventionally performed is a linear stability analysis. Infinitesimal perturbations are given to all variables, to simulate the effects of irregularities in the flow rates, reservoir heterogeneities, etc. which give rise to fluctuations in the basic flow quantities. These perturbed variables are governed by force and material balance equations and can be solved for a perturbation of a given wave number. The solution shows if this perturbation will grow in time. Since the wave number is selected arbitrarily, the results show disturbances of which wave numbers grow in time. If none of them grow, then the process is stable, and any growth indicates instabilities. Since the solutions are obtained for disturbances of arbitrary wave numbers, these can be combined to represent the results of all arbitrary but small disturbances by using Fourier integrals.[12,20] In water flooding,[12] the rate of growth is obtained as

$$\beta = \frac{\alpha V_{wA}}{\phi(S_{wA} - S_{wB})} \frac{(M - 1)}{(M + 1)} \tag{15}$$

where M is the mobility ratio $(k_{RwA}/\mu_w)(\mu_o/k_{RoB})$ where the subscripts A and B indicate the regions immediately behind or ahead of the front. The wavelength of disturbance is given by $2\pi/\alpha$ where α is the wave number. The ratio $M > 1$ since oil is more viscous. Consequently, $\beta > 0$; this signifies positive growth rates and viscous fingering takes place as shown schematically in Figure 1. The rate of growth also can be shown to be positive if the porous medium is inhomogeneous, or if the front slightly slants.[12] The latter case develops into the gravity override situation shown in Figure 1. Stability analyses also have been conducted for steam[16] and combustion drives.[21]

Two points of interest that emerge from these analyses are noted below. First, the stability analyses often show that the disturbances of large wavelength are unstable. When these displacements are studied in the laboratories — "core tests" — the front sizes are too small to accommodate the disturbances of large wavelengths and the observed results show that the displacements are "stable". In oil fields the fronts are larger, of the order of miles, where large-scale disturbances can occur and the drives become unstable. This result is in

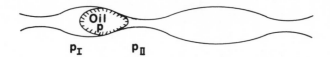

p_I p_{II}

FIGURE 3.   A trapped oil ganglion in a pore with changing cross-sectional areas.

contrast to that predicted by the labscale experiments.[16] Secondly, the mobility ratio M is always expected to be adverse, that is, M > 1. Consequently, instability is a bothersome and persistent problem in most drives. To remedy this situation, a slug of polymer solution is injected following the injection of the displacing medium. This high viscosity slug helps to restore favorable mobility ratio, M < 1.[22] The polymers are usually water-soluble bio-polymers and have a problem in that they degrade at ~150°F, typical oil field temperatures. At the present time, foams have been found to be a good material to control mobility.[23]

III. CAPILLARITY

The effects of capillarity are reviewed in Reference 24. Though more modern research and newer concepts have been reported since then which provide different perspectives (and probably a better chance at quantitizing these effects), the more conventional views in Reference 24 provide at least a basic grasp of the physical aspects of capillarity in porous media.

During the formation of sandstone, and certainly limestone, water was present and these rocks are thus assumed to be water wet. Cleaned oil-bearing rocks are also found to be water wet. However, if the oil contains metal ions, then they combine with asphaltic materials to form surfactants. The polar parts of the surfactants adsorb on the walls and as the hydrocarbon parts are then exposed, the rocks become oil wet. The immediate effect of wettability is that it affects the capillary pressures, the relative permeabilities, connate water, residual oil, etc. The most important of these for present purposes is the residual oil. As mentioned previously, at this stage the relative permeability of oil $k_{Ro} = 0$, and the oil becomes immobile, trapped by the capillary action in the form of disconnected ganglia.

The mechanisms of capillary action in trapping oil are discussed next. The first is demonstrated in Figure 3, where a single pore with changing cross-section is shown. The pore is assumed to be water wet. It is possible to use the Laplace equation to write the pressure differences on the two sides of a trapped oil drop as

$$p - p_I = \frac{2\gamma_{ow}}{R_I} \tag{16a}$$

$$p - p_{II} = \frac{2\gamma_{ow}}{R_{II}} \tag{16b}$$

and combining the two the relation

$$p_I - p_{II} = 2\gamma_{ow}\left(\frac{1}{R_{II}} - \frac{1}{R_I}\right) \tag{16c}$$

is obtained, where R_I and R_{II} are the two radii of curvature at the two interfaces. As $R_I > R_{II}$, $p_I - p_{II} > 0$, that is, in spite of a driving force on the oil drop it is at equilibrium. If $p_I - p_{II}$ is increased slightly over the value on the right-hand side in Equation 16c, the drop

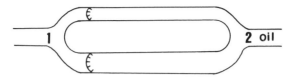

FIGURE 4.   Scheme for displacement of oil in a pore-doublet.

will move quite rapidly until it is trapped again in a similar position. This rapid movement is known as the Haines jump. This jump can also be demonstrated for a stack of tori or spherical packings. The latter cases also demonstrate the role played by pendular rings, where the metastable states are formed by the droplets of the wetting liquid wrapping themselves around the points of contact.

With some adaptations, the geometry of Figure 3 can also be used to demonstrate the trapping of oil when the rock is oil wet. In general, Equations 16a—c are modified by the wettability, i.e., the equilibrium contact angle λ. Since the advancing contact angles are different from the receding contact angles,[25] it can be seen that the entrapment will change with imbibition or drainage. Whether or not this method traps a significant amount of oil obviously depends on whether the lithologic rock type contains many constrictions. The most important observation to make here is that the oil cannot be trapped if γ_{ow} becomes zero. It is noteworthy that γ_{ow} can become zero if the two fluids become miscible.

A second kind of trapping action is demonstrated with the help of a doublet, shown in Figure 4. The doublet has two pores in parallel containing oil. One of the pores has a smaller radius than the other. In this case, the rock is assumed to be water wet. Because of the capillary forces, water is imbibed into both pores (see the development of Equations 4 and 5). However, since the capillary driving force in the upper pore with the smaller radius is higher, water will move into that pore faster, and when it reaches point 2 the oil in the larger pore becomes isolated. Obviously the reverse is true when the rock is oil wet, when the oil is trapped into the pore with a smaller radius. The complicating factor here is the viscous effect. Since the viscous resistance to flow is much stronger in small pores, the capillary pressure may not be able to overcome this. Further, the effect of a pressure drop $\Delta p = p_1 - p_2$ in the flooding process needs to be accounted for as well. Detailed force and material balances[26,27] lead to the conclusions that both in oil wet and water wet rocks oil is trapped in the smaller pore, except for some cases in water wet rocks where the oil can be trapped in the large pore. Whereas the amount of trapped oil can be reduced by decreasing the oil-water interfacial tension to ultralow values in a water wet rock, this cannot be achieved if the oil is trapped in the smaller pores in an oil wet rock. Thus, in most cases, the amount of oil trapped can be reduced by reducing the interfacial tensions to ultralow values, but this method is expected to work well only in water wet or partially water wet rocks where an order of magnitude estimate can be used to show that the interfacial tension between the displacing medium and oil will have to be below 10^{-2} mN/m.[28]

The third model used to depict oil entrapment and mobilization is based on the performance of a ganglion or ganglia in model or random pore networks. The fluid mechanics in a segregated porous medium have been studied[29] as well as fluid mechanics with mass transfer,[30] both for a single ganglion. More interesting results are obtained, which can be used to explain residual saturations and relative permeabilities, in the studies on regular or random network structures.[31-33] In one case it was found that a single large ganglion disintegrates rather easily into daughter ganglia which get trapped, unless the oil-water interfacial tension is very low. The results of these studies are rich in details, but it is quite likely that they are affected by the assumed nature of the pore structures. A detailed discussion of what random pore structures could be and their impact on the properties of two-phase flow in porous media has been given by Larson et al.[19]

It is fair to say that tertiary oil recovery processes are those in which the attempt to improve the oil recovery is made by decreasing the interfacial tensions, and at times by improving the rock wettability to water. Whereas the first controls the amount of residual oil in water floods, the second determines where and how such oil is trapped. The simplest of these processes is the NaOH flood. It improves the water wettability of the rock, and in the presence of NaCl can attain interfacial tensions of less than 10^{-2} mN/m.[34] Consequently, in other processes a NaOH pre-flush is often employed. Carbon dioxide flood in another process of considerable interest and a very promising one. Although by itself it is not miscible with oil, on continued contact sufficient carbon dioxide can dissolve in the oil and sufficient oil can evaporate into the carbon dioxide phase and attain miscibility.[35] In this case, the interfacial tension becomes zero. The problem of instability here is quite serious, but the recent attempts at delivering carbon dioxide with foams[23] seem promising.

The obvious interest here is in the use of surfactants to attain ultralow interfacial tensions and improve oil recovery. These surfactants in the first generation processes are sulfonated crude oil, neutralized with caustic. The delivery of these surfactants has been proposed in various forms, low concentration micellar solutions, high concentration solutions, foams, and even vesicles, and in the form of microemulsions.[36-38] The knowledge of how these processes work is incomplete, with the exception of microemulsion flooding where at least the concept is simple and direct even though much of the details are not known.

IV. MICROEMULSION FLOODING

In the previous sections, it has been discussed that there are two basic problems facing improved oil recovery. The first is that the floods are usually unstable, and hence have reduced sweep efficiencies. The second is that even if the floods are stable, the capillary effects overwhelm the viscous forces and a good portion of oil cannot be recovered. The solution is to provide a flood medium with a favorable mobility ratio $M < 1$, which also has an ultralow tension with oil. The microemulsion flood is discussed below from these viewpoints.

The physical chemistry, thermodynamics, and phase equilibrium of oil-water-surfactant systems which are capable of producing ultralow interfacial tension are the subjects of detailed discussion in this volume. Only a brief summary, particularly of the features of interest in oil recovery, is given later. Many phases are seen in such systems, phases which mainly differ from one another in their microstructures. Some of these are micelles, reverse micelles, oil-in-water (O/W) and water-in-oil (W/O) microemulsions, lamellar liquid crystals, hexagonal or cubic cylindrical liquid crystals and their inverses, etc. A system of importance to oil recovery would have brine as a single pseudocomponent instead of water, since the connate water in the reservoir is brine and the water available for flooding is often brine. Strong electrolytes have a significant influence on the ionic surfactants under consideration. In addition to the effects of NaCl, one needs to study the effects of cosurfactants in the system. The need[39] and the efficacy of the cosurfactants,[40] which are usually short chain alcohols, to produce systems with ultralow surface tensions with the ionic surfactants have long been known. The oil is usually modeled as a single pseudocomponent: *n*-hexane, *n*-heptane, *p*-xylene, etc. Consequently, the phase equilibria in such systems are shown on pseudoternary diagrams. Here, oil forms one component and the remaining ones are surfactant-cosurfactant and brine as two pseudocomponents. The microemulsions and the middle phase are found on the low surfactant-cosurfactant side of the phase diagram, which is very significant as the oil recovery process with microemulsions, etc. will involve low surfactant concentrations and consequently low surfactant consumption.

The O/W microemulsion has a very low surface tension against excess oil and similarly W/O microemulsion has a low surface tension against excess water. The O/W microemulsions

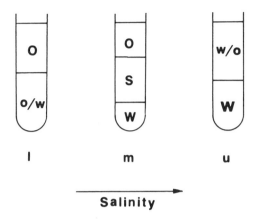

FIGURE 5. Schematically shown *l*, *m*, and *u* microemulsion types in a test tube.

occur at low salt concentrations. When the salt concentration is increased, an isotropic middle phase (also known as the surfactant phase) appears in equilibrium with oil and water. With increasing salt concentrations, W/O microemulsion results in equilibrium with excess water. Because of the way these phases arrange themselves in a test tube, they are known as *l*, *m*, and *u* phase microemulsions, respectively, as shown in Figure 5. The notations stand for lower, middle, and upper phases. The interfaces in *l*, *m*, and *u* systems have ultralow surface tensions and have been shown for a real system in Figure 6. It is obvious that there are two interfaces, an oil interface which vanishes with increasing salinity and a water interface which appears with increasing salinity. Such transitions can also be effected by changing the amounts or natures of the cosurfactants[41] or the nature of the oil, specifically its aromaticity.[42] The details of the phase equilibria itself and the description of where and under what conditions microemulsions are to be found are very complex.[43,44] It will suffice to state here that to obtain microemulsions it is necessary to have a surfactant film at brine oil interface that has an ultralow surface tension.[45] From practical considerations, it is necessary to have a knowledge of the salinity at which the two interfaces in Figure 6 have equal surface tensions. This salinity is known as the optimal salinity where the tensions of both water and oil interfaces are ultralow and, as shown also in Figure 6, the amount of either water or oil solubilized by the surfactant is very high. The optimum salinity has been correlated with the structures of oil, cosurfactants, and even extended to mixtures of surfactants.[46,47] Consequently, given the salinity (from concentration in the connate water or brine flush), a surfactant-cosurfactant system can be designed so that the optimum conditions result, viz., high solubilization and ultralow interfacial tensions.

Following Reference 28, a simple form for microemulsion flooding is given below. Figure 7 shows a part of the phase diagram where oil is in equilibrium with O/W microemulsions. When the microemulsion phase (M) is contacted with a brine and oil mixture (X), diffusion of components among phases results. The diffusion paths, in the first approximation, are given by straight lines,[48] MX in this case. Whereas the segment MQ lies in the single-phase region, QX lies in the two-phase region. According to the average composition of Q′, the average between Q and X, two phases appear, K and oil connected by a tie-line. The diffusion path moves to MK, where K is in equilibrium with oil. In general, equations of conservation of individual species need to be solved to obtain the point K, the locus of which can lie on the brine-K-Q curve. From this diffusion path it is possible to construct the layout of the flood process as shown in Figure 8.

The microemulsion slug of composition M moves into regions of composition Q to K which are in equilibrium with oil. Because of its ultralow surface tension against oil, the

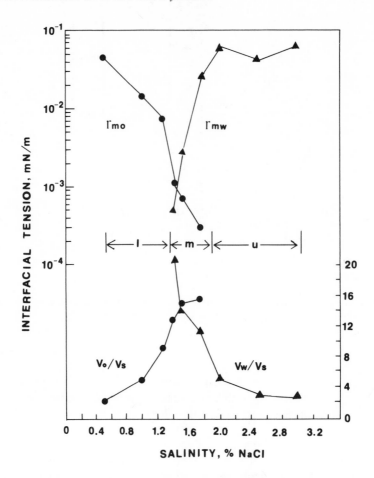

FIGURE 6.   Interfacial tensions γ_{mo} (oil-microemulsion), γ_{mw} (water-micro-emulsion), and volume of oil solubilized per unit volume of the surfactant V_o/V_s and volume of water solubilized per unit volume of the surfactant V_w/V_s. The chemistry of the components is given in Reference 28. (From Reed, R. L. and Healy, R. N., *Improved Oil Recovery by Surfactant and Polymer Flooding*, Shah, D. O. and Schechter, R. S., Eds., Academic Press, New York, 1977, 383. With permission.)

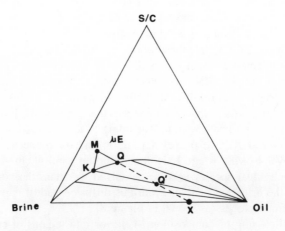

FIGURE 7.   Contacting process between microemulsion (μE) of composition M and oil and water mixture at X.

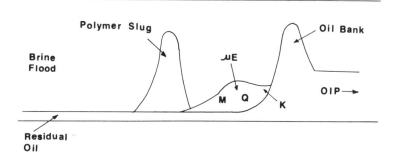

FIGURE 8. The displacement process has been shown schematically. OIP is the oil-in-place and μE is the microemulsion phase.

trapped oil ganglia become mobile and coalesce to form an oil bank. As the oil bank moves forward, it coalesces with the isolated ganglia which are thus mobilized. Some residual oil is still left behind. The microemulsion flood is backed by a polymer flood which imparts to the process the favorable mobility ratio $M < 1$.

Since in an actual system the diffusivity of surfactant is different from that of the cosurfactant and the diffusivity of water and NaCl are different, the diffusion paths cannot be represented on a single pseudoternary diagram. Indeed, they need to be traced on a pentenary diagram of a oil-water-brine-surfactant-cosurfactant system. When, for instance, variations in the brine concentrations occur, different phase equilibria (on the pseudoternary diagrams) are encountered at different points, which affect the oil recovery.[30] Similar effects may be found with the variation of the cosurfactant.[41] The very interesting observation made in these cases is the formation of liquid crystals, sometimes seen in between the microemulsion phase and the oil phase.[49,50] Such liquid crystals are transient in nature as the pseudoternary phase equilibria diagrams show. However, when the very viscous liquid crystals occur between the microemulsion phase and the oil phase,[50] the mass transfer is reduced and an unfavorable mobility ratio $M > 1$ is established in the drive. When the liquid crystals occur behind the microemulsion region, mass transfer resistance there becomes irrelevant and the favorable mobility ratio $M < 1$ is imposed.[51] In systems in which a micellar solution is contacted with oil,[52] the formation of liquid crystals appears to be almost certain since the liquid crystals are to be found in equilibrium with micellar solution.

The mechanism of how oil ganglia become mobile is an important one. The original diffusion path on contacting is MQX in Figure 7. As stated earlier, the path QX is eventually split into R and oil. Actually this path is established by diffusion, but as it falls in the two-phase region it denotes supersaturation. The process by which phase separation takes place in the supersaturated solution to give rise to two saturated liquids K and oil is called spontaneous emulsification,[48] which is often suggested as the mechanism under which an oil ganglion is broken up or becomes mobile. Some investigations suggest that the swelling of the ganglion takes place when the surfactant dissolves in oil.[30] This results in the ganglion being squeezed out of a narrow pore into a more favorable position. Of course this implies that the right-hand corner of Figure 7 is a surfactant-oil solution, and in fact inverse micelles are often found in that region.[53]

Ultimately the loosened oil ganglia must coalesce to form the oil bank, a feature which has been studied as well.[54-56]

The difficulty with microemulsion floods is the adverse mobility ratio $M > 1$, to rectify which a polymer slug is introduced following the microemulsion slug. The polymer solutions have viscosities higher than oil and hence provide the process with a favorable mobility ratio, $M < 1$. As mentioned previously, the polymers are water soluble, and one difficulty with them is that they degrade at higher temperatures. One peculiarity of polymers when they flow through a porous medium is the "chromatographic effect", in which the large polymer molecules prefer to enter into pores of larger diameters and hence elute faster than anticipated under Darcy's law. A polymer slug may thus overtake a slug of micellar solution under proper conditions. Other methods for improving the delivery systems have been suggested in the form of vesicles.[38] In carbon dioxide flooding, foams have been used advantageously.[23] It is to be noted that at high pressures the foam morphology changes, giving rise to very small cell sizes[57,58] and in that form the system may not be very different from vesicles. The use of foams in surfactant flooding has not seen detailed investigation.

Micellar or microemulsion floods can fail if there are cations of higher valences in the system. The surfactants with divalent (Ca^{2+}, Mg^{2+}, etc.) or trivalent (Al^{3+}, etc.) cations, produced by ion-exchange process in the reservoir, have low solubilities and precipitate. By using branched chain alcohols it is possible to obtain surfactant films with a Ca^{2+}:Na^+ ratio found in sea water with which microemulsions can be formed.[59]

The question of how classical thermodynamics (particularly of phase equilibria) and fluid flow are affected by the special conditions in a reservoir demands attention. Under the conditions of high temperatures, most nonionic surfactants cannot be used. Temperature effects in usual anionic surfactants are not known in detail. The high pressure in the reservoir is known to change phase equilibria[60] but not much is known in a quantitative form. Whereas classical thermodynamics deal with systems in "container-less walls", the wall effects are very important here. At present, a rigorous theory of wall effects is available for only simple fluids.[61] In nonequilibrium studies, mass transfer is very important, the fundamentals of which are yet to be established in a formal way. The little that is known[50-52,62-64] provides a variety of mechanisms. The diffusion process is also known to change inside pores or other restrictive geometries.[65] In fluid flow, where it is known that at large velocities when one fluid displaces another in a straight capillary tube, the displaced fluid is left behind as a lubricating layer next to the walls;[66] in contrast in capillaries of diameters in the submicron range, the lubricating layer is that of the wetting liquid,[67] hence overriding some of the conventional hydrodynamics in favor of surface effects. Further, the question of flow resistance of microstructures through very narrow pores remains to be investigated.

Finally, the question of feasibility has to be dealt with. From a technical point of view, these processes have been tested in oil fields. Reference 34, together with many publications in the petroleum engineering journals and publications of the U.S. Department of Energy, lists the performance in terms of pore volumes of oil produced and water to oil ratio, at the production well against pore volumes of surfactant solution injected. A significant rise in the production level is often observed, especially over water flood. When the process does not work, precipitation of surfactant is often the important cause. An overview of the process, in terms of the expected ranges of oil recovery, is not yet known. If the area of drive in an oil field of 25% porosity is 10 mi², and microemulsion bank containing 15% surfactant is 1-ft wide, it would require about 3×10^8 kg of surfactant. Thus when the process goes into production, the amount of surfactant. Thus when the process goes into production, the amount of surfactant needed would swamp the current production levels or demand. To cut the cost, the surfactant is to be produced at site with crude oil as mentioned earlier. Lastly, the economic feasibility will determine when the process would go into practice, the economic factors being the cost of oil so produced vs. the cost of oil produced by other means, demand, supply, and scarcity of oil, many of which can be affected by global economic and political factors as well.

ACKNOWLEDGMENT

This work was supported by DOE Grant #DOE-ACO2 83ER13083.

REFERENCES

1. **Craig, F. F., Jr.,** *The Reservoir Engineering Aspects of Waterflooding,* Vol. 3, Society of Petroleum Engineers of AIME, Dallas, 1971.
2. **Scheiddeger, A. F.,** *The Physics of Flow through Porous Media,* 3rd ed., University of Toronto Press, 1974.
3. **Bear, J.,** *Dynamics of Fluids in Porous Media,* Elsevier, Amsterdam, 1972.
4. **Wooding, R. A. and Morel-Seytoux, N. J.,** *Annu. Rev. Fluid Mech.,* 8, 233, 1976.
5. **Craft, B. C. and Hawkins, M. F.,** *Applied Petroleum Reservoir Engineering,* Prentice-Hall, Englewood Cliffs, N.J., 1959.
6. **Muskat, M.,** *Physical Principles of Oil Production,* McGraw-Hill, New York, 1949.
7. **Cole, F. W.,** *Reservoir Engineering Manual,* Gulf Publishing Co., Houston, 1969.
8. **Amyx, J. W., Bass, D. M., Jr., and Whiting, R. L.,** *Petroleum Reservoir Engineering, Physical Properties,* McGraw-Hill, New York, 1960.
9. **Nehring, R.,** *Giant Oil Fields and World Resources,* Rand, Santa Monica, Calif., 1978.
10. **Schumacher, M. M., Ed.,** *Enhanced Recovery of Residual and Heavy Oils,* 2nd ed., Noyes Data Corp., Parkridge, N.J., 1980.
11. **Van Poollen, H. K. and Associates, Inc.,** *Fundamentals of Enhanced Oil Recovery,* Penn Well Books, Tulsa, 1980.
12. **Miller, C. A.,** Underground Processing (lecture notes); see also *Chem. Eng. Educ.,* 15, 198, 1981.
13. **Claridge, E. L. and Bondor, P. L.,** *Soc. Pet. Eng. J.,* 14, 609, 1974.
14. **Saffman, P. G. and Taylor, G. I.,** *Proc. R. Soc. London, Ser. A,* 245, 312, 1958.
15. **Schechter, R. S. and Gidley, J. L.,** *AIChE J.,* 15, 339, 1969; **Grim, J. A. and Schechter, R. S.,** *Soc. Pet. Eng. J.,* 11, 390, 1971; **Lund, K., Fogler, H. S., and McCune, C. C.,** *Chem. Eng. Sci.,* 28, 691, 1973; **Lund, K., Fogler, H. S., McCune, C. C., Conwell, C., and Ault, J. W.,** *Chem. Eng. Sci.,* 30, 825, 1975; **Fogler, H. S., Lund, K., and McCune, C. C.,** *Chem. Eng. Sci.,* 30, 1325, 1975; **Lund, K., Fogler, H. W., McCune, C. C., and Ault, J. W.,** *Chem. Eng. Sci.,* 31, 373, 1976; **Lund, K. and Fogler, H. S.,** *Chem. Eng. Sci.,* 31, 381, 1976; **McCune, C. C., Fogler, H. S., Lund, K., Cunningham, J. R., and Ault, J. W.,** *Soc. Pet. Eng. J.,* 15, 361, 1975.
16. **Miller, C. A.,** *AIChE J.,* 21, 474, 1975.
17. **Jaycock, M. J. and Parfitt, G. D., Eds.,** *Chemistry of Interfaces,* Ellis Horwood, New York, 1980.
18. **Childs, E. C.,** *The Physical Basis of Soil Water Phenomena,* Wiley-Interscience, London, 1969.
19. **Larson, R. G., Scriven, L. E., and Davis, H. T.,** *Chem. Eng. Sci.,* 36, 57, 1981; **Larson, R. G., Davis, H. T., and Scriven, L. E.,** *Chem. Eng. Sci.,* 36, 75, 1981.
20. **Miller, C. A.,** *Surface and Colloid Science,* Vol. 10, Matijevic, E., Ed., Plenum Press, New York, 1978.
21. **Armento, M. E. and Miller, C. A.,** *Soc. Pet. Eng. J.* 17, 423, 1977.
22. **Sandiford, B. B., Willhite, G. P., Dominguez, J. G., and Trushenski, S. P.,** *Improved Oil Recovery by Surfactant and Polymer Flooding,* Shah, D. O. and Schechter, R. S., Eds., Academic Press, New York, 1977.
23. **Dellinger, S. E., Patton, J. T., and Holbrook, S. T.,** *Soc. Pet. Eng. J.,* 24, 191, 1984.
24. **Stegemeier, G. L.,** *Improved Oil Recovery by Surfactant and Polymer Flooding,* Shah, D. O. and Schechter, R. S., Eds., Academic Press, New York, 1977, 55.
25. **Zisman, W. A.,** *Contact Angle, Wettability, and Adhesion,* Adv. in Chem. Ser. 43, ACS, Washington, D.C., 1964, 1.
26. **Slattery, J. C.,** *AIChE J.,* 20, 1145, 1974.
27. **Oh, S. G. and Slattery, J. C.,** *Soc. Pet. Eng. J.,* 19, 83, 1979.
28. **Reed, R. L. and Healy, R. N.,** in *Improved Oil Recovery by Surfactant and Polymer Flooding,* Shah, D. O. and Schechter, R. S., Eds., Academic Press, New York, 1977, 383.
29. **Ng, K., Davis, H. T., and Scriven, L. E.,** *Chem. Eng. Sci.,* 33, 1009, 1978.
30. **Lam. A. C., Schechter, R. S., and Wade, W. H.,** *Soc. Pet. Eng. J.,* 23, 781, 1983.
31. **Ng, K. M. and Payatakes, A. C.,** *AIChE J.,* 26, 419, 1980.
32. **Payatakes, A. C., Ng, K. M., and Flumerfelt, R. W.,** *AIChE J.,* 26, 430, 1980.
33. **Lin, C.-Y. and Slattery, J. C.,** *AIChE J.,* 28, 311, 1982.
34. **Wilson, L. A., Jr.,** *Improved Oil Recovery by Surfactant and Polymer Flooding,* Shah, D. O. and Schechter, R. S., Eds., Academic Press, New York, 1977.
35. **Hutchinson, C. A. and Braun, P. H.,** *AIChE J.,* 7, 64, 1961.
36. **Bansal, V. K. and Shah, D. O.,** *Micellization, Solubilization, and Microemulsions,* Vol. 1, Mital, K. L., Ed., Plenum Press, New York, 1977, 87; *Microemulsions: Theory and Practice,* Prince, L. M., Ed., Academic Press, New York, 1977, 149.
37. **Gogarty, W. B.,** *Improved Oil Recovery by Surfactant and Polymer Flooding,* Shah, D. O. and Schechter, R. S., Eds., Academic Press, New York, 1977, 27.

38. **Puig, J. E., Franses, E. I., Tahmon, Y., Davis, H. T., Miller, W. G., and Scriven, L. E.,** *Soc. Pet. Eng. J.,* 22, 37, 1982.

39. **Prince, L. M.,** *Microemulsions: Theory and Practice,* Academic Press, New York, 1977, 91.

40. **Salter, S. J.,** The Influence of Type and Amount of Alcohol on Surfactant-Oil-Brine Phase Behavior and Properties, presented at Annu. Fall Meet., SPE preprint 6843, Denver, 1977.

41. **Benton, W. J., Natoli, J., Qutubuddin, S., Mukherjee, S., Miller, C. A., and Fort, T., Jr.,** *Soc. Pet. Eng. J.,* 22, 53, 1982.

42. **Robbins, M. L.,** *Micellization, Solubilization, and Microemulsions,* Vol. 2, Mittal, K. L., Ed., Plenum Press, New York, 1977, 713.

43. **Ekwall, P.,** *Advances in Liquid Crystals,* Vol. 1, Brown, G. H., Ed., Academic Press, New York, 1974, 1.

44. **Bellocq, A. M., Biais, J., Bothorel, P., Clin, B., Fourche, G., Lalanne, P., Lamaire, B., Lemanceau, B., and Roux, D.,** *Adv. Colloid Interface Sci.,* 20, 167, 1984.

45. **Ruckenstein, E. and Chi, J. C.,** *J. Chem. Soc. Faraday Trans. 2,* 81, 1690, 1975.

46. **Salager, J. L., Dourrel, M., Schechter, R. S., and Wade, W. H.,** *Soc. Pet. Eng. J.,* 19, 271, 1979.

47. **Cash, L., Cayias, J. L., Fournier, G., MacAllister, D., Shares, T., Schecter, R. S., and Wade, W. H.,** *J. Colloid Interface Sci.,* 59, 39, 1977; **Cayias, J. L., Schechter, R. S., and Wade, W. H.,** *Soc. Pet. Eng. J.,* 16, 351, 1976; **Puerto, M. C. and Reed, R. L.,** *Soc. Pet. Eng. J.,* 23, 699, 1983.

48. **Ruschak, K. J. and Miller, C. A.,** *Ind. Eng. Chem. Fundam.,* 11, 534, 1972.

49. **Miller, C. A., Mukherjee, S., Benton, W. J., Natoli, J., Qutubuddin, S., and Fort, T., Jr.,** *Interfacial Phenomena In Enhanced Oil Recovery,* Wasan, D. and Payatakes, A., Eds., AIChE Symp. Ser. 212, New York, 1982, 28.

50. **Ghose, O. and Miller, C. A.,** *J. Colloid Interface Sci.,* 100, 444, 1985.

51. **Friberg, S. E., Podzimek, M., and Neogi, P.,** *J. Dispersion Sci. Tech.,* 7, 57, 1986.

52. **Friberg, S. E., Podzimek, M., Mortenson, M., and Neogi, P.,** *Sep. Sci. Tech.,* 20, 285, 1985.

53. **Friberg, S. E.,** *Interfacial Phenomena in Apolar Media,* Parfitt, G. D. and Eicke, H., Ed., in preparation.

54. **Wasan, D. T. and Mohan, V.,** *Improved Oil Recovery by Surfactant and Polymer Flooding,* Shah, D. O. and Schechter, R. S., Eds., Academic Press, New York, 1977, 161.

55. **Wasan, D. T., Shah, S. M., Chan, M., Sampath, K., and Shah, R.,** *Chemistry of Oil Recovery,* Johansen, R. T. and Berg, R. L., Eds., ACS Symp. Ser. 91, Washington, D.C., 1979, 115.

56. **Wasan, D. T., Milos, F. S., and DiNardo, P. E.,** *Interfacial Phenomena and Enhanced Oil Recovery,* Wasan, D. T. and Payatakes, A., Eds., AIChE Symp. Ser. 212, New York, 1982, 105.

57. **Holcomb, D. L., Callaway, E., and Curry, L. L.,** *Soc. Pet. Eng. J.,* 21, 410, 1981.

58. **Rand, P. B. and Kraynik, A. M.,** *Soc. Pet. Eng. J.,* 23, 152, 1983.

59. **Shinoda, K., Kunieda, H., Arai, T., and Saija, H.,** *J. Phys. Chem.,* 88, 5126, 1984.

60. **Rossen, W. R. and Kohn, J. P.,** *Soc. Pet. Eng. J.,* 24, 537, 1984.

61. **Teletzke, G. F., Scriven, L. E., and Davis, H. T.,** *J. Colloid Interface Sci.,* 87, 550, 1982.

62. **Neogi, P., Kim, M., and Friberg, S. E.,** *Sep. Sci. Tech.,* 20, 613, 1986.

63. **Benton, W. J., Miller, C. A., and Fort, T., Jr.,** *J. Dispersion Sci. Technol.,* 3, 1, 1982.

64. **Raney, K. H., Benton, W. H., and Miller, C. A.,** in *Macro- and Microemulsions,* Shah, D. O., Ed., American Chemical Society, Washington, D.C., 1985, 193.

65. **Malone, D. M. and Anderson, J. L.,** *Chem. Eng. Sci.,* 33, 1429, 1978.

66. **Miller, C. A. and Neogi, P.,** *Interfacial Phenomena: Equilibrium and Dynamic Effects,* Marcel Dekker, New York, 1985, 304.

67. **Templeton, C. C. and Rushing, S. S., Jr.,** *Trans. AIME,* 201, 211, 1956.

INDEX

A

Acetamide, 80
Additives, effects of, 94, 98, 104—107
Aerosol OT, 137—140, 156
Alcohol, 2—4, 94, 104, 158, 159
Alcohol partition between M′ and O′, K_m, 11—12
Alcohol self-association, 26
Amplitude distortion, 131
Anisotropic nematic phases, 34
Antonov relation, 182
Antonov rule, 190
Aqueous microdomains, 4
Aqueous nonionic systems, 97—99
Area per polar group, 110—111
Association constants of equilibria, 9
Association mode, 7
Attractive droplet structure, 135
Azeotropic-like states, 75

B

Biaxial nematic phase, 34
Bicontinuous microemulsions, 95, 109, 186
Bicontinuous models, 192—193
Bicontinuous structure, 134, 140, 148, 168
Bilayer lamellas, 110
Bilayers, 34
Binary mixtures, 34—35
Birefringent microemulsions, 160
Boltzmann's constants, 121
Bound state, 157
Bragg spacings, 111
Brine salinity, 105
Buckley-Leverett scheme, 203
Butanol-octane-SDS-water, 12
n-Butanol octane SDS water, 11

C

Capillarity, 199, 204—206
Capillary pressure, 200—201
Capillary tube method, 131
CEP P_{CE}^3, 63—65
Chain configuration, 111
Chaotic movements, 95
Chemical potentials, 6—8, 53
Chemical relaxation, 159
Classic concentration-gradient relaxation, 133
Cloud point, 102—103, 105
Cloud point curve, 97, 98, 103
Coagulation-fragmentation, 155, 161—162, 165, 168
Coexistence curve, 37, 42, 61, 97
Coexistence surface, 62
Coexisting phases in multiphase equilibria, 45—52

Conductivity, 108
Consolute temperature, 99—100
Constant composition, critical behavior at, 68—70
Constant value, 53
Continuous solubility region, 107
Cosolubilized structure, 168
Cosurfactant partitioning, 142
Cosurfactants, see also specific types, 140—142, 158—159
Counterions, 158
Critical behavior, 168—169, 193—194
  CEP P_{CE}^3, 63—65
  constant composition, 68—70
  fixed temperature, 65—68
  Line P_c^1, study in several points of, 65—70
  nonaqueous microemulsions, 87
  quaternary microemulsion system, 60—77
  sodium dodecyl sulfate systems, 181—189
Critical end points (CEP), 34—35, 60, 75, 194
  sodium dodecyl sulfate systems, 182—183
Critical exponent, 184
Critical-like behavior, 167—169
Critical micellar concentration (CMC), 24—25, 96—98, 178
Critical points, 34—35, 60, 95
  aqueous systems, 97
  hydrocarbon content, 87
  light scattering results, 87, 90
  nonhydrocarbon system, 87
Critical solution, 87
Critical surface, 62
Critical temperature, 99
Cubic liquid crystals, 125—126
Cubic phases, 34
Cylinders, 34

D

Darcy's law, 199, 210
Densities, 35
Density-like variable, 61
Density space, 35
Diaphragm-cell method, 131
Dielectric constant, 160
Dielectric studies, 110
Diffusion, 119—152
  methods for measuring, 127—134
  theory of, 120
Diffusional constraints, 125
Diffusion-controlled process, 159
Diffusion paths, 209
Diffusion rate, 120
Dilute micelles, 107
Dilution procedures, 143
  surfactant layer properties, 180—181
Dimensionality of order parameter, 60
Dimensionality of space, 60

Q

R

S

X

ntally acceptable alternative. Tests on ebonite, a hard rubber
l, and ASR, a waste product of automotive shredders, indicate
tu gas can be produced in short residence times. Considerable
 to be done to optimize operating conditions and to determine
est with heavy metals, chlorine, and sulfur compounds.

ted

R. J. *Automobile Shredder Residue - The Problem and Potential*
 CMP Report No. 90-1; Center for Materials Production:
, PA, 1990; pp iii, 1-1.
V. S.; Most, I. G.; Wolman, M. R. *Investigation of the Energy
1utomobile Shredder Residue*; USDOE Contract No. DE-AC07-
1; EnerGroup, Inc.: Portland, ME, 1987; pp VI-8-8, VI-8-14, VI-

mical Engineers' Handbook, 6th Edition; Perry, R. H.; Green, D.
1ey, J. O., Eds.; McGraw-Hill Book Co.: New York, NY, 1973;
3-103.

22, 1992

and at 900°C is 2 minutes, whereas conversion time at 900°C without a catalyst is 18 minutes. Residence times for 50 percent and 100 percent conversion were found graphically. The point for 100 percent conversion was found by extrapolating the linear portion of the conversion line. The reaction appears to be zeroth order with respect to carbon. As conversion approaches 100 percent, the reaction is no longer strictly zeroth order because unreactable material (ash) limits access to carbon, but the order goes up only to approximately 0.2, introducing a very small error into the calculated time for total conversion.

Ebonite Continuous Fluid-Bed Reactor Gasification Tests. Bench-scale testing was performed on ebonite in a 1- to 4-lb/hr continuous fluid-bed reactor (CFBR) system, shown in Figure 6. Preheated gas and steam are introduced into the bottom of a 3-inch-diameter reactor. The lower section of the reactor, which is attached to the coal feed system, is made of 3-inch pipe and is 33 inches in length. The freeboard section is made of 4-inch pipe and is 18.75 inches in length. Solids remain in the bed until, through weight loss from gasification, they reach the top of the 3-inch section and fall out through the top bed drain leg, where they are collected in an accumulation vessel. Unreacted fines and some ash particles are entrained and separated from the gas stream by a 3-inch cyclone. Liquids are condensed in one of two parallel, indirect-cooled condensation trains. Gas is then metered and sampled by an on-line mass spectrometer.

Carbon conversion for the ebonite was found to be approximately 90 percent at 900°C, with most of the unreacted ebonite found in the condensation train, indicating that fines blew out of the bed before having sufficient residence time for complete conversion. A narrower particle size for the feed, a lower fluidization velocity, or a larger diameter freeboard section would most likely raise this conversion by reducing fines entrainment. Alternatively, a reactor/cyclone recycle system that is designed for this particular feedstock would also produce higher conversions. Comparing the amount of material in the bed with the feed rate indicates that the residence time for the test was less than one hour. The residence time is extremely dependent on temperature and heatup rate. Ebonite agglomerates at temperatures below approximately 800°C. If the reactor is not above 850°C and at a high heatup rate, the ebonite will agglomerate, greatly reducing the reaction rate and the overall conversion.

Gas produced from gasification and from water-gas shift reactions is between 220 to 280 lbs per 100 lbs of moisture- and ash-free ebonite feed material. The average composition of the product gas is shown in Table IX. Gas produced has a Btu content of approximately 260 Btu/scf. This number does not include nitrogen used in fluidization. Btu content will be lower when inert gas is included, but since the amount of inert gas is process-specific, Btu content of gas produced only is given.

Water conversion was found to be 1.5-2.0 mole water/mole fixed carbon based on material balance data. Trace element analysis showed considerable loss of lead from the ebonite, going from 660 ppm in the feed to 257 ppm in

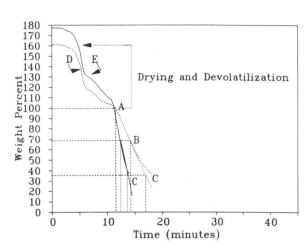

Figure 5. Ebonite - With K_2CO_3 - 800° and 900°C

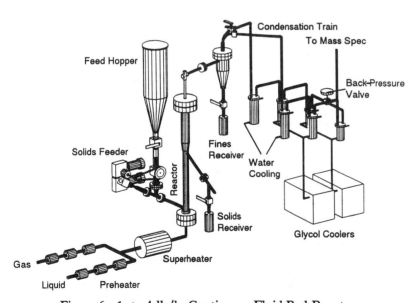

Figure 6. 1- to 4-lb/hr Continuous Fluid-Bed Reactor

Table IX. Ebonite

H_2
CO_2
H_2S
CH_4
CO

[a] Without nitrogen fluidizing gas.

the product char. Antimony also decreas
ppm and ending up at 129 ppm. Chlorine
149 ppm.

Environmental Impacts of Gasifying Wast

Gasification of waste materials offers not o
but also decreases waste volume that ne
density of the feed material is approximate
material (top bed drain) is approximately (
12.8 m³ out of 100 m³ fed will be left over
conversion). If 100 percent conversion is
of feed will be left over to landfill, resu
percent. ASR density is approximately 0.3
is estimated to be about 1.6 g/cm³, resultin
percent.
    Wastewater from the process may co
lead and antimony. Acid leaching the el
desirable to eliminate as much of the hea
operations. Additionally, leachability studi
unconverted material. Gas cleanup proble
sulfur- and chlorine-containing compounds
stream, as well as trace metals that
particulates. The ebonite feed material is
received sulfur content of 3.9 percent.
removal in reduced forms may be one of th
for using a gasifier instead of a combustor.
more soluble in H_2O than Cl_2 (3).

Summary

The future of waste disposal appears to be
incineration and toward recycling and usi
energy. New technologies in pollution
techniques, such as gasification, make ener

an environ
waste mate
that a high
work rema
how to dea

Literature

1. Schmitt
   Solutio
   Pittsbu
2. Hubble
   Value
   84ID12
   8-23.
3. *Perry's*
   W.; M
   pp 3-9

RECEIVED Ju

SEWAGE AND INDUSTRIAL SLUDGE

Chapter 12

Estimating the Heating Value of Sewage Sludge

A New Correlation

M. Rashid Khan, Matthew A. McMahon, and S. J. DeCanio

Research & Development Department, Texaco, Inc.,
P.O. Box 509, Beacon, NY 12508

On a dry-basis, the heating value of sewage sludge is greater than that of oil shale or tar sand. The volatile matter content of dry sludge can be higher than that of the high volatile bituminous coal. Available correlations in the literature, developed for coals, were applied for predicting the experimentally determined heating values. In addition, the sludge compositional data (C, H, S, and ash) were used to develop a new correlation specifically for raw sewage sludge. Compared to the models tested, the new correlation developed in this study for sewage sludge provided a better fit between the measured and predicted values. Sewage sludge is composed of organic and inorganic materials. The organic portion of the sludge is predominantly composed of C, H, N, and S.

A variety of organic and inorganic materials can be found in a waste-water treatment plant (1). Treatment plants receive tremendous quantities of waste-water containing dissolved and suspended solids from numerous sources including domestic, industrial and urban-offs as well as from storm drainage.

The solid residue or sludge, the principal product of primary and secondary treatments, has been traditionally ocean dumped, incinerated or landfilled. However, current federal regulations restrict such traditional practices. The option to dispose of such materials by landfilling also suffers from psychological (e.g., "not-in-my-backyard" syndrome) and genuine environmental concerns (e.g., contamination of ground water or agricultural products and leaching). A recent survey of compositional characteristics of domestic sludges indicate that most sludges can be

0097–6156/93/0515–0144$06.00/0

classified as "hazardous," and consequently not suitable for disposal by landfilling (2). The incineration technology can suffer from emission and public perception problems. Keeping the limitations of these alternatives in mind, conversion of sewage sludge to clean fuels by, for example, gasification (which readily converts essentially all the organic constituents) to synthesis gas (CO and H_2) for power generation or as chemical feedstock, provides an excellent avenue to utilize this renewable resource (3).

The processing and the utilization of sewage sludge requires a better understanding of its physical and chemical properties. In particular, the ability to estimate its calorific value would indeed be of great importance keeping in mind that the measured heating values of sludge are generally not readily available and the reported data can suffer from a relatively large experimental variation (partly due to possible biological/chemical degradation of samples during various treatments). Correlations are important for justification and modeling of the conversion processes now being developed.

The correlations between the coal composition and heating value were reported as early as 1940. Over 20 different equations are reported in the literature which enable one to calculate the heating value of coal based on the ultimate/proximate analyses (4-9). However, essentially nothing could be found in the literature that could be readily applied to specifically estimate the heating value of dewatered sludge.

To assess the utility of existing correlations (developed for coal), the most widely used equations were tested for dry sewage sludge. Mott and Spooner (1940) claimed that their equation will yield heating values agreeing within 200 btu for the whole range of fuels, from peat to anthracite (4). We, however, were much less successful with this equation for dewatered sewage sludge.

Mason and Ghandi (1980) developed a correlation based on coal samples from the Pennsylvania State University coal data base (6). A comparison of the experimental results and the predicted values (based on Mason and Ghandi's equation) was made. Compared to the equation by Mott and Spooner, this equation (termed Data Base [DB] Equation) did a better job in estimating the heating value of sludge. Various equations reported for coal in the literature were attempted for dewatered sewage sludge. However, like Mott and Spooner's equation, these equations (10, 11) appeared to be inappropriate for dewatered sewage sludge.

Experimental

Dewatered sewage sludge samples (originating in various treatment plants of the country) were dried in a lab vacuum oven under N_2. The dry samples were characterized by monitoring the following: ultimate analysis (C, H, S, N), ash content and high heating value. A selected set of

samples were characterized in multiple laboratories which
included the following: Huffmann Laboratories, Inc. (Golden,
CO), Institute of Gas Technology (IGT, Chicago, IL), and
Texaco Research & Development (Beacon, NY) to ensure that
analyses in various laboratories provide comparable results.
In general, the data obtained from various labs were within
the variation allowed by the conventional ASTM guidelines
for each analyses. All analyses for a given sample were
completed relatively rapidly to minimize degradation of
samples due to bacterial growth.

The data (30 observations in total) were analyzed by
using the Statistical Analytical System (SAS) package
developed by SAS Institute (12). The regression program
available in this package was applied to develop an
empirical model.

**The Heating Value of Sewage Sludge Compared to the Various
Fossil Fuels.** The mean heating value (gross) of sludge
(based on 30 observations) compared to various fossil fuels
is shown in Figure 1. The heating value of oil shale (Green
River formation of Mahogeny zone; Colorado; 33 gal/ton,
described by Khan, and Khan and others [13, 14]) was 3200
Btu/lb. The heating value of eastern Kentucky shale can be
significantly lower than the western shale considered in
this study. The heating value of the Asphalt Ridge basin tar
sand was less than 2000 Btu/lb (13, 14). By contrast, the
heating value of an average sewage sludge is considerably
higher (6400 btu/lb). The heating value of an industrial
biosludge observed in this study to be greater than 9000
btu/lb. However, no industrial sludges were included in the
data base aimed at developing the new correlation.

Compared to essentially all fossil fuels (excluding
petroleum based fuels), sewage sludge has a higher H/C
(atomic) ratio (Figure 2). The mean H/C ratio of sewage
sludge was 1.65 (based on 30 observations), considerably
higher than that of the bituminous coals (Pitt#8) with H/C
ratio of 0.89 or a sub-bituminous coal (H/C of 0.96 for
Wyodak coal). The H/C of the sewage sludge is comparable to
tar sand bitumen (with H/C of 1.5).

In addition to the elements described above,
significant amounts of chlorine and various volatile metals
can be present in sewage sludge. For example, the chlorine
content of one sludge was as high as 0.6% (dry basis). Other
volatile inorganics identified in the this sludge included
the following: Beryllium (less than 0.02 ppm), Vanadium
(less than 1 ppm), and Manganese (900 ppm). However, this
study did not consider the role of chlorine or vaporizable
metals on the heating value of sludge.

Variations in the Sewage Sludge Composition. The mean,
standard deviation, minimum and maximum values for the
compositional analyses are presented in Figures 3 and 4. The
variations in the C, H, N and S content in different samples
are shown in Figure 3. The sulfur content for various
sludges ranged between 0.18 and 3.61 percent with a mean

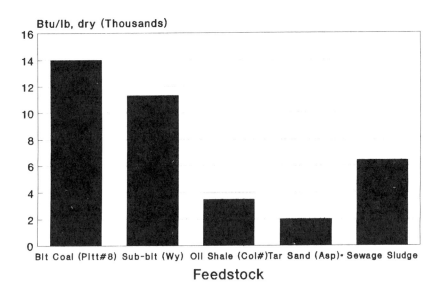

Figure 1. Comparison of Heating Values of Various Feedstocks (Dry-basis)

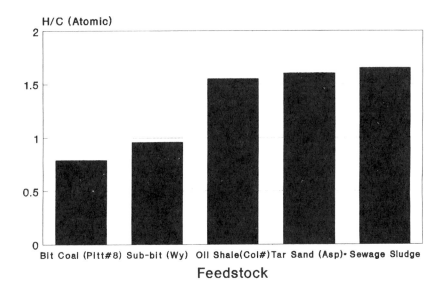

Figure 2. Comparison of H/C of (Atomic) of Various Feedstocks (Dry-basis)

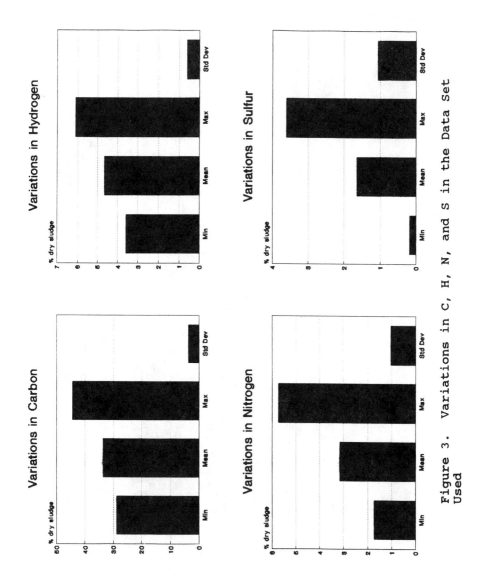

Figure 3. Variations in C, H, N, and S in the Data Set Used

Variations in Sludge Heating Value

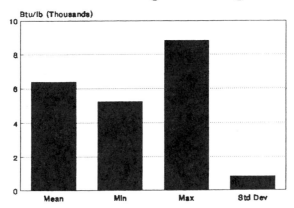

Variations in Ash Content

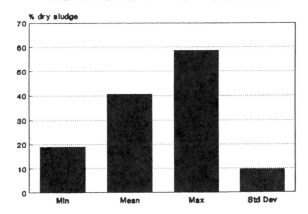

Figure 4. Variations in the Heating Value and Ash Content in the Data Set Used

value of 1.71 (with a standard deviation of 1.05 about the
mean). The oxygen content of sludge ranges between 3.5 and
27.8% with a mean of 16.5 (and a standard deviation of
6.3%).

The volatile matter content for a given sludge ranged
between 45 and 62% (dry basis). These values are
significantly higher than the volatile matter content of a
high volatile bituminous coal (with a volatile matter
content of 35%, dry basis). The H/C ratio (atomic) for the
data set used ranges between 1.44 and 1.86 with a mean of
1.65 with a standard deviation of 0.106.

Figure 4 shows that the mean heating value of the
sludge was 6409 with a standard deviation of 816 (based on
30 observations). The measured values for the sludge ranged
between 5261 and 8811 Btu/lb. The heating value and the
compositional characteristics of sludge are dependent on the
nature of sludge as well as on the degree of digestion (or
pretreatment) a sludge has undergone. The minimum ash
content for the sludge was 18.9% while the maximum value for
the sludge was 58.68% (the mean was 40.5%). The higher ash
content generally reflects that the sludge has either been
digested or heat-treated to convert a large portion of the
organic constituents.

The sludge composition is dependent on the nature of
pretreatment a given sludge has experienced. For example,
the sludge conditioned by a wet oxidation process
(intermediate pressure, 300-400 psi; oxidizing atmosphere;
temperature of 250-375 F) has a significantly different
composition and a lower heating value (Figure 5) compared to
the untreated sludge (lower C, H but a higher oxygen content
compared to the untreated materials; all data on dry basis).

The compositional differences between various sludges
can be significant. For example the differences in the
compositional characteristics between the Los Angeles (CA)
and Passaic (NJ) are shown in Figure 6. It is interesting to
note that the pyritic sulfur is the dominant sulfur type for
several sludges (Figure 7). The presence of this large
concentration of pyrite is not typical of domestic sludges
but suggests the formation of pyritic sulfur from organic
sulfur by bacterial action. Such conversion of organic to
inorganic sulfur has been reported in the coal literature.

Evaluation of the New Correlation. The variations (between
the measured and predicted values) for the three models
evaluated in this study are summarized in Figure 8
(discussed in details in the following section). Figure 9
compares the measured and predicted heating values based on
the Data Base Model. The disagreement between the predicted
and measured values in Figure 9 is much larger than those
shown in Figure 10 (based on the new correlation developed
in this study).

Figure 10 compares the measured and predicted heating
values calculated based on the correlation developed in this
study. The following equation describes this model:

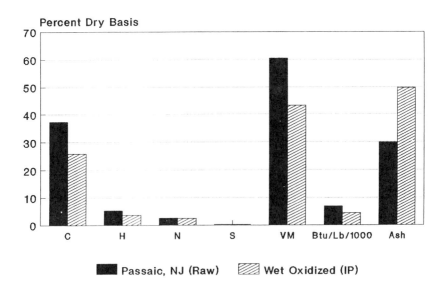

Figure 5. Effect of Wet Oxidation on Sewage Sludge Composition (Passaic Sewage Sludge); Raw and Wet Oxidized (@ Intermediate Pressure)

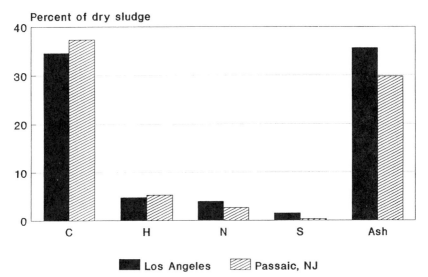

Figure 6. Comparison of LA versus Passaic Sludge Composition

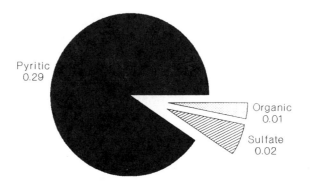

Figure 7. Types of Sulfur Present in the Passaic Sludge
(Total Sulfur Content 0.32%)

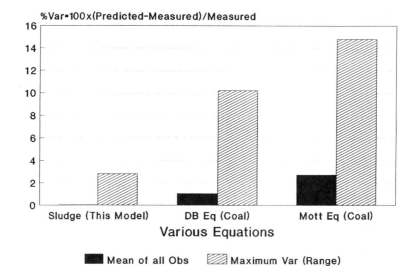

Figure 8. Composition of Various Models Tested for the
Data Set

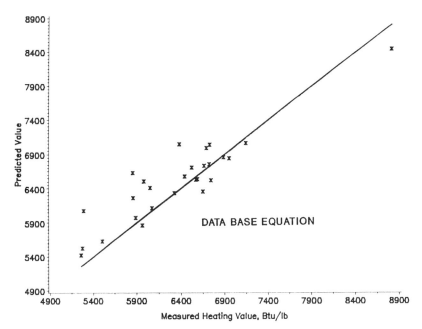

Figure 9. Predicted and Measured Heating Values for Sewage Sludge (Based on the Data Base Equation, i.e., Literature Model)

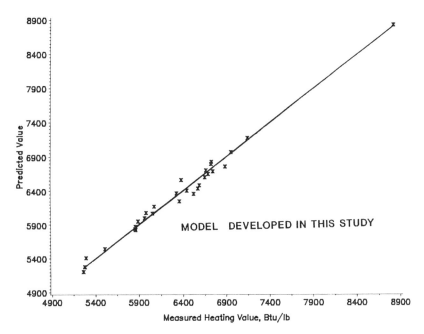

Figure 10. Predicted and Measured Heating Value for Sludge Based on the New Sludge Model

Q (Btu/lb) = 241.89*C + 264.26*H + 236.2*S + 20.99*Ash - 4174.68

R^2 for the model is 0.99. The parameter measures the proportion of total variations explained by the regression. It is calculated by dividing the sum of squares due to regression by the total sum of squares. R^2 is related to correlation coefficient, r, by the following in simple linear regression: r = square root of R^2. In addition, r, has the same sign as the slope of the computed regression.

The F value for the model was 650.6. The F ratio is the ratio produced by dividing the mean square for the model by the mean square of error. It tests how well the model as a whole (after adjusting for the mean) accounts for the behavior of the independent variable.

The P value for the model was 0.0001. P defines the "observed level of significance." In statistical terms, the level of significance, alpha, of a test is defined as the probability of rejecting the null hypothesis (i.e., no linear relationship between the dependent and independent variables), given the null hypothesis is true. The P-value gives us the largest value of alpha that would lead to the acceptance of the null hypothesis. In other words, from statistical standpoint, the correlation developed in this study for estimating the heating value of sludge is highly significant.

Comparison of Various Correlations. Attempts were made to estimate the heating value of sludge using correlations widely reported in the literature applicable for coal (and oil shale). In particular, the equation by Mott & Spooner and the Data Base equations were compared with the newly developed correlation.

The percent variation between the measured and the predicted values were calculated for each model by the following equation:

% Variation = 100 x (Predicted-Measured)/Measured

The variations (between the measured and predicted values) for the three models are summarized in Figure 8. The model developed in this study provides a <u>mean variation</u> of 0.019% between the predicted and the measured values. The <u>maximum variation</u> between the measured and predicted values was never greater than 2.83%, based on the new model. In contrast, the Data Base Equation provides a maximum variation of 10.2% while the equation by Mott & Spooner yielded a maximum difference of 14.8% between the measured and predicted values.

Summary & Conclusions

The following conclusions are derived based on this study:

o   The heating value of dry municipal sewage sludge is considerably higher than tar sand or oil shale but lower than that of bituminous coal. The atomic H/C ratio of sewage sludge, however, is higher than that of bituminous coal, but comparable to the H/C ratio of oil shale. Some industrial biosludges can have heating value comparable to that of a low rank coal.

o   The volatile matter content of sludge is higher than that of coal, oil shale or tar sand.

o   The compositional characteristics (C, H, N, S and ash) of sludge can vary widely among sewage sludge of different origin. Wet oxidation of sewage sludge significantly reduces its heating value as well as its C and H content.

o   The conventional equations developed for coal are not readily applicable for sewage sludge. The equation developed in this study serve reasonably well for estimating the heating value of sludge of various origin based on its analysis (C, H, S, and ash).

Literature Cited

1. V. T. Chow, R. Eliassen and R.K. Linsley, Editors, "Wastewater Engineering: Treatment, Disposal & Reuse", McGraw-Hill, 1979.
2. Federal Register, "National Sewage Sludge Survey: Availability of Information and Data, and Anticipated Impacts on Proposed Regulations; Proposed Rule," Vol 55., No. 218, 40 CFR Part 503, Nov 9, 1990.
3. Willson, W. G. et. al, "Low-Rank Coal Water Slurries for Gasification," Final Report by UNDEERC, EPRI Report No. AP-4262, Nov., 1985.
4. Mott, R. A. and Spooner, C. E., "The Calorific Value of Carbon in Coal," Fuel 19, 226-31, 242-51 (1940).
5. Selvig, W. A. and Gibson, F. H., "Calorific Value of Coal," in Lowry, H. H. ed., Chemistry of Coal Utilization 1, 139, New York, 1945.
6. Mason, D. M. and Ghandi, K., "Formulas for Calculating the Heating Value of Coal and Char: Development, Tests and Uses," Am. Chem. Soc. Div. Fuel Chem Preprints, 1980, 25(3), 235.
7. Boie, W., "Fuel Technology Calculations," Energietechnic 3, 309-16, 1953.
8. Grummel, E.S., and Davis, I. A., "A New Method of Calculating the Calorific Value of a Fuel From Its Ultimate Analysis," Fuel 12, 199-203, 1933.
9. Neavel, R. C., S. E. Smith, E. J. Hippo, and R. N. Miller, "Interrelationships Between Coal Compositional Parameters," Fuel 65, 3, pp. 312-320, 1986.
10. Wilson, D. L., "Prediction of Heat of Combustion of Solid Wastes from Ultimate Analysis," Environmental Science Technology 6, No. 13, 1119-1121, 1972.

11. Metcalf and Eddy, "Wastewater Engineering:Treatment
     Disposal and Reuse," pp. 6:13, McGraw Hill, New York,
     1979.
12. SAS User's Guide: Basic, Version 5 Edition; SAS
     Institute; Cary, NC, 1985.
13. Khan, M. R., K. S. Seshadri, and T. E. Kowalski,
     "Comparative Study on the Compositional Characterisitcs
     of Pyrolysis Liquids Derived from Coal, Oil Shale, Tar
     Sand and Heavy Residue," Energy & Fuels, 3, 412-420,
     1989.
14. Khan, M. R., "Correlations Between Physical and
     Chemical Properties of Pyrolysis Liquids Derived from
     Coal, Oil Shale, and Tar Sand," Energy & Fuels, 2, 834-
     842, 1988.

RECEIVED August 5, 1992

Chapter 13

Preparing Pumpable Mixtures of Sewage Sludge and Coal for Gasification

Matthew A. McMahon and M. Rashid Khan

Research & Development Department, Texaco, Inc.,
P.O. Box 509, Beacon, NY 12508

The flow behavior of coal water slurries is
significantly degraded when untreated sludge is
mixed with coal at a sufficient concentration.
Various methods of treating sludge were
evaluated in an effort to make coal slurries
containing more than 25 per cent sludge solids
more fluid so that they could be pumped through
pipes and nozzles into a pressurized gasifier.
Drying sludge in commercial dryers at
temperatures ranging from 180 deg F to 400 deg F
significantly improved its slurrying characteristics with coal. The fluidity
characteristics could also be improved by
removing water under vacuum, filtering with high
intensity filter presses and subjecting the
sludge to shearing stresses. Slurry viscosity
measurements were made at 70 to 212 deg F in
novel viscometers.

Over 26 billion gallons of waste water are treated by more
than 15,000 publicly owned treatment works in the United
States serving over 70% of the population (1). This
treatment results in the production of 7 million metric
tons per year of sewage sludge. Over 40% of this is
applied to the land while about 14% is incinerated and
another 5% is dumped into the ocean (2). (See Figure 1)
The recent ban on ocean dumping--with penalties for
continued dumping reaching $600 per dry ton in 1992--along
with a decreasing number of landfills and other
environmental concerns, have created a need for
environmentally sound sewage disposal alternatives.
The Texaco Coal Gasification Process, (TCGP) which
has operated satisfactorily in large scale facilities for
several years, appears to offer attractive features as

0097–6156/93/0515–0157$06.00/0

such an alternative. Figure 2 shows a slurry of coal being converted into electricity by means of gasification and combined cycle electricity generation. In this process coal is combusted in a deficiency of oxygen to produce synthesis gas--a mixture of carbon monoxide and hydrogen. This mixture can be used for several purposes including use as a fuel for a combined cycle power process. In this process, high temperature synthesis gas is heat exchanged with water to produce steam, which is converted into power in steam turbines, and, after cleanup, the cooled synthesis gas is burned in gas turbines to produce additional power.

Coal/water slurries containing about 60% coal are a usual feed for the TCGP. On the other hand, concentrated sludge slurries in the form of sludge filter or centrifuge cakes containing 70 to 80% water are a common product of water treatment plants. Sludge in this form is a low quality fuel with an insufficient Btu content to be gasified alone in the process. It must therefore be mixed with an auxilliary higher quality fuel such as coal, oil or gas, as shown in Figure 3, to form a satisfactory feed for the process. This study examines only the use of coal as the auxilliary fuel.

In addition to having a satisfactory heat content, slurry mixtures which are suitable feeds for the process must be pumpable at high solids concentrations and contain sufficient sludge to justify the incremental cost of handling it. This paper describes the results of our efforts to characterize the fluidity properties of sludge/coal slurries and to identify a treatment process that would enable sludge concentrations to be increased to practical levels in pumpable slurries with coal.

The hydraulic transport of particulate solids has recently been reviewed (3). Campbell and Crescuolo have examined the rheological characteristics of dilute sludge slurries (4), and Beshore and Giampa have reported on the rheological properties of concentrated coal slurries containing minor amounts of sludge (5). No detailed studies of the rheological characteristics of coal slurries containing high concentrations of raw or thermally treated sludge have been reported. A fuel comprised of raw (non-dewatered) sludge and coal has also been claimed to be pumpable and useful as a boiler fuel (6).

Experimental

The viscometer used for this work was developed in Texaco Research and Development facilities and calibrated with oils of known viscosity. Usually, apparent viscosity vs solids concentration curves were obtained from which the total solids that could be included in a slurry at a given viscosity were determined. About 80 grams of slurry was required for each measurement. Slurries were prepared by mixing the desired amounts of sludge, coal and water to a measurable consistency in the measuring cup and noting the torque at a stirrer speed of 600 rpm. Measurements were

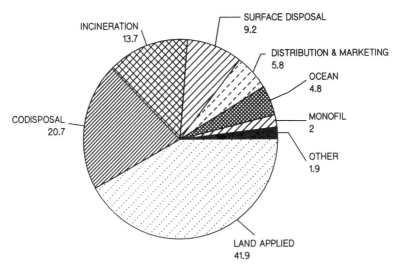

Figure 1.   Methods of disposing of sewage sludge in the United States-1990.

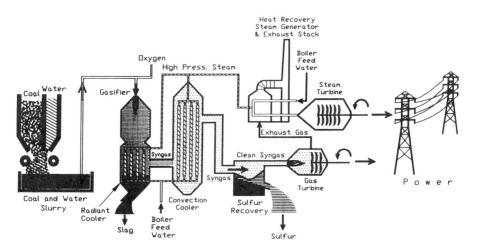

Figure 2.   Texaco Gasification Process-combined cycle mode of operation.

then repeated as incremental amounts of water were
subsequently added. Torques were related to viscosities
by measurements on oils of known viscosity. Typical
data are included in Figure 4. Replicate measurements
indicated that the standard deviation of measurements
using this technique was 0.92 for a sludge sample which
contained 13.0% solids at 1000 cp. Most of the error in
this measurement originates from sampling the raw sludge
because its accuracy is dependent on the accuracy of the %
solids determination in the original sample. The standard
deviations calculated for solids content measurements in
raw sludge samples were between 0.7 and 0.94 for samples
containing 23% solids.

A Haake viscometer, RV-100, was also used for
rheological measurements.

To establish the patterns of sludge behavior,
dewatering of sludge was achieved by an advanced
dewatering technique, high intensity press (HIP). The
pressing action of the HIP was simulated on a laboratory
scale by means of a device designed and constructed by
Andritz Corporation, manufacturer of the commercial HIP
apparatus. Dewatered cake was distributed on a 4"x4"
piece of filter fabric which was supported by a specially
designed perforated square metal tray. This was
surrounded by a square metal box. A similar piece of
fabric was placed over the sample followed by a square
upper tray. Pneumatic piston pressure was then applied to
the upper tray forcing out the entrapped water. The
applied pressure was changed with various retention times
to simulate the actual HIP zone pressures. Upon
completion of the pressure cycles, the pneumatic lever was
pushed up and the pressure box quickly removed. The solid
content was determined by actual measurement and the
throughput calculated by empirical equations developed by
Andritz.

Sludge Characteristics

The digested sludge used for most of the measurements made
in this study was obtained from water treatment plants in
Los Angeles County, Los Angeles City and San Bernadino
County in California. The as-received centrifuge cakes
were amorphous, fibrous materials containing 20 to 30%
total solids. Polymeric flocculating agents were employed
in their preparation at the water treatment plants. These
materials were not pumpable but could be made so by
diluting to a slurry containing about 15% solids. Their
composition is compared with coal and peat in Table I.
Digested sludge solids generally contain: 30-60% volatile
solids, 5-20% grease and fats, 5-20% protein, 10-20%
silica and 8-15% cellulose (7).

As-Received Sludge/Coal Slurries

One of the main purposes of this work was to determine the
maximum amount of sludge that could be incorporated into a
pumpable slurry with coal. Economics dictated that
commercially viable sludge containing feeds should include

Table I
Typical Analysis of Sewage Sludge and Other Solid Fuels

Fuel	Moisture %	Ash	C	H	N	O By Diff	S	BTu/lb (Dry)
Sewage Sludge	80	36	31	4.8	3.9	22.1	1.7	6400
Peat	83.7	3.4	47.1	5.4	1.4	42.6	0.1	
Lignite, TX	29.3	21.5	55.7	4.5	1.0	15.8	1.4	9788
Subituminous C, Wyoming	28	7.8	68.1	4.9	1.1	17.2	0.6	11840
Bituminous Pittsburgh 8	0.8	8.6	76.5	5.1	1.4	5.8	2.5	13765

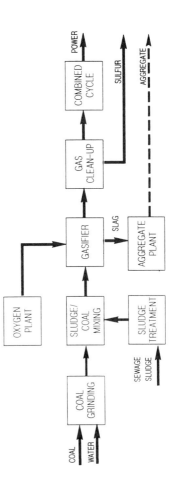

Figure 3. Flow diagram of a conceptual sewage sludge gasification process.

at least 25% sludge solids and have a total solids content
of above 50%.   Experience with coal slurries indicates
that slurries having apparent viscosities of about 1000 cp
are pumpable.   Results of viscosity measurements on a
number of sludge/coal slurries containing varying amounts
of Los Angeles sludge in Utah-Sufco bituminous coal are
presented in Figure 4. These results, which are typical of
many, showed that the amount of total solids which can be
incorporated   into   a   pumpable   slurry   decreases   with
increasing sludge content.   It is also apparent that in
the 1000 to 2000 cp range, small increases in total solids
content of slurries effect large increases in viscosity.
Presenting these data in another way as in Figure 5 shows
that plots of sludge content vs total solids content are
linear at constant viscosity. None of these mixtures was
considered a satisfactory fuel because either their sludge
or total solids content was too low at the 1000cP pumpable
viscosity for economic operation.

The   rheological   properties   of   sludge   vary
considerably with its treatment history as well as the
nature and amount of industrial components in it.   As can
be seen in Figure 6, the sludge from City A has poor
fluidity properties.   This is attributable to its containing over 10% short paper fibers from nearby paper
recycling plants.   City D's sludge appeared to be quite
fibrous also making it quite viscous relative to sludges
from several cities compared in this figure.   Among the
factors   influencing   viscosities   are   the   degree   of
digestion, the relative amounts of sludge from primary and
secondary treatment and the solids content of the filter
or centrifuge cakes produced from water treatment plants
(Figure 7).   Lime treatment, because it removes   water
from the slurry, also seriously degrades the fluidity of
sludge.

Thermal Treatment

Previous work indicated that that thermal treatment of
sludge   would   improve   its   slurrying   characteristics.
Innumerable studies (7,8)   and a recent review (9)
indicate that heat treatment of sludge coagulates the
solids and breaks down the colloids and cells. Protein
material is also denatured and microorganisms are killed.
These changes combine to irreversibly reduce the water
affinity of the sludge solids.   Heat treated sludge is
readily dewatered on filters to solids concentrations of
30 to 50%, while unheated sludge is usually dewatered to
about 20 to 30%   and only with the aid of polymeric or
inorganic conditioning agents.

To examine the effects of thermal treatment on the
slurrying characteristics of sludge solids, we obtained
samples of sludge that had been thermally treated on a
commercial scale in different types of dryers and thermal
treatment units throughout the country.   Dilute sludge
slurries containing 5 to 25% solids are usually fed to

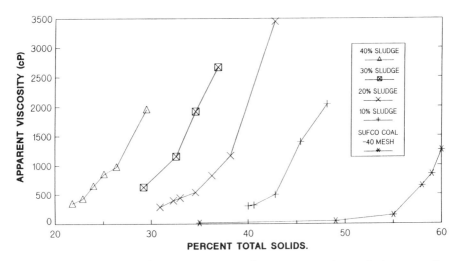

Figure 4. Viscosity concentration curves for mixtures of Los Angeles sludge and Utah Sufco coal.

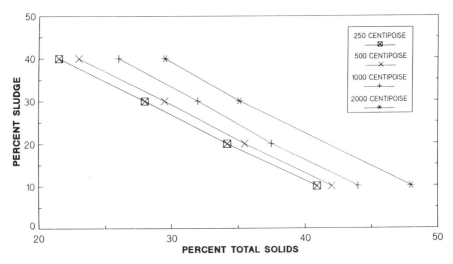

Figure 5. Plots of total solids content vs percentage sludge solids in slurries at selected constant viscosities.

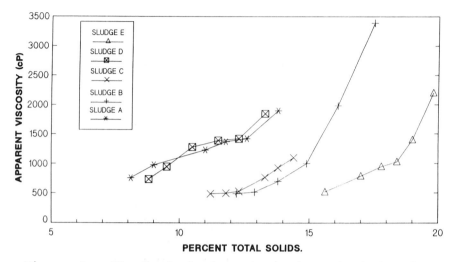

Figure 6.   Rheological characteristics of sludge from different water treatment plants.

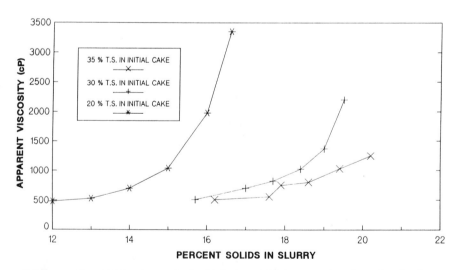

Figure 7. Effects of initial solids content of sludge cake on the apparent viscosities of slurries formed upon dilution.

these dryers. Thermal treatment over the wide temperature range encompassed by these processes did in general improve slurrying properties (See Table II). The products of various treatment processes are physically and compositionally different from raw sludge (Table III). Ash, carbon and oxygen contents vary with treating conditions. Some of the products are dry homogeneous powders somewhat coal-like in appearance while others are quite fibrous containing about 60% moisture. The moist products could not be ground into a powder satisfactorily for slurry testing without drying them first. After drying, they slurried very well.

Results of viscosity tests on slurries prepared from these materials are summarized in Table II. The slurrying characteristics of the as-received sludges and of mixtures containing 30% sludge and 70% coal (dry basis) were routinely measured. Clearly, all of these materials demonstrated slurrying characteristics far superior to those of untreated sludge. Compare the results in Figure 4 with those in Table II, for example. Some treated sludges were coal-like in slurry behavior hardly affecting the fluidity of the coal at low concentrations, while the other materials degraded the fluidity of the slurries to varying extents. For example, about 60% solids can be included in a 1000 cp pumpable slurry of coal alone. But the solids contents of mixtures containing 30% sludge and 70% coal having viscosities of 1000 cp ranged from 45 to 58.5%. This behavior is not unexpected since these materials have not only been heated at different temperatures but also under different conditions. In one process, sludge is heated while suspended in oil allowing oil soluble compounds to be extracted from it, while in others, organic components in sludge are oxidized or simply volatilized.

Thermally dried sludge could also be slurried in oil but the viscosity of the slurrying oil determined to some extent the amount of slurry solids that could be included in a pumpable mixture.

Overall, the rheological characteristics of raw and treated sludge were found to be very consistent. Plotting the total solids in a 1000 cp slurry of as-received or treated sludge against the maximum total solids of the same material that could be incorporated into a 1000 cp 30% sludge/70% coal slurry afforded the linear relationship shown in Figure 8.

Effect of Shear

The rheology of a thermally treated sludge as a function of shear rate is shown in Figure 9. These results, which were obtained with a Haake RV-100 viscometer, show that the viscosity of dried sludge is dependent upon shear rate and temperature--with the viscosity decreasing, for example, from about 900 cp to a little over 400 cp as the temperature is increased from 30 to 90 degC. Basically, dried sludge particles are very friable with no real structure--and break quite readily under shearing stress.

Table II
Slurrying Characteristics of Commercially Available
Thermally Treated Sludges

Process	Max Temp, F	% Ash	Slurry Composition Sludge %	Coal %	Total Solids at 1000 CP
A	250	50.9	30 / 100	70 / -	51.0 / 36
B	1200	32.29	30 / 100	70 / -	58.5 / 48.2
C	365	33.8	30 / 100	70 / -	45 / 30
D	358	44.0	30 / 100	70 / -	62 / 53

Table III
Analyses of Dried Sludge from Various Processes

	Los Angeles as Received	Process A	Process B	Process C	Process D
% Moisture	80.20	3.03	7.93	2.79	9.20
% Ash	36.30	50.90	32.29	31.38	44.00
% C	31.30	30.47	34.70	33.80	28.90
% H	4.83	4.48	4.99	5.32	3.80
% N	3.92	3.95	5.51	2.57	3.20
% S	1.65	1.60	0.71	0.63	1.10
% O (By Diff)	22.00	8.60	21.80	26.30	19.00
BTU Value Lb/Dry	6140	5856	6797	6304	4886

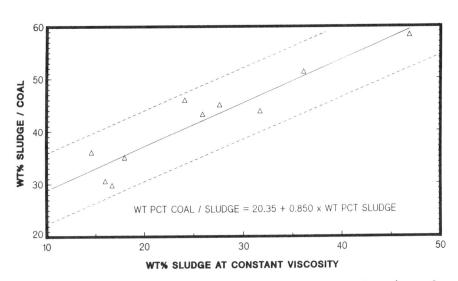

Figure 8. Plots of solids contents of 1000cp slurries of raw and treated sludge vs solids contents of 1000cp slurries of 30/70 sludge/coal slurries.

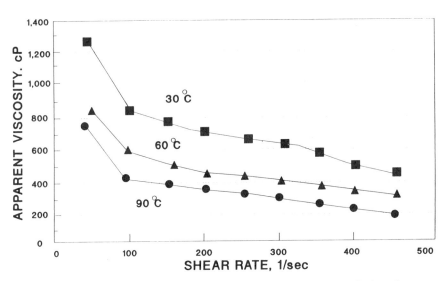

Figure 9. Apparent viscosity vs shear rate plots for a 38wt% solids slurry of dried sludge in water.

Their breakdown is probably non-reversible although no attempts were made to prove it.

Raw sludge is also quite unstable to shearing stress. Hatfield (10) many years ago and others more recently (4) reported that shearing reduces the viscosity of sludge. Results reported here using sludge containing high concentrations of solids are consistent with the previous reports. As would be expected, more of this sheared product could be incorporated into a pumpable slurry than the "as-received" sludge.

A picture of what may be occurring when sludge is sheared is shown in Figure 10. The sludge being sheared consists of coagulated and flocculated colloidal sludge particles. The particles are coagulated with the aid of positively charged polymers, usually polyamides, which neutralize the negatively charged colloidal particles. The observed viscosity reduction is no doubt attributable to at least two factors: the first of these is the simple shearing of the cellulosic and polyamide flocculating polymers in the sludge. The second is a consequence of the shearing stresses on the flocculated colloidal particles. The drag forces and unfolding of the flocculated particles probably release trapped water and make it available as a carrier fluid with a consequent decrease in viscosity.

High Intensity Press Treatment

To further establish the effects of water removal on the patterns of sludge behavior, dewatering of sludge was achieved by an advanced dewatering technique: high intensity pressing (HIP). A schematic of how a HIP operates is depicted in Figure 11. Quite simply, filter cake from a conventional belt filter, which contains 15 to 25% solids, is fed to a high intensity press where it is subjected to increased mechanical pressure to remove water. HIP applied a pressure of 125 psi compared to only 25 psi for a conventional belt filter press. As described in the experimental section, we simulated the HIP treatment in the laboratory.

This treatment improved the rheological properties of sludge--increasing the amount of solids that can be included in a 1000cp pumpable slurry from about 12 to 16% (Figure 12). Worth noting is that high intensity pressing is a way of applying a high shear to sludge.

Vacuum Drying

Results of the shearing and thermal treatment experiments indicated that the most important aspect of improving the slurryability of sludge was to free the water trapped in the sludge. In the shearing case, the trapped water becomes available as part of the carrier fluid while thermal treatment simply volatilizes the water--apparently somewhat irreversibly. This suggested that removing water

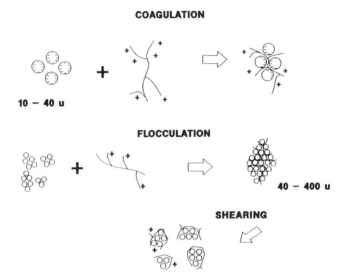

Figure 10.   Conceptual diagram showing the formation of
flocculated sludge particles from colloidal sludge
particles using cationic polymers and their breakup by
shearing stress.   Adapted from reference 11.

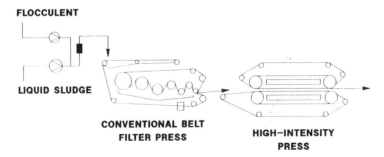

Figure 11.   High intensity press flow scheme.

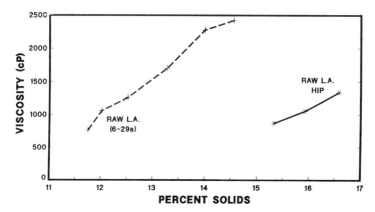

Figure 12. Effect of HIP treatment on the rheological properties of Los Angeles sludge.

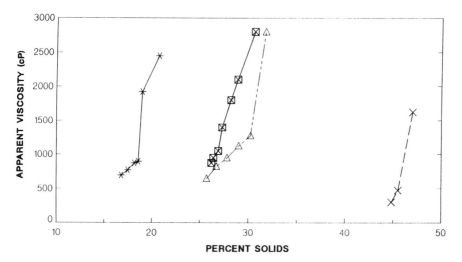

Figure 13. Comparison of the rheological characteristics of sludge which has been dewatered by different methods. ✳ Raw sludge, 23.8% solids. ✕ Raw sludge dried to 98.4% solids. ⊠ Dewatered to 43% solids by vacuum drying. △ Dewatered to 45.3% solids by high intensity pressing.

by simple air or vacuum drying might improve the slurrying characteristics of sludge. And indeed this was found to be true as can be seen in the results presented in Figure 13. Also included in this figure are the results of another experiment in which the same amount of water was removed by vacuum drying and by high intensity pressing. These techniques both afforded products having similar rheolgical properties

Summary

In an effort to explore various means for treating sewage sludge in order to improve its slurrying characteristics with coal, it has been found that virtually all means of removing water trapped in the raw sludge centrifuge filter cake including thermal treatment, shearing, vacuum drying and simple air drying will improve the slurrying characteristics.

Acknowledgments

We would like to express our appreciation to Ronald J. McKeon and Christine A. Albert for their contributions to the work which made this paper possible.

Literature Cited

1.   Bastion, R.K, <u>Proceedings of the National Conference on Sewage Treatment Plant Sludge Management</u>, May, 1987.
2. <u>Federal Register</u>, Nov 7,1990, pp 179-181.
3. Shamlou, P.A.; <u>Handling of Bulk Solids, Theory and Practice</u>; Butterworth and Company, Ltd.,1988.
4. Campbell, H.W.; Crescuolo, P.J., <u>Water Science and Technology</u>, **1982**, vol 14, pp 475-489.
5. Beshore, D.G.; Giampi, V.M., U.S.Patent 4,762,527, 1988.
6. Rodriguez,L.A.; Ashworth, R.A.; Armstead, R.; Aristedes, P.A.; Spake, N.B., U.S. Patent 4,405,332, 1983.
7. <u>Wastewater Engineering: Treatment Disposal Reuse</u>; Cerra, F.J.; Maisel,J.W., Eds.; McGraw-Hill Series in Water Resources and Environmental Engineering, McGraw Hill, 1979.
8. Nicholls, T.P.; Lester, J.N.; Perry, R., <u>The Science of the Total Environment</u>, **1980**, 14, pp 19-30.
9. Reimann, D. Umwelt, 1987, 6 , pp 332.
10. Hatfield, W.D. <u>Sewage Work Journal</u>, **1938**, 10(1),pp 3-30.
11. Doyle, C.L.; Haight, D.M. <u>Proceedings of the National Conference on Municipal Treatment Plant Sludge Management</u>, **1986**, pp103.

RECEIVED May 1, 1992

METALS EMISSIONS CHARACTERIZATION AND CONTROL TECHNOLOGIES

Chapter 14

Metal Emissions Control Technologies for Waste Incineration

James R. Donnelly

Davy Environmental, San Ramon, CA 94583

Control of metals emissions from municipal, hospital, and hazardous waste incinerators has recently been mandated in regulations proposed by the U.S. Environmental Protection Agency. An understanding of regulatory requirements and control technologies is needed to select the most cost-effective system for site-specific applications. This chapter presents a review of the current U.S. regulations covering incinerator emissions and describes technologies used for their control. Typical emission levels and control efficiencies achievable are presented.

A major issue facing industrialized nations is the environmentally sound disposal of municipal solid wastes, hospital wastes, and industrial hazardous wastes. The amounts of these wastes generated have grown annually over the past several decades (1), and improper disposal has resulted in numerous environmental problems. Incineration in properly designed combustion systems has been demonstrated as a method of achieving a very high degree of destruction and control of these wastes. Incineration is often combined with heat recovery systems to simultaneously recover energy in the form of steam or electricity. A wide variety of incinerator types, boilers, and industrial furnaces are used for destroying these wastes.

Incineration of municipal and hazardous waste has the potential for increasing air pollution due to emissions of constituents contained in these waste streams and products of their combustion. These wastes are likely to contain sulfur and chlorine compounds, as well as numerous toxic metals (e.g., arsenic, beryllium, cadmium, chromium, lead, mercury, and silver). Highly chlorinated hydrocarbons and polynuclear aromatic compounds such as dioxins and furans may also be present. During combustion, sulfur and chlorine compounds are converted to the acid gases SO_2 and HCl; toxic metals are converted to their oxide or chloride forms. The high combustion

0097–6156/93/0515–0174$06.00/0

temperatures employed in modern incinerators will cause many of the metal compounds present to volatilize and be carried out of the incinerator in the hot flue gas stream. These compounds can then condense out as fine particulate matter or in some instances leave the system while still in vapor form. High combustion temperatures and residence times are used, as they have been shown to effectively destroy complex organic compounds which may be in the waste stream.

The increase in waste incineration has been accompanied by increased public concern over air pollution and an increase in local, state, and federal regulations. The USEPA recently revised federal regulations to further limit incinerator emissions. This increased regulatory climate has resulted in an increase in the complexity and efficiency of air pollution controls employed for emissions control.

This paper presents a review of the current U.S. regulations covering incinerator emissions and describes technologies used for their control. Typical emission levels and control efficiencies achievable for various metals are presented.

Air Pollution Regulations

Air pollution regulations governing incinerator flue gas emissions vary widely in the compounds controlled, emission levels allowed, removal efficiencies required, averaging times used, and testing requirements. On the national level, municipal waste incinerators are regulated under Clean Air Act (CAA) provisions, whereas hazardous waste incinerators are regulated under the Resource Conservation and Recovery Act (RCRA). In addition to national regulations, local or state permitting agencies may require more stringent emissions controls or control of additional pollutants as part of a facility's operating permit. The EPA has recently been active in setting standards for municipal waste incinerators, hazardous waste incinerators and boilers, and industrial furnaces which burn hazardous wastes. The EPA is currently in the process of setting regulations for hospital waste incinerators.

Municipal Waste Incinerators. The EPA promulgated "New Source Performance Standards and Emissions Guidelines for Existing Facilities" for Municipal Waste Combustors (MWCs) in February 1991 *(2)*. These standards are summarized in Table I.

In setting these standards, EPA recognized differences in facility size, type of incineration (mass burn fired versus refuse derived fuel fired), and new sources versus existing sources. The facility capacity refers to the total burn rate for all refuse combustors at a single site. EPA selected total particulate matter emission limits as a way of controlling trace toxic metal emissions. EPA will add emission limits for mercury, cadmium, and lead in the coming year based on applying Maximum Achievable Control Technology (MACT). EPA has until late 1992 to establish comparable emission standards for smaller combustors, those less than or equal to 250 tons per day per train.

Opacity limits are set at 10 percent and must be continuously monitored. Opacity is used as an indication of particulate matter emissions.

Table I. USEPA Municipal Waste Combustion Emission Standards°

	New Source Performance Standards	Emission Guidelines For Existing Facilities	
Capacity (tons/day)	Unit	Unit	Facility
	>250	>250 ≤ 1100	>1100
Particulate Matter (gr/dscf)	0.015	0.030	0.015
Opacity (%)	10	10	10
Organic Emissions (ng/dscm) Total Chlorinated PCDD Plus PCDF*			
-Mass burn units	30	125	60
-RDF fired units	30	250	60
Acid Gas Control % Reduction or Emissions (ppm)			
HCl	95 (25)	50 (25)	90 (25)
SO_2	80 (30)	50 (30)	70 (30)
NO_x	(180)	None	None
Carbon Monoxide (ppm)	50-150**	50-250**	50-250**

°   All emissions limits are referenced to dry gas conditions at 20°C and 7% oxygen.

*   PCDD - Polychlorinated Dibenzodioxins; PCDF - Polychlorinated Dibenzofurans.

**   Range of values reflect differing types of MWC's.

Cumulative emission limits have been established for polychlorinated dibenzodioxins (PCDD) plus polychlorinated dibenzofurans (PCDF). These compounds were selected as surrogates for organic emissions because of their potentially adverse health effects. In addition, EPA has established carbon monoxide (CO) emission limits as a measure of "good combustion practices" to limit the formation of PCDD, PCDF, and their key precursors. CO emission limits vary from 50 to 150 ppm (7% O_2 dry gas conditions) depending on the type of combustion.

Acid gas emissions limits (HCl and SO_2) are based on either a percent reduction or a maximum stack emission level, whichever is least stringent. Nitrogen oxide (NO_x) emissions levels are proposed only for large new sources.

Hazardous Waste Incinerators. In April 1990, the EPA published a proposed rule and requests for comments in the Federal Register for Standards for Owners and Operators of Hazardous Waste Incinerators and Burning of Hazardous Wastes in Boilers and Industrial Furnaces *(3)*. The final rules for "Burning Hazardous Waste in Boilers and Industrial Furnaces" were published in the Federal Register in February 1991 *(4)*. Key provisions of these regulations are presented in Table II.

Table II. USEPA Proposed Hazardous Waste Incineration Standards

Destruction and Removal Efficiency (DRE)	99.9999% Dioxin-Listed Wastes 99.99% All Other Wastes
Particulate Matter	0.08 gr/dscf @ 7% O_2
Carbon Monoxide (Tier I)	100 ppmv (d) @ 7% O_2
Hydrocarbons (Tier II)	20 ppmv (d) @ 7% O_2
Continuous Emissions Monitoring	CO, O_2, HC

Tier III Reference Air Concentrations (annual limits, μ g/m³)			
Hydrogen Chloride	0.7	Free Chlorine	0.4
Carcinogenic Metals		Non-Carcinogenic Metals	
Arsenic	2.3×10^{-3}	Antimony	0.3
Beryllium	4.1×10^{-3}	Barium	50
Cadmium	5.5×10^{-3}	Lead	0.09
Chromium	8.3×10^{-4}	Mercury	0.3
		Silver	3
		Thallium	0.3

EPA proposed extending current emissions limits covering Destruction and Removal Efficiencies for organic constituents and for particulate matter. EPA also proposed establishing risk-based emission limits for individual toxic metals, hydrogen chloride, and organic compounds. EPA added limits for chlorine when they published their final rule for boilers and industrial furnaces (BIFs) *(4)*. Reference Air Concentrations (RACs) were proposed for maximum modeled annual average ground concentrations of these pollutants. RACs for the carcinogenic metals were set at levels which would result in an increased cancer risk for a Maximum Exposed Individual of less than 1 in 100,000. RACs for the non-carcinogenic metals and chlorine were set at 25 percent of the Reference Dose (RfD) with the exception of lead, which was set at ten percent of the National Ambient Air quality level. The RAC for HCl is based directly on inhalation studies.   RfD's are estimates of a maximum daily exposure (via injection) for the human population that is not likely to cause deleterious effects.

In setting these standards, EPA established a three tiered approach for demonstrating compliance.   The tiers are arranged from the easiest to demonstrate and most conservative to the more complex to demonstrate and less conservative. Compliance with any tier is considered to prove compliance with these regulations.

Tier I   EPA established conservative maximum feed rates (lb/hr) for each constituent as a function of effective stack height, terrain and land use. In setting these limits, EPA assumed no partitioning in the incinerator, no removal in an air pollution control system, and reasonable worst case dispersion. Demonstration of compliance is through monitoring of feed composition. Two examples of Tier I screening limits are 2.4×10^{-4} to 4.1×10^{-3} for arsenic and 9.4×10^{-3} to 1.6 pounds per hour for lead, depending on stack height, terrain, and land use.

Tier II   EPA established conservative emission rate limits for each constituent as a function of effective stack height, terrain, land use and assumed reasonable worst case dispersion. Demonstration of compliance is through periodic stack emission testing and continuous emission monitoring of carbon monoxide, hydrocarbons and oxygen. Two examples of Tier II screening limits are 3.1×10^{-5} to 5.3×10^{-3} grams per second for arsenic and 1.2×10^{-3} to 2.0×10^{-1} grams per second for lead.

Tier III   EPA established RACs which must be met for each component. Demonstration of compliance is through periodic emissions testing and site specific dispersion modeling to demonstrate actual (measured) emissions do not exceed RACs. For the carcinogenic metals, the ratios of each metal's measured value to its RACs are added to give a cumulative value which must be below 1.0 (risk of 1 in 100,000). Tier III RACs for all metals are shown in Table II.

The standards will be implemented through limits on specific incinerator and air pollution control system operating parameters. In addition, emissions testing of all dioxin/furan tetra-octa congeners, calculation of toxic equivalents, dispersion modeling, and health risk assessments will be required for incinerators equipped with dry particulate control devices (electrostatic precipitators or fabric filters) operating at an inlet temperature between 450° and 750°F, or if hydrocarbon emission levels exceed 20 ppmv (d) *(4)*.

EPA also established a requirement for continuous emissions monitoring of stack carbon monoxide (CO), oxygen (O_2), and total hydrocarbons (HC). Compliance with Products of Incomplete Combustion (PICs) limitations will be demonstrated through monitoring data that meet specified rolling hourly averages for CO and HC. If these emissions levels are not achieved, then the incinerator operator must demonstrate PIC compliance through dioxin/furan emissions measurements, toxic equivalent calculations, and health-risk assessments.

Air Pollution Controls

The major fraction of toxic metals found in incinerator flue gases exists as fine particulate matter, although a significant fraction of some metals can exist in the vapor phase at typical incinerator exit flue gas conditions. The fraction of toxic metals present in the vapor phase is a function of the specific metal and its chemical form, the combustion conditions, and the flue gas temperature.

Control of the particulate fraction is achieved by utilizing traditional particulate control devices. Control of the vapor phase fraction is achieved through cooling of the flue gas and collection of the fine particulate thus formed. Table III lists the types of controls typically employed to control toxic metals.

Table III. Toxic Metal Controls

Fraction	Control Device
Particulate	Electrostatic Precipitators Fabric Filters Wet Scrubbers
Vapor Phase	Spray Dryer Absorbers Wet Scrubbers Condensing Wet Scrubbers

Particulate Metals Fraction Control. Traditionally, particulate matter control from incinerator flue gases was achieved utilizing either a wet venturi scrubber or an electrostatic precipitator (ESP). As emissions regulations have become more stringent and control of the fine particulate fraction has become more important, there has been a shift to using fabric filters and combinations of ESP with wet scrubbers.

Electrostatic Precipitators. Electrostatic precipitators collect particulate matter along with toxic metals and trace organics condensed on it by introducing a strong electrical field in the flue gas, which imparts a charge to the particles present. These charged particles are then collected on large plates which have an opposite charge applied to them. The collected particulate is periodically removed by rapping the collection plates. The agglomerated particles fall into a hopper, where they are removed. Key design parameters for ESP's include: particulate composition, density, and resistivity; flue gas temperature and moisture content; inlet particulate loading and collection efficiency; specific collection area (SCA = square feet of collecting surface per 1,000 cubic feet of flue gas); number of fields; flue gas velocity; collector plate spacing; rapping frequency and intensity; and transformer rectifier power levels.

Properly sized ESP's applied to waste incinerators can reduce particulate matter emissions to the range of 0.01 - 0.015 gr/dscf and achieve collection efficiencies of greater than 95 percent of fine particulate. These ESP's would typically have 3 to 5 collecting fields, specific collecting areas of 400 to 550, flue gas velocities of 3 to 3.5 feet per second, and transformer rectifiers capable of supplying secondary power levels of 35-55 KV at 30-50 milliamps per 100 square feet of collecting area. ESP sizing for hazardous waste incinerator application presents a challenge because of the wide range of materials which may be incinerated.

Fabric Filters. Pulse-jet-type fabric filters are typically used for particulate, toxic metal, and trace organics emissions control for waste incinerators. Fabric filters are always installed downstream of a quenching device because of temperature concerns and to minimize the chance of burning embers reaching the fabric. Fabric filters achieve particulate emission levels between 0.01 and 0.015 gr/dscf and fine particulate control greater than 99 percent.

Fabric filters offer some significant advantages over ESP's for waste incinerator applications. Fabric filters are relatively insensitive to ash properties, flue gas conditions, or operating loads. Fabric filters are more efficient chemical reactors when they are applied as a component of a spray dryer absorption system. They appear to achieve higher control efficiencies for toxic metal and trace organic compounds.

Fabric filters also have some limitations when compared to ESPs. They have a narrower range of operating temperatures. Under some operating conditions, the fabric is subject to blinding, with an associated high-pressure drop and shortened bag life. Fabric filters are more prone to corrosion problems due to their inherent design and operation. Selection of either an ESP or fabric filter should be based on an overall evaluation of site-specific conditions and requirements.

Wet Scrubbers. Wet scrubbers are generally used for particulate control in waste incineration applications that do not require very low emissions of total and fine (< 10 micron) particulate, or where the inlet loading is not too high. Wet scrubbers used for particulate control include: venturi scrubbers, wet ionizing scrubbers, and condensation scrubbers. Venturi scrubbers are used as a primary particulate control device for flue gas with high particulate loadings. Ionizing wet scrubbers and condensation scrubbers are normally used as a part of a total control system where inlet particulate loadings are low. Figure 1 shows a schematic of a common venturi scrubber design used for particulate and acid gas control.

Flue gas enters the venturi, where it is contacted with water sprays and is accelerated to a high velocity (100-150 feet per second) through the venturi. The primary mechanism for particulate removal is impaction with water droplets, although some condensation also takes place. From the venturi, the flue gas stream enters a second stage, where liquid de-entrainment takes place. The second chamber may contain sprays or packing where an alkaline reagent is introduced to remove acid gases present. The flue gas may then pass to another scrubber type for additional particulate and toxic metal control.

Vapor Phase Metals Control. Vapor phase toxic metals are removed from the flue gas stream through condensation and collection of the particulate formed. This is typically accomplished in a two-step process. The most commonly employed systems are spray dryer absorption systems. Also employed are two-stage processes utilizing an ESP followed by a wet scrubber designed for fine particulate removal.

Spray Dryer Absorption Systems. Spray dryer absorption (SDA) has been widely applied for waste incinerator emissions control and has demonstrated high collection efficiencies for most toxic metals present in the flue gas. SDA has been specified as best available control technology (BACT) in a number of municipal waste incinerator air permits.

Figure 2 shows a simplified process flow diagram of the SDA process. The SDA system is comprised of a reagent preparation system, a spray dryer absorber, and a dust collector. Incinerator flue gas enters the spray dryer, where it is contacted by a cloud of finely atomized droplets of reagent (typically hydrated lime slurry). The flue gas temperature decreases and the humidity increases as the reagent slurry simultaneously reacts with acid gases and evaporates to dryness. In some systems a portion of the dried product is removed from the bottom of the spray dryer, while in others it is carried over to the dust collector. Collected reaction products are sometimes recycled to the feed system to reduce reagent consumption.

Several different spray dryer designs have been employed for incinerator SDA applications. These include: single rotary, multiple rotary, and multiple dual fluid nozzle atomization; downflow, upflow, and upflow with a cyclone pre-collector; and single and multiple gas inlets. Flue gas retention times range from 10 to 18 seconds, and flue gas temperatures leaving the spray dryers range from 230° to 400°F *(5)*.

Toxic metals removal in the dust collector is enhanced by cooling of the incoming flue gas (from 2,000° to 450°F) as it passes through the spray dryer. Subsequent to the cooling, some vaporized metals condense to form fine particulates, which grow through impaction and agglomeration with the very

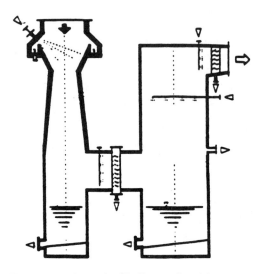

Figure 1. Venturi with Spray Scrubber

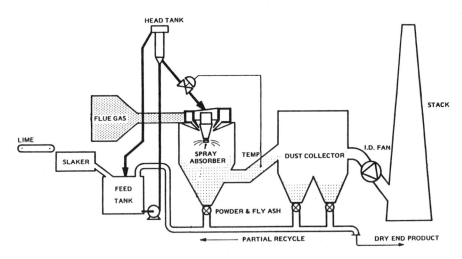

Figure 2. Spray Dryer Absorption Process

high number of lime droplets produced by the atomization devices. These agglomerated particles are then easily removed *(6)*.

Generally, the lower the spray dryer outlet temperature, the more efficient is the acid gas absorption and vaporized toxic metals removal. The minimum reliable operating outlet temperature depends on the spray dryer and dust collector design and on the composition of the dry fly ash reaction product. The spray dryer outlet temperature must be maintained high enough to ensure complete reagent evaporation and the production of a free-flowing product. Low outlet temperature operation requires efficient reagent atomization, good gas dispersion and mixing, adequate residence time for drying, and dust collector design that minimizes heat loss and air in-leakage.

The dust collector downstream from the spray dryer may be an electrostatic precipitator, a reverse-air baghouse, or a pulse-jet type baghouse. The selection of dust collector type is dependent on site specific factors such as particulate emission limits, overall acid gas removal requirements, and project economics. Each of the dust collection devices offers process advantages and disadvantages that need to be evaluated on a site specific basis. When strict acid gas control is required, (95+ % HCl, 85+ % SO$_2$), baghouses are generally utilized, as they are better chemical reactors than electrostatic precipitators. Toxic metals control efficiencies achievable with a SDA system are quite high (99+ %) except for the relatively volatile mercury. Mercury emissions, however, can be controlled at greater than 90 percent efficiency levels through the use of additives such as sodium sulfide or activated carbon *(1,7)*.

Wet Scrubbers. Wet scrubbers control vapor phase emissions through gas cooling and collection of the resulting condensed fine toxic metal particulate. The most commonly used wet scrubbers for this type of service are the ionizing or electrostatically enhanced wet scrubbers and the condensing wet scrubber.

Electrostatically-enhanced scrubbers capture the condensed fine particulate by imparting a charge to the incoming particulate and then collecting these charged particles on neutral packing material or on negatively-charged collecting electrodes. Prior to entering the charged section of the scrubbers, the flue gas typically passes through a presaturation stage, where it is cooled to its adiabatic saturation temperature (140° - 180°F). The cooling causes a significant fraction of vapor phase metals to condense out either as fine particulate on the surfaces of existing particulate matter or on the surfaces of water droplets. These scrubbers also remove acid gases by reaction with an alkali that is added to the scrubber water. Typically, caustic or soda ash is used as the alkali.

Condensing wet scrubbers sub-cool the incoming flue gas to below its adiabatic saturation temperature. This causes a larger fraction of the vapor phase metals to condense and in addition causes water vapor to condense forming a large number of droplets to aid in the collection of the fine toxic metal particulate. Figure 3 shows major components of a condensing wet scrubber.

In this system flue gas enters a quench section where it is first cooled to its saturation temperature. The gas is then ducted to a condenser/absorber where it is contacted by a cooled reagent stream and further cooled to about 80° - 90°F. The flue gas (now containing condensed toxic metal particulate and water droplets) is passed to a collision scrubber, where the fine droplets impinge on a flat surface. Here the fine particulate and water droplets interact and agglomerate, resulting in particulate capture. The flue gas then passes through an entrainment separator for droplet removal and is discharged through the stack.

The reagent streams from the various scrubbing stages are combined and pumped to a cooling tower (or refrigeration system) for cooling prior to being recycled to the scrubbing system. Condensing wet scrubbers have achieved very high removal efficiencies for acid gases, toxic metal, and fine particulate matter.

Metals Emissions

Metals emissions can be effectively controlled from municipal, hospital, and hazardous waste incinerators. The types of controls typically employed vary with the type of incinerator and local regulatory requirements. Table IV presents data on typical emissions levels for municipal waste incinerators.

Table IV. Typical Refuse Incinerator Uncontrolled and Controlled Emissions

Pollutant	Uncontrolled Emissions	Controlled Emissions	Percent Reduction
Particulate Matter, gr/dscf	0.5-4.0	0.002-0.015	99.5+
Acid Gases ppmdv			
HCl	400-100	10-50	90-99+
SO$_2$	150-600	5-50	65-90+
HF	10-0	1-2	90-95+
NO$_x$	120-300	60-180	30-65*
Toxic Metals mg/nm^3			
Arsenic	<0.1-1	<0.01-0.1	90-99+
Cadmium	1-5	<0.01-0.5	90-99+
Lead	20-100	<0.1-1	90-99+
Mercury	<0.1-1	<0.1-0.7	10-90+
Total PCDD/PCDF ng/nm^3	20-500	<1-10	80-99

* Reference conditions - Dry Gas 20°C and 12% carbon dioxide.

SOURCE: Adapted from refs. 1 and 4.

The controlled emissions levels are typical of modern large municipal waste incinerators equipped with spray dryer adsorption pollution control systems.

Table V presents controlled emissions estimates for the major toxic metals found in hazardous waste incinerator flue gas.

Table V. Hazardous Waste Incinerator Emissions Estimates

	EPA* Conservative Estimated Efficiencies	Typical Actual Control Efficiencies	Typical Range of Emissions Rates
Particulate matter	99+ %	99.9+ %	0.005-0.02 gr/dscf
Arsenic	95	99.9+	1-5 $\mu g/m^3$
Beryllium	99	99.9	<0.01-0.1
Cadmium	95	99.7	0.1-5
Chromium	99	99.5	2-10
Antimony	95	99.5	20-50
Barium	99	99.9	10-25
Lead	95	99.8	10-100
Mercury	85-90	40-95+	10-200
Silver	99	99.9+	1-10
Thallium	95	99+	10-100

*   Based on spray dryer fabric filter system or 4-field electrostatic precipitator followed by a wet scrubber.

SOURCE: Adapted from refs. 5–9.

The control efficiencies and emissions rates presented are based on the application of a spray dryer absorption system equipped with a fabric filter dust collector or a four-field electrostatic precipitator followed by a wet scrubber designed for fine particulate removal. All of the toxic metals can be controlled at greater than 99 percent efficiency except for mercury. High-efficiency mercury removal (95 percent) requires the use of an adsorption enhancer in the spray dryer absorption system or a condensing wet scrubber following a dust collector.

Conclusions

The increased use of incineration for control and destruction of municipal and hazardous wastes has led to increasingly stringent air pollution control

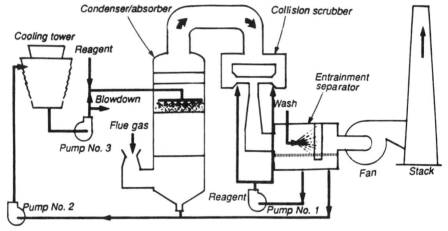

Figure 3. Condensing Flue-Gas Cleaning System

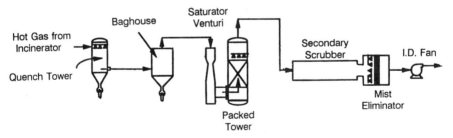

Figure 4. Combined Systems Approach - Emissions Control Scheme

regulations.   EPA has recently promulgated New Source Performance Standards for municipal waste combustors which require health risk based emissions limits for specific metals to be established within the next year.   EPA has also proposed hazardous waste incinerator emissions limits which include health risk based emissions limits for ten toxic metals.   EPA is currently preparing new standards for hospital waste incinerators.

Spray dryer absorption is considered to represent BACT for many municipal waste incinerator applications and is capable of achieving high collection efficiencies for the metals of concern. Spray dryer absorption is also used for control of emissions from hazardous waste incinerators.   Dust collectors with downstream wet scrubbers are also commonly used to control metals emissions from hazardous waste incinerators.   Both systems have demonstrated the ability to achieve high collection efficiencies for the ten toxic metals proposed for regulation.   Emissions rates of medium to large incinerators equipped with properly designed air pollution control systems can comply with current regulations.

Emissions levels of toxic metals from incinerators equipped with modern air pollution control systems are orders of magnitude lower than for 1980 levels. Current regulations are based on health risk assessments and set emissions limits that are highly protective of human health and the environment. Continued public concern regarding incinerator emissions, however, may result in additional and more complex control systems being required in the future. Figure 4 shows a combined systems approach to control incinerator emissions.

The quench tower and fabric filter would result in capture of the major fraction of toxic metals and greater than 99 percent removal of particulate matter.   The venturi and packed tower scrubber stages would achieve additional toxic metal removal and act as acid gas control devices.   The secondary scrubber would be a fine particulate control device and remove additional toxic metals. The subsequent mist eliminator might be followed by an activated carbon absorber for removal of any residual vapor-phase metals and toxic organic compound present in the flue gas.

Cleanup levels of greater than 99 percent for all major pollutants can be achieved by employing multiple control devices but at significant additional cost and loss of overall system reliability without significant increases in protection of human health or the environment. Industry, the public, and regulators must balance all factors in setting reasonable future emissions limits.

Literature Cited

1.   J.R. Donnelly, "Overview of Air Pollution Controls for Municipal Waste Combustors," Second International Conference on Municipal Waste Combustion, Tampa, April 1991.
2.   USEPA, "New Source Performance Standards and Emissions Guidelines for Existing Municipal Waste Combustors," 40 CFR Part 60 Ca, Federal Register, Volume 56, No. 28, February 11, 1991.
3.   USEPA, "Standards for Owners and Operators of Hazardous Waste Incinerators and Burning of Hazardous Wastes in Boilers and Industrial

Furnaces; Proposed and Supplemental Proposed Rule, Technical Corrections and Request for Comments," 40 CFR Parts 260, 261, 264, and 270, Federal Register, Volume 55, No. 82, April 27, 1990.

4. USEPA, "Burning of Hazardous Waste in Boilers and Industrial Furnaces: Final Rule," 40 CFR Part 260, et al, Federal Register, Volume 56, No. 35, February 21, 1991.

5. T.G. Brna and J.D. Kilgroe, "The Impact of Particulate Emissions Control of the Control of Other MWC Air Emissions," Journal of the Air and Waste Management Association, Volume 40, No. 9, Pages 1324-1330, September 1990.

6. J.R. Donnelly and S.K. Hansen, "Joy/Niro Spray Dryer Acid Gas Removal System for Hazardous Waste and Sewage Sludge Incineration," HMCRI-Sludge 87, Boston, May 1987.

7. J.R. Donnelly, S.K. Hansen, M.T. Quach, "Joy Niro Control Technology Concepts for Hazardous Waste Incinerators," 1986, AWMA Annual Meeting, Minneapolis, June 1986.

8. EPA/530-SW-91-004, "Metal Control Efficiency Test at a Dry Scrubber and Baghouse Equipped Hazardous Waste Incinerator," September 1990.

9. "E.I. Dupont Company Trial Burn Test Program, Emission Test Results, Sabine River Works, Final Report," Alliance Technologies Corporation, August 1990.

RECEIVED July 31, 1992

Chapter 15

Metal Behavior During Medical Waste Incineration

C. C. Lee and G. L. Huffman

Risk Reduction Engineering Laboratory, U.S. Environmental Protection Agency, Cincinnati, OH 45268

Toxic metals such as lead, cadmium, and mercury are contained in medical waste. Consequently, the incineration of medical waste may result in the emissions of trace metals into the environment, if incinerators are not properly designed and operated. EPA's Risk Reduction Engineering Laboratory initiated a study in 1988 to document what is known about medical waste treatment, particularly in the area of medical waste incineration. This paper is to summarize the findings from this study regarding the behavior of metals in incineration processes. Highlights of these findings are as follows: (1) Lead and cadmium are the two most-often-found metals in medical waste; (2) Metals can partition into different phases (gas, liquid or solid) but cannot be destroyed during incineration; (3) There are several potential pathways that metals follow to reach the environment. They exit incinerators with siftings, bottom ash, fly ash, scrubber waste, and flue gas; (4) Data on the capture efficiency of metals by air pollution control equipment used at medical waste incinerators is very limited; and (5) Wet scrubbers generally capture cadmium moderately well but normally perform poorly in removing chromium and lead. Fabric filter systems efficiently capture all metals.

Medical waste contains toxic metals such as lead, cadmium, and mercury. These metals will only change forms (chemical and physical states) but will not be destroyed during incineration. They can be emitted from incinerators on small particles capable of penetrating deep into human lungs. Thus, the emission of trace amounts of heavy metals from medical waste incinerators is one of the major concerns to those who are involved in medical waste management. A clear understanding of metals behavior in medical waste incinerators is critically needed.

EPA's Risk Reduction Engineering Laboratory initiated a study in 1988 to document what is known about medical waste treatment, particularly in the area of medical waste incineration. Potential toxic metal emissions from medical waste incineration was one of main subjects studied. This paper is to summarize the findings of that study.

Metal Sources

University of California at Davis researchers conducted a study to identify the sources of toxic metals in medical wastes [1]. The research effort focused on lead and cadmium because they were the two most-often-found metals in medical waste. They concluded that plastics in the waste contributed most to the presence of these two metals. Cadmium is a component in common dyes and thermo- and photo-stabilizers used in plastics. Lead was found in many materials including plastics, paper, inks, and electrical cable insulation. However, the primary source of lead appeared to be plastics. Like cadmium, lead is used to make dyes and stabilizers which protect plastics from thermal and photo-degradation. It is ironic to note that the dyes made from lead and cadmium are used to color plastic bags. Thus, part of the lead and cadmium emissions could be due simply to the "red bags" that infectious waste is placed in.

Under the authority of the hazardous waste program required by the Resource Conservation and Recovery Act, EPA has identified ten (10) metals of most concern from 40 CFR 261 Appendix VIII. Four of the ten metals are classified as carcinogenic and the other six metals are considered to be toxic. The EPA's Carcinogen Assessment Group has estimated the carcinogenic potency for humans exposed to low levels of carcinogens. An assigned "Unit Risk" indicates the relative health threat of the metals. Unit Risk (UR) is the incremental risk of developing cancer to an individual exposed for a lifetime to ambient air containing one microgram of the compound per cubic meter of air. Inhalation is the only exposure pathway considered in determining UR.

Data on toxicity are used to define concentrations for the six toxic metals below which they are not considered dangerous. Ambient concentrations should not exceed this concentration. The EPA has defined the maximum toxic concentration, or Reference Air Concentration (RAC), for each metal. If ground level concentrations of any of these metals exceeds its RAC, adverse health effects are likely. The Unit Risk of the four carcinogenic metals and the RAC of the six toxic metals are listed in Tables I and II.

Table I. Unit Risk (UR) Values for Four Carcinogenic Metals

Metals species	Unit risk
Arsenic (As)	0.0043
Beryllium (Be)	0.0025
Cadmium (Cd)	0.0017
Chromium (Cr^{+6})	0.012

UR: incremental lifetime cancer risk from exposure to 1 μg/cubic meter

Table II. Reference Air Concentrations (RACs) for Six Toxic Metals

Metals species	RAC $(\mu g/m^3)$
Antimony (Sb)	0.025
Barium (Ba)	50.00
Lead (Pb)	0.09
Mercury (Hg)	1.70
Silver (Ag)	5.00
Thallium (Tl)	500.00

Emission Pathways

A majority of metal emissions is in the form of solid particulate matter and a minority is in vapor form. It was generally concluded that particulate emissions from the incineration of medical wastes are determined by three major factors:

(1)   Suspension of noncombustible inorganic materials;

(2)   Incomplete combustion of combustible materials (these materials can be organic or inorganic matter); and

(3)   Condensation of vaporous materials (these materials are mostly inorganic matter).

The ash content of the waste feed materials is a measure of the noncombustible portion of the waste feed and represents those materials which do not burn under any condition in an incinerator. Emissions of noncombustible materials result from the suspension or entrainment of ash by the combustion air added to the primary chamber of an incinerator. The more air added, the more likely that noncombustibles become entrained. Particulate emissions from incomplete combustion of combustible materials result from improper combustion control of the incinerator. Condensation of vaporous materials results from noncombustible substances that volatilize at primary combustion chamber temperatures with subsequent cooling in the flue gas. These materials usually condense on the surface of other fine particles.

The transformation of mineral matter during combustion of metals-containing waste is shown in Figure 1. The Figure is self-explanatory. There are several potential pathways to the environment that metals may follow. Most metals remain in the bottom ash. A small fraction of the ash (on a weight basis) is entrained by the combustion gases and carried out of the primary chamber as fly ash. Volatile metals may vaporize in the primary combustion chamber and leave the bottom ash. These metals recondense to form very small particles as the combustion gases cool. Some of the entrained ash and condensed metals are captured

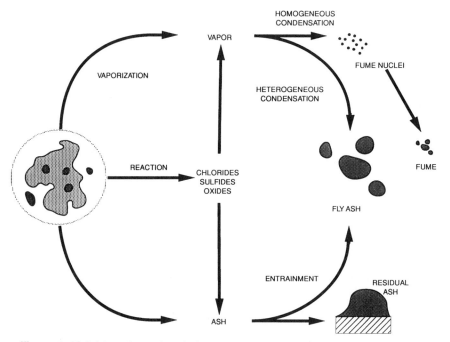

Figure 1. Metal transformation during incineration. Reproduced from ref. 2.

in the air pollution control equipment (APCE). The rest enters the atmosphere. Four key variables affecting the vaporization of metals are *(2)*:

● Chlorine concentration in the waste;

● Temperature profiles in the incinerator;

● Metal species concentration in the waste; and

● Local oxygen concentration.

Current Control Practice

Typically, two strategies are used to minimize metals emissions: (1) The primary chamber is operated at conditions which do not promote vaporization or entrainment of metals; and (2) Any metals which do escape can be captured in the APCE, if present. The parameters usually used to control the escape of metals from the primary chamber are the primary chamber temperature and gas velocity. The key APCE parameters used are specific to the device which is utilized.

(1)   Combustion control: Most operating medical waste incinerators are simple single-chamber units with an afterburner located in the stack. The ability of batch incinerators to control metals emissions is limited because only the temperature in the stack is usually monitored.

Most new incinerators are starved-air units. The primary chamber is designed to operate at low temperatures and low gas flow rates. This minimizes the amount of materials entrained or vaporized.

To ensure that metal emissions are minimized, operators must maintain the primary chamber at the temperatures and gas flow rates for which it was designed. Usually the only parameter that system operators can directly control is feed rate. High feed rates can lead to high temperatures and high gas velocities. Thus, many operators carefully control the feed rate. The feed rate is reduced when primary temperatures increase.

(2)   APCE control: When metals reach the APCE, they are present in one of three forms. Non-volatile metals are on large entrained particles. Metals which have vaporized and recondensed are usually present on fly ash particles with diameters less than 1 micron. Extremely volatile metals are present as vapors. Table III summarizes the ability of common APCE to control these different metal forms. The Table is based on data and worst case predictions. Wet scrubbers are often used to minimize the temperature of the

flue gases. Use of low temperatures ensure that all metal
vapors have condensed. As indicated in Table III, vapors
are much more difficult to capture than particles *(2)*.

TABLE III. Typical APCE Control Efficiencies

Control Efficiency (%)

APCE	Particulate	Fume	Vapor
Venturi scrubber 20" pressure drop	90	85	60
Venturi scrubber 60" pressure drop	98	97	90
Fabric filter	95	90	50
Spray drier/ fabric filter	99	95	90

Emission Data

Figure 2 compares the concentration of arsenic (chosen merely for
illustrative purposes) in flue gases before any APCE, and in emitted
gases for a variety of incinerators. As shown, a wide variety of
flue gas cleaning equipment is used. The Figure indicates the
effectiveness of the various types of APCE. Arsenic is predicted to
be relatively volatile, compared to other metals. Significant
amounts of arsenic are therefore expected to vaporize in an
incinerator. Figures 3 and 4 present similar data for the two most
common metals found in medical waste, lead and cadmium *(2)*.

Conclusion

Some metals and metal species found in medical waste are volatile
and will vaporize at the conditions found in medical waste
incinerators. The vapors are carried away from the waste by the
exhaust gas and they recondense as the gas cools. The vapors
condense both homogeneously to form new particles and
heterogeneously on the surfaces of existing fly ash particles. To
control metal emissions, metals which are of a highly volatile
nature are of main concern in terms of installing the proper APCE.
Because there are many APCE sizes and types, it is very important to
fully understand metal emissions characteristics, combustion control
and operating possibilities, and expected APCE performance so that
metal emissions can be minimized from medical waste incinerators.

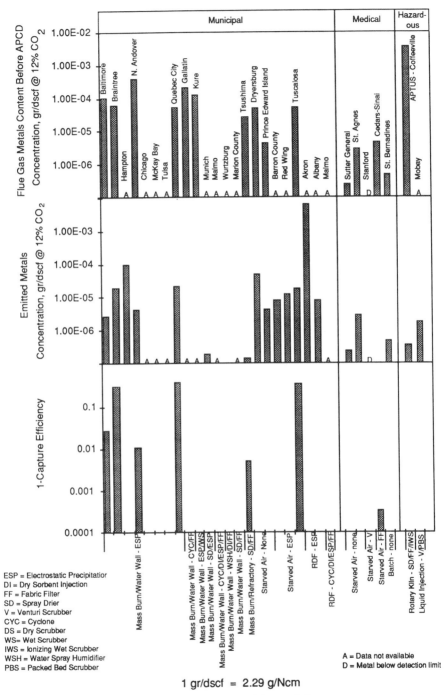

1 gr/dscf = 2.29 g/Ncm

Figure 2. Arsenic emissions. Reproduced from ref. 2.

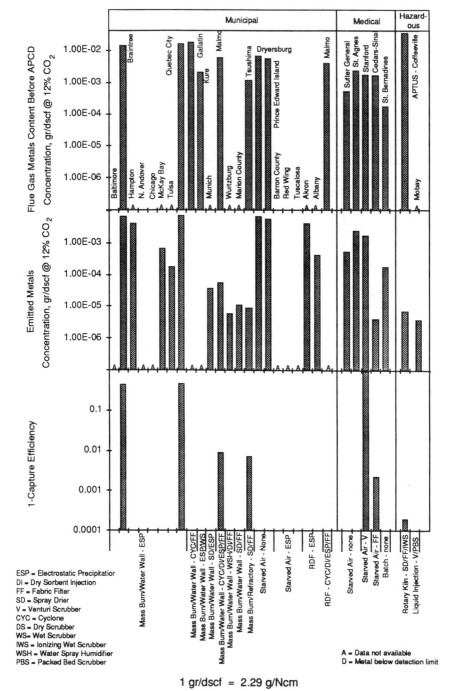

1 gr/dscf = 2.29 g/Ncm

Figure 3. Lead emissions. Reproduced from ref. 2.

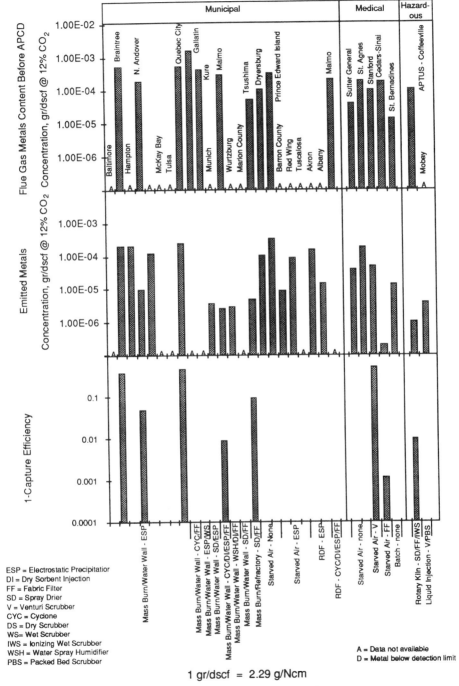

Figure 4. Cadmium emissions. Reproduced from ref. 2.

Literature Cited

1. Hickman, D. C., "Cadmium and Lead in Bio-Medical Waste Incinerators," Master of Science Thesis, University of California, Davis, 1987.
2. Energy and Environmental Research Corporation, "State-of-the-Art Assessment of Medical Waste Thermal Treatment," A Draft Report to EPA's Risk Reduction Engineering Laboratory, April 1991.

RECEIVED September 15, 1992

Chapter 16

Trace Metal Analysis of Fly Ash from Combustion of Densified Refuse-Derived Fuel and Coal

Bassam S. Attili, Kevin D. Ingram, Chia-Hui Tai, and Kenneth E. Daugherty

Department of Chemistry, University of North Texas, Denton, TX 76203

Analysis of the trace metals in fly ash produced from the combustion of quicklime binder enhanced densified refuse derived fuel (bdRDF) with coal is discussed.

In 1987 a full-scale cofiring of bdRDF and high sulfur coal was conducted at Argonne National Laboratories. About 567 tons of bdRDF pellets were cofired with coal at 0 to 50 percent bdRDF by Btu content and 0, 4, and 8 percent binder.

Analysis has continued on the samples acquired at Argonne. The fly ash was dissolved in a mixture of aqua-regia and hydrofluoric acid in a Parr bomb using a microwave dissolution method . The solution was then analyzed by Inductively Coupled Plasma (ICP) for As, Ba, Be, Cd, Cr, Cu, Hg, Ni, Pb, Sb, Se, Tl, V, and Zn. Results indicated that some trace elements decreased in fly ash with the increase in dRDF percentage while others increased.

The disposal of refuse is an increasing concern of municipalities and state governments throughout the U.S. Ten years ago there were approximately 10,000 sanitary landfills in the country. Currently there are less than 5,000. By the year 2000, many existing landfills will become filled to capacity, and new landfills will be more costly to site (1-3). The NIMBY syndrome, or Not IN My Back Yard, dominates peoples minds when it comes to siting new landfills. The development of an attractive disposal method is critical to overcome these problems.
There are three ways to dispose of Municipal Solid

0097–6156/93/0515–0199$06.00/0

Waste (MSW). It can be landfilled, dumped at sea or burned. Incineration is the major potential solution to the landfill problem (4-6). The conventional way to conduct incineration of MSW is called mass burn, in which the entire incoming garbage stream is run through a high temperature kiln. There are several potential problems with mass burn. Since all of the materials that are in the garbage stream are incinerated, there are high levels of metals in the ash. Because of the nature of the materials, there are also high levels of residual ash which can be up to 25 percent by weight and up to 10 percent by volume of the incoming MSW (5). Also, there are concerns about air emissions if the kiln temperatures are not kept at proper levels.

An increasingly attractive option is to separate the metals, aluminum, high density plastic, corrugated cardboard and glass from the incoming MSW so that they can be recycled. Then any remaining non-combustible materials are separated from the combustible materials that remain. These remaining combustible materials (approximately 50% of the MSW by weight) consists largely of paper. This can be ground and turned into an alternative energy source called Refuse Derived Fuel (RDF) which can then be co-fired with coal. There is obviously a need to densify the material so that it can be transported to end users. Also, in order to reduce any chemical and biological degradation that might occur during storage, a binder material might need to be incorporated into the densified material.

Approximately 150 potential binders were tested at the University of North Texas (UNT), to be used as binding agents with RDF. This initial study took into account the cost and environmental acceptability of these materials to determine the best candidates to be used in pelleting trials. A commercial test of pelletizing RDF was conducted at Jacksonville, Florida, Naval Air Station in the summer of 1985 with the binder candidates that were identified earlier. Durability tests along with analysis of the effectiveness of the binder to impede degradation were conducted. The results of the test proved that quicklime ($Ca(OH)_2$) is the best binder.

The material was then ready for full scale plant demonstration. This cofiring of RDF with coal was conducted at Argonne National Laboratories (ANL) in the summer of 1987 and involved the combustion of over five hundred tons of the binder enhanced dRDF with a high sulfur coal over a six week period. Over 1500 emission samples were collected from the combustion test and included flue gas emissions, fly ash, bottom ash, and feed stock samples.

There are two types of ash that result due the combustion of any material. Fly ash consists of fine particulate matter that escapes in the flue gas. It is collected by electrostatic precipitators or baghouses, although approximately 1-2 percent of fly ash does escape to the atmosphere (3) even with the pollution control devices on-line. There is also bottom ash which is a courser material that drops through the grates in the furnace.

Fly ash is the major by-product of burning MSW (7,8). There are about 35,000 tons of fly ash produced for each million ton of waste incinerated (9,10). Fly ash consists of 70-95% inorganic matter and 5-30% organics (3). There are many constructive ways to use fly ash, including as an additive to improve the performance of Portland cement and as a soil stabilizer (11).

Since fly ash is the major by-product of incineration and incineration is the main attractive solution to landfills the physical and chemical characteristics of fly ash is important in determining its method of disposal or use (12). Trace metals are important because of their relationship to regulatory criteria under the Resource Conservation and Recovery Act (RCRA) regarding toxicity (13). Trace metals play a major role in whether a fly ash is hazardous and is disposed of or if it can be used in a productive way.

The most toxic elements of concern in ash are arsenic, barium, beryllium, cadmium, chromium, copper, mercury, nickel, lead, antimony, selenium, thallium, vanadium and zinc. Inductively Coupled Plasma Atomic Emission Spectroscopy (ICP-AES), was selected over other methods for determining these toxic elements because of the trace nature of the elements involved, the need for quantitation and the ability to determine all of the elements in the samples sequentially.

A microwave oven dissolution method was used to dissolve the ash in a mixture of aqua-regia and hydrofluoric acid using a Parr bomb. The solution was then analyzed by ICP.

Methodology

Fuel Preparation. The binder enhanced dRDF pellets for the 1987 ANL study were supplied by two facilities, one a 40 ton per day plant located at Thief River Falls, Minnesota (Future Fuel Inc.) and the other a 470 ton per day plant at Eden Prairie, Minnesota (Reuter Inc.). The dRDF was made with 0, 4, and 8 percent $Ca(OH)_2$ binder.

Before each test run, dRDF pellets and coal were blended together using a front-end loader until the material appeared approximately homogenous. Due to differences in bulk densities and energy values of the materials, to produce a blend close to 10 percent dRDF by Btu content it takes a mixture of three volumes of a high-

sulfur Kentucky coal and one volume of dRDF . The blend
was moved by front-end loader to the coal pit and
transported by conveyor to coal bunker prior to use in the
ANL stoker fired boiler. Nine tons per hour of the fuel
mixture on average were burned.

Sampling Plan. A total of 567 tons of dRDF pellets were
cofired with 2,041 tons of sulfur-rich coal in 12 separate
test runs. The runs were classified according to the
different Btu contents of dRDF in the fuel and different
binder content of dRDF (Table I). Runs 1 and 12 used coal
alone in order to establish base line data. To avoid
cross-contamination between the different runs, coal only
runs were also performed between the other runs to
cleanout the dRDF from the coal pit and to reduce any
memory effects that might occur in the boiler due to the
inclusion of the calcium binder from previous runs.

Sample Collection. Over 1,500 samples of flue gas
emissions, fly ash, bottom ash, and feedstock were
collected during the 12 runs. A total of 190 bottom ash
samples were collected from under the grate and through
the traveling grate in the boiler. A total of 176 fly ash
samples were collected from the multi-cyclone and from the
economizer. Random ash samples were taken every eight
hours. The samples were collected either by one of the
UNT research teams or one of the ANL operators at the
required times. Aluminum containers were used to collect
the ash samples. After the samples cooled they were
transferred into ziplock bags which were then labeled with
the date, run number, and the time the sample was
collected. The ash samples were then packed and
transported to UNT where they were arranged according to
the run number, date, and time of collection.

Equipment

Parr Bombs. Parr Teflon acid bombs were obtained from
Parr Instrument Company. The bomb is made of a microwave
transparent polymer. A compression relief disc is built
into the closure to release excessive pressure if the bomb
reaches an internal pressure of over 1500 psi. In most
cases all parts of the bomb were reusable except for the
O-ring.

Microwave Oven. Microwave digestion is becoming
increasingly accepted as a fast and reliable alternative
to the traditional hot plate method for the digestion of
samples before elemental analysis. A Kenmore commercial
microwave oven was used to facilitate this work. The oven
has a variable timing cycle from 1 second to 100 minutes

Table I. Coal/dRDF Test Run Schedule

Run #	Date	Composition	%Binder
1	1-5 June	Coal	xxxxx
2	5-8 June	Coal, 10% dRDF	0
3	8-12 June	Coal, 10% dRDF	4
4	12-15 June	Coal, 10% dRDF	8
1	15-18 June	Coal	xxxxx
5	18-23 June	Coal, 20% dRDF	0
7	23 June	Coal, 20% dRDF	4
6	23-26 June	Coal, 30% dRDF	8
7	26-28 June	Coal, 20% dRDF	4
8	28 June-1 July	Coal, 20% dRDF	8
12	1-4 July	Coal	xxxxx
11	4-5 July	Coal, 50% dRDF	4
12	5-6 July	Coal	xxxxx
9*	6-7 July	Coal, 30% dRDF	0
10*	7-8 July	Coal, 30% dRDF	4
12	8 July	Coal	xxxxx

* reduced plastic content dRDF pellets

and a variable heating cycle based on its' power settings from 70 watts through 700 watts at full power.

Inductively Coupled Plasma Atomic Emission Spectrometry ICP-AES. A Perkin-Elmer ICP-5500 Atomic Emission Spectrometer with a 27.12-MHz RF generator was used in this analysis.

The performance characteristics of ICP, namely its versatility, wide applicability, and ease of use are almost unparalleled among other methods of elemental analyses (14). ICP is theoretically capable of determining any element in the sample matrix except argon, which is used to form the plasma (15).

The application of ICP to simultaneous determination of major, minor, and trace level elements in various matrices has been well documented (16-19). ICP offers several advantages as an alternative approach for the analysis of geochemical and environmental samples (20-24). ICP permits the determination of a large number of elements with high sensitivity and precision and with relative freedom from chemical interferences (25-27).

Sample Analysis

After the samples were returned to the laboratory, they were arranged according to the dates and times they were collected. To determine the trace metal content of each sample, about 10 grams of the ash was ground to pass a 75 mesh sieve. A 400 mg sample was placed in a teflon container and treated with 1 mL of hydrofluoric acid and 3 mL of aqua regia. The teflon container was then placed in the Parr bomb and the bomb was tightly capped. The bomb was placed in the microwave oven and heated for 4 minutes and left for several hours to cool. After cooling, the teflon container was uncapped and 2 mL of saturated boric acid was quickly added. The container was then recapped, returned to the microwave oven and reheated for 1 minute, then cooled again (28-30).

At this stage some uncombusted carbon remained, so the solution was filtered to remove the residue. The residue was washed with deionized water and the filtrate was diluted to 50 mL in a polyethylene volumetric flask.

The solution was finally analyzed by ICP using a blank and a standard solutions containing the same amounts of acids. Standards with varied concentrations of As, Ba, Be, Cd, Cr, Cu, Hg, Ni, Pb, Sb, Se, Tl, V, and Zn were used for the analysis.

Results and Discussion

The chemical composition of the coal and dRDF ash depends on many factors. The geological and geographic factors

related to the coal deposits have a major effect on the
initial composition of the fuelstock. Also, the
combustion temperatures, residence time in the combustion
zone and air flow rate in the boiler along with the
efficiency of air pollution control devices have an effect
on the elements present in the fly ash.

The fly ash samples which were investigated in this
study by ICP were the from the economizer which comes
after the multicyclone in the pollution control system.
The results from this study are summarized in Table II.
Fourteen metals were analyzed in this study. They were
arsenic, barium, beryllium, cadmium, chromium, copper,
mercury, nickel, lead, antimony, selenium, thallium,
vanadium, and zinc. The metals arsenic, cadmium, mercury,
lead, antimony, selenium and thallium are not included in
these results because their concentrations were too low
to be detected by ICP. Table III summarizes the ICP
detection limits under the conditions of this study of all
of the elements examined.

Effect of dRDF content on trace metals. The processing
of MSW to RDF removes much of the unwanted trace metals
since many of the metal containing pieces of waste are
separated before incineration to be recycled. The metal
content of the coal and RDF blend ash is expected to be
affected by the different percentages of RDF. Elements
such as barium, cadmium, chromium, copper, mercury, lead
and zinc are known to be enriched in RDF related to coal.
However, the levels of cadmium, mercury and lead observed
were below detection limits for all of the coal and bdRDF
mixes examined in this study.
Table II and Figures 1-7 show the amount of each
metal in the fly ash. Each figure illustrates the variance
of the level of the given metal based upon the percent
replacement by Btu content of RDF for coal and upon the
percent binder used in conjunction with the bdRDF.
The graphs show an increase in copper and zinc
concentration with an increase in bdRDF concentration.
Since these elements are generally higher in RDF, this
was expected. The elements barium and chromium varied
greatly but showed a general increase with an increase in
bdRDF concentration at different binder percentages.
These increases are also due to these elements being more
enriched in RDF ash than in coal ash.
The elements vanadium, beryllium and nickel
generally were decreased in concentration with an increase
in bdRDF percentage. Those elements are believed to be
at approximately the same concentration or slightly lower
in RDF than coal.
Obviously the amount of all of these metals will
deviate as their concentrations vary in the incoming waste

**Table II. Summary of Toxic Metals Concentration
in Economizer Fly Ash (ug/g)**

Run#	Ba	Be	Cr	Cu	Ni	V	Zn	Btu% dRDF	Binder %
1.	158.2	25.3	105.2	152.7	130.0	223.4	324.8	0	–
2.	240.1	27.7	111.3	199.5	135.3	234.1	338.5	10	0
3.	202.2	19.6	100.3	151.8	122.2	177.9	293.7	10	4
4.	144.4	14.7	94.9	143.7	100.8	160.7	390.8	10	8
5.	227.7	20.2	143.8	243.4	137.4	231.0	404.6	20	0
6.	155.7	11.3	108.3	208.5	149.0	187.1	478.2	20	4
7.	182.8	13.3	127.6	243.6	121.3	193.4	466.7	20	8
8.	160.1	11.4	114.4	360.6	92.7	181.3	443.6	30	0
9.	158.3	10.5	115.9	207.9	75.6	149.6	455.9	30	4
10.	190.2	10.5	112.2	227.6	130.5	161.3	470.8	30	8
11.	228.5	14.9	126.1	353.5	97.7	179.8	372.1	50	4
12.	177.4	16.6	93.5	217.1	97.8	171.6	240.3	0	–

Table III. Detection Limits of ICP (ug/g)

Detection Limits of ICP (ug/g)	
As	62.50
Ba	12.50
Be	6.25
Cd	6.25
Cr	6.25
Cu	6.25
Hg	125.00
Ni	12.50
Pb	125.00
Sb	125.00
Se	62.50
Tl	125.00
Zn	6.25

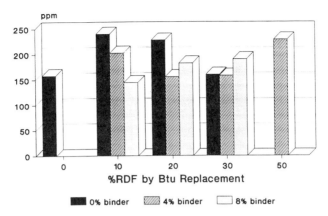

Figure 1.   Barium concentrations in the fly ash.

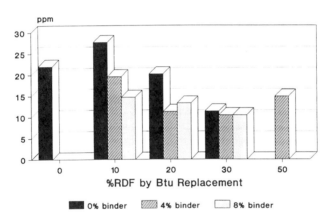

Figure 2.   Beryllium concentrations in the fly ash.

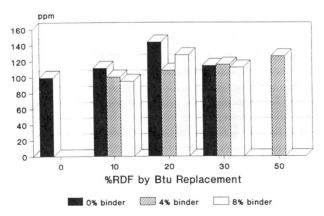

Figure 3.   Chromium concentrations in the fly ash.

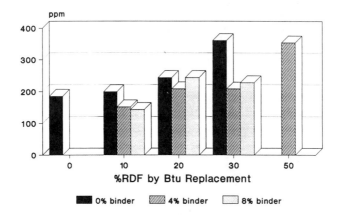

Figure 4.   Copper concentrations in the fly ash.

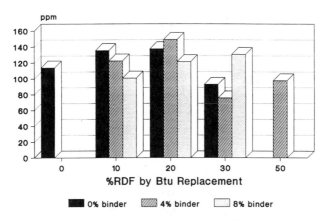

Figure 5.   Nickel concentrations in the fly ash.

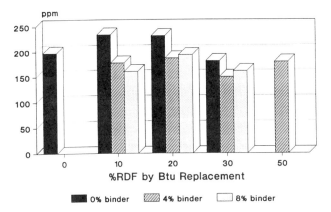

Figure 6. Vanadium concentrations in the fly ash.

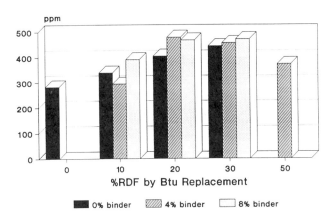

Figure 7. Zinc concentrations in the fly ash.

stream. Care has to be taken in the processing of the MSW
to remove all of the possible contaminants. However, the
production of RDF is very dependent upon what individuals
throw away as garbage.

The most promising aspect of this study is that the
metal emissions vary little from that of coal burned
alone. Only copper showed a large increase in its
presence in the fly ash. As recycling gains popularity
in this country, the amount of copper in the countries
waste streams should decrease and in turn help decrease
the amount of copper in the fly ash of the combustion of
refuse derived fuel. However, it should be reiterated
that nothing in this study indicates that the ash from the
combustion of refuse derived fuel must be landfilled as
a hazardous waste and not used in productive activities.

Conclusion

The binder-enhanced dRDF is a promising technique for the
future to be used as fuel or as a substitute for coal. It
is an economical way of disposal of MSW in the sense that
it reduces the heavy cost of landfilling, and this
technique generates extra income if sold as fuel.
According to the results presented here it shows promise
in reducing emissions, especially the trace heavy metals
emissions, which also makes it safer to use the fly ash
on a large scale, including for many construction
projects.

Literature Cited

1.   Ohlsson, O., presentation at Resource Recovery for
Small Communities in Panama City, Florida, 1988.
2.   Ohlsson, O., Daugherty, K.E., presentation at Air
and Waste Management Association Forum 90 in Pittsburgh,
Pennsylvania, 1990.
3.   Attili, B.S., Daugherty, K.E., Kester, A.S.,
Shanghai International Conference on the Utilization of Fly
Ash and Other Coal Products, 1991, pp. 18.1-18.12.
4.   Daugherty, K. E., Phase 1, Final Report; U.S.
Department of Energy, ANL/CNSV-TM-194, 1988.
5.   Rogoff, M.J., Noyes Publications: Park Ridge, New
Jersey, 1987, pp. 39-45.
6.   Gershman, Brickner, and Bratton, Inc., "Small-Scale
Municipal Solid Waste Energy Recovery Systems", Van
Nostrand Reinhold Company, NY., 1986, pp. 4-20.
7.   Eiceman, G.A., Clement, R.E., Karasek, F.W.,
Analytical Chemistry, 1979, 51, pp. 2343-2350.
8.   Karasek, F.W., Gharbonneau, G.M., Revel, G.J., Tong,
H.Y., Analytical Chemistry, 1987, 59, pp. 1027-1031.
9.   Huheey, James E., Inorganic Chemistry, Harper and
Row Publishers, New York, Third Edition, 1983, pp. 851-
936.

10. Roy, W.R., Thiery, R.G., Schuller, R.M., Environmental Geology Notes 96, State Geological Survey Division, April, **1981**.
11. Hecht, N., "Design Principles in Resource Recovery Engineering", Butterworth Publishers: Boston, **1983**, pp. 23-33.
12. Poslusny, M., Daugherty, K., Moore, P., American Institute of Chemical Engineers Symposium Series, **1988**, 265, pp. 94-106.
13. Attili, B. S., Ph.D. Dissertation, University of North Texas, Denton, Texas, **1991**.
14. McLaren, J.W., Berman, S.S., Boyko, V.J., Russel, D.S., Analytical Chemistry, **1981**, 53, pp. 1802-1806.
15. Tikkanen, M.W., Neimczyk, T.M., Analytical Chemistry, **1986**, 52, pp. 366-370.
16. Nygaard, D.D., Analytical Chemistry, **1979**, 51, pp. 881-884.
17. Ward, A.F., Marciello, L.F., Analytical Chemistry, **1979**, 51, pp. 2264-2272.
18. McQuaker, N.R., Brown, P.D., Chany, G.N., Analytical Chemistry, **1979**, 51, pp. 888-895.
19. Nadkarni, R.A., Analytical Chemistry, **1980**, 52, pp. 929-935.
20. Fassel, V.A., Analytical Chemistry, **1979**, 51, pp. 1290A-1308A.
21. Bennet, H., Analyst, **1977**, 102, pp. 153-158.
22. Greenfield, S., Jones, I.L., McGreachin, H.M., Smith, P.B., Analytica Chimica Acta, **1975**, 74, pp. 225-245.
23. Robinson, A., Science, **1978**, 199, p. 1324-1329.
24. Uchida, H., Uchida, T., Iida, C., Analytical Chimica Acta, **1979**, 108, pp. 87-95.
25. Cox, X.B., Bryan, S.R., Linton, R.W., Analytical Chemistry, **1987**, 59, pp. 2018-2023.
26. Sugimae, A., Barnes, R., Analytical Chemistry, **1986**, 58, pp. 785-789.
27. Beckwith, P.M., Mullins, R.L., Coleman, D.M., Analytical Chemistry, **1987**, 59, pp. 163-167.
28. Nadkarni, R. A., Analytical Chemistry, **1984**, 56, pp. 2233-2237.
29. Bettinelli, M., Baroni, U., Pastorelli, N., Journal of Analytical Atomic Spectroscopy, **1987**, pp. 485-490.
30. ASTM standard method 200.7 ICP/AES.

RECEIVED October 1, 1992

METALS EMISSIONS CONTROL USING SORBENTS

Chapter 17

Aluminosilicates as Potential Sorbents for Controlling Metal Emissions

Mohit Uberoi[1] and Farhang Shadman

Department of Chemical Engineering, University of Arizona,
Tucson, AZ 85721

Various aluminosilicate type material were evaluated as potential
sorbents for removal of toxic metal compounds such as lead
and cadmium chloride and alkali metal compounds such as sodium
and potassium chloride. The overall sorption process is not just
physical adsorption, but rather a complex combination of
adsorption and chemical reaction. The sorbents show high
capacities and fast kinetics for metal vapor removal. The sorption
mechanism and nature of the final products formed have
important implications in the way each sorbent may be used.
The overall rate of metal capture can be increased by facilitating
intraphase and interphase transport of metal vapors by pore
structure optimization and by using different catalyst and reactor
geometries.

Emissions of metal compounds during combustion and incineration is becoming
an increasingly important problem. Coal contains various compounds of alkali
and heavy metals in different forms and concentration. These compounds are
released to the gas phase during combustion and gasification of coal. Trace metal
compounds could also be present in municipal and industrial wastes. Incineration
of these wastes results in emission of toxic metal compounds, causing various
environmental problems. Alkali vapors generated during the processing of coal
or waste are considered to be the major cause of fouling and slagging on equipment
surfaces. Emissions of heavy metal compounds of lead, mercury, cadmium and
arsenic etc. could lead to severe environmental problems due to the toxic nature
of these compounds. For efficient processing of coal and wastes, the concentration
of these metal compounds has to be reduced to levels that are acceptable for
both efficient operations and a non-polluting environment.

Removal of metal compounds from the feed stock prior to its high temperature
processing is found to be relatively inefficient and expensive. Controlling the
release of metal vapors during high temperature processing is very difficult. A
promising technique for the removal of metal vapors after they have vaporized

[1]Current address: W. R. Grace & Co., 7379 Route 32, Columbia, MD 21044

0097–6156/93/0515–0214$06.00/0

is through the use of solid sorbents to capture the metal compounds by a combination of reaction and adsorption .
The sorbent can be used in two ways:
   a. It could be injected as a powder (similar to lime injection) for in-situ removal of metal compounds.
   b. The metal containing flue gas could be passed through a fixed or fluidized bed of sorbent. The sorbent could be used in the form of pellets, beads, or monoliths (for high dust applications).

Previous studies by us and other investigators indicate that solid sorbents can be very effective in removing alkali and lead vapors from hot flue gases (1-4). In the present work, a number of potential sorbents were screened and compared for their effectiveness in removing cadmium compounds from hot flue gases. Details of the sorption mechanism were investigated for the selected sorbents.

Cadmium compounds are considered to be among the most toxic trace elements emitted into the environment during fuels combustion and waste incineration. Cadmium and cadmium compounds are primarily used in the fabrication of corrosion resistant metals. Cadmium is also used as a stabilizer in poly-vinyl chlorides, as electrodes in batteries and other electrochemical cells, and for numerous applications in the semiconductor industry (5). Due to this wide range of applications, cadmium is present in many municipal and industrial wastes. Cadmium is also present in coal in trace quantities (6). Consequently, emission of cadmium compounds is a problem in many waste incinerators and coal combustors. The chemical form and concentration of these compounds depend on a number of factors including feed composition and operating conditions (7).

The increased use and disposal of cadmium compounds, combined with their persistence in the environment and relatively rapid uptake and accumulation in the living organisms contribute to their serious environmental hazards. The present technology is inadequate to meet the expected cadmium emission standards. Therefore, new and effective methods need to be developed and investigated for controlling the emission of cadmium and other toxic metals in combustors and incinerators.

Experimental

Materials. In the first part of this study, several model compounds and naturally available materials were evaluated as potential sorbents for removal of gaseous cadmium compounds from hot flue gases. The model compounds included silica (MCB grade 12 silica gel) and alpha alumina (Du Pont Baymal colloidal alumina, technical grade). The naturally available materials included kaolinite (52% SiO_2, 45% Al_2O_3, 2.2% TiO_2, 0.8% Fe_2O_3), bauxite (11% SiO_2, 84% Al_2O_3, 5% Fe_2O_3), emathlite (73% SiO_2, 14% Al_2O_3, 5% CaO, 2.6% MgO, 3.4% Fe_2O_3, 1.2% K_2O) and lime (97% CaO). Cadmium chloride was used as the cadmium source. For the screening experiments, the sorbents were used in the form of particles, 60 -

80 mesh in size. For the kinetic and mechanistic study the sorbents were used in the form of thin flakes (disks). The flake geometry is easy to model and characterize using analytical techniques. All the sorbents were calcined at 900°C for two hours and stored under vacuum until used. All the experiments were conducted in a simulated flue gas atmosphere containing 15% CO_2, 3% O_2, 80% N_2 and 2% H_2O.

Equipment and Procedures. The equipment and procedures used for the screening and mechanistic study are described below.

Screening Experiments. The main components of the experimental system were a Cahn recording microbalance, a quartz reactor, a movable furnace and analyzers for determining the composition of the gaseous products. This system has been previously used for screening of sorbents for removal of lead compounds. Therefore, only the salient features of the system are described here. Details can be found in a previous publication (1). The cadmium source was suspended by a platinum wire from the microbalance, which monitored the weight change during the experiments. A fixed bed of the sorbent particles was made by placing 100 mg of the sorbent particles on a 100 mesh stainless-steel screen in a quartz insert. All experiments in this study were performed with the source at 560°C and the sorbent at 800°C. This method ensured that the concentration of cadmium vapors around the sorbent was much below saturation, thereby preventing any physical condensation on the outer surface of the sorbent.

Heating of the cadmium source resulted in vaporization of $CdCl_2$ which was carried by the flue gas through the sorbent fixed bed. The percentage of cadmium adsorbed was determined from the amount of cadmium delivered (microbalance measurement) and the cadmium content of the sorbent at the end of the experiment. The cadmium content of the sorbent was determined by dissolving the samples in a $H_2O/HF/HNO_3$ (2/1/1 proportion by volume) mixture and subsequently analyzing the solution by atomic absorption spectroscopy. Separate water leaching experiments were performed to determine the water soluble fraction of adsorbed cadmium. The leaching of cadmium was conducted at 40°C in an ultrasonic bath for two hours. The cadmium content of the solution was subsequently determined by atomic absorption spectroscopy.

Study of Sorption Details. A microbalance reactor system was used for studying the sorption details. The sorbent flakes were suspended by a platinum wire from the microbalance. The weight of the sorbent was continuously monitored during the experiments by the microbalance. The cadmium source was placed in the horizontal arm of the reactor. The simulated flue gas from the gas preparation section was split in two parts. One part entered the reactor from the inlet below the balance and the remaining gas entered through the horizontal arm in which the cadmium source was placed. The cadmium source was heated by a heating tape and a furnace heated the sorbent flakes. All experiments were conducted with the source temperature of 560°C and the sorbent temperature of 800°C. When steady flow rates and cadmium concentrations were achieved, the sorbent was exposed to the cadmium containing flue gas. The sorbent weight was continuously recorded as it captured the cadmium vapors. The microbalance was continuously purged with ultra high purity nitrogen gas.

Results and Discussion

Screening Experiments. The results obtained from the screening experiments are given in Figure 1. Multiple experiments were conducted for most of the sorbents and good reproducibility was obtained. Since all experimental parameters except the sorbent type were kept constant, the amount of cadmium adsorbed is a good indication of the sorbent effectiveness (rate) for cadmium removal from hot flue gases. The most obvious feature of these results is the difference in the ability of the sorbents to capture cadmium from the flue gas passing through them. Alumina and bauxite had the highest cadmium capturing efficiencies. A large fraction of the cadmium captured by these two sorbent was water insoluble. Silica and kaolinite were not effective for removal of cadmium. Since cadmium chloride has a high water solubility, the formation of water insoluble compounds on sorption by alumina and bauxite leads to the conclusion that chemical reaction is the dominant mechanism of cadmium capture by these sorbents. Lime, a sorbent used for removal of sulfur compounds, did not have a high cadmium capturing efficiency. Since most of the cadmium captured by lime was water soluble, physical condensation is the dominant mechanism of cadmium capture for lime. From the screening experiments, it seems that compounds containing aluminum oxide have a high cadmium capturing efficiency. Bauxite was therefore further studied to determine the mechanism of cadmium capture. Since previous studies have indicated that kaolinite is a good sorbent for removal of lead and alkali compounds, it was also included in the mechanistic study.

Details of the Sorption Process. In the first part of this study, the cadmium sorption capacity of kaolinite and bauxite was investigated at 800°C. In these experiments the sorbent flakes were exposed to cadmium vapors until no further mass change was observed. The profiles for sorption for both kaolinite and bauxite are shown in Figure 2. The rate of sorption decreases with time and a final limit is achieved beyond which no further cadmium sorption takes place. The observed initial rate for cadmium capture by kaolinite was much slower than that for bauxite. This is consistent with the results from the screening experiments, where kaolinite captured much less cadmium compared to bauxite. Also, the final saturation limit for kaolinite (18%) was found to be lower than that for bauxite (30%) (Figure 2). When the concentration of cadmium in the flue gas wa reduced to zero, no desorption from either sorbent was observed indicating that reversible physical adsorption was not the dominant sorption mechanism.

X-ray diffraction (XRD) analysis was used to identify the final products formed by sorption of cadmium chloride on kaolinite and bauxite. Analysis of kaolinite flakes exposed to cadmium vapors indicated the formation of a cadmium aluminum silicate compound, $CdAl_2Si_2O_8$, which is water insoluble.

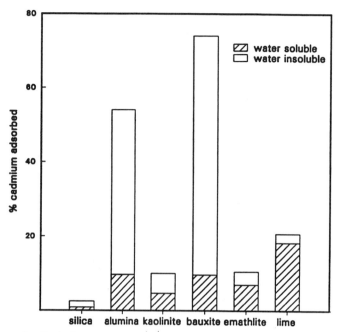

Figure 1. Cadmium removal efficiencies of various sorbents. Mass of sorbent used = 100 mg. T=800°C. Amount of cadmium vaporized in each experiment = 6.5 mg.

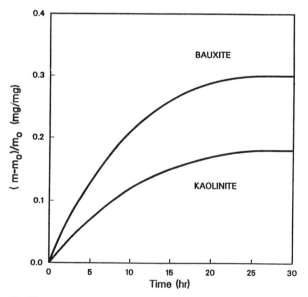

Figure 2. Temporal profiles of cadmium sorption on kaolinite and bauxite. T=800°C. Flow rate = 200 cc/min. Mass of flake = 15.1 mg (bauxite); 20.4 mg (kaolinite). Flake thickness = 0.71 mm (bauxite); 1.1 mm (kaolinite).

Based on XRD results, the following reaction scheme is proposed for capture of cadmium :

$$Al_2O_3.2SiO_2 + CdCl_2 + H_2O \ ---- \ CdO.Al_2O_3.2SiO_2 + 2HCl \quad (1)$$
metakaolinite

where metakaolinite is the dehydration product of kaolinite. Holland et al. (8) also observed the formation of this compound when a solid mixture of cadmium carbonate and kaolinite was heated to 800°C for twenty hours. Based on the stoichiometry of the overall reaction, 1 kg of kaolinite can capture 0.51 kg of cadmium, forming a product which is water insoluble and therefore safely disposable. The maximum weight gain postulated from this reaction (58% by weight) is much higher than that obtained experimentally in the microbalance reactor setup (18 wt%).

To further understand the reasons for low sorbent utilization, a kaolinite flake which had captured cadmium to its maximum capacity was mounted in epoxy and analyzed by SEM and EDX analysis. A cadmium map of the flake shows that cadmium is concentrated on the kaolinite edge (Figure 3). An EDX line scan on the flake surface indicated that the concentration of cadmium varied from 49 wt% at the edge to 2 wt% at the center. The concentration of cadmium at the edge (49%) is close to the value calculated from the postulated reaction mechanism assuming complete conversion (51%). This indicates that the surface of kaolinite was completely converted by reaction to form a cadmium aluminosilicate. Silicon and aluminum maps on the kaolinite surface indicate complete uniformity of distribution of these elements on the kaolinite surface. Since this flake had captured cadmium to its maximum capacity (microbalance data, Figure 2), the non-uniform distribution of cadmium indicates incomplete sorbent utilization. Since the volume of the cadmium aluminosilicate phase is higher than that of the aluminum silicate phase, the formation of cadmium aluminum silicate at the outer surface probably blocks the sorbent pores, resulting in incomplete sorbent utilization.

XRD analysis of bauxite particles exposed to cadmium vapors indicated the formation of two crystalline compounds: a cadmium aluminum silicate and a cadmium aluminate. Since the cadmium aluminate has a higher water solubility compared to that of the aluminosilicate, bauxite has a larger fraction of water soluble cadmium in the screening experiments. The amount of SiO_2 present in bauxite is not enough to combine with all Al_2O_3 to form an aluminosilicate compound. Based on the stoichiometry of Reaction 1 and the amount of SiO_2 present, 1 kg of bauxite can capture 0.10 kg of cadmium to form a cadmium aluminosilicate. Based on the remaining alumina, 1 kg of bauxite can capture 0.94 kg of cadmium oxide forming a cadmium aluminate according to the following reaction mechanism:

$$Al_2O_3 + CdCl_2 + H_2O \ ---- \ CdO.Al_2O_3 + 2HCl \quad (2)$$

If all the alumina and silica reacted with kaolinite to form cadmium aluminum silicate and cadmium aluminate (Reactions 1 and 2), the maximum weight gain possible is 1.06 kg per kg of bauxite. This is higher than the value obtained in the microbalance experiments conducted to completion (0.30 kg/kg bauxite).

To further understand the sorption mechanism, a bauxite flake exposed to cadmium vapors to the point of "no further weight change" was mounted in epoxy and analyzed by SEM and EDX analysis. The cadmium map on the bauxite surface indicated a non-uniform cadmium distribution (Figure 4). Comparison of the cadmium map with the aluminum and silicon maps on the flake surface indicated that regions high in both alumina and silica have high concentrations of cadmium. As confirmed by the XRD analysis, the cadmium present in the alumina phase reacts to form a cadmium aluminate. Regions rich in both alumina and silica combine with cadmium to form a cadmium aluminum silicate. EDX analysis of regions high in both aluminum and silicon indicated that the cadmium concentration was 26 wt%. EDX analysis of regions high in alumina indicated that the cadmium concentration in this phase (11 wt%) was much lower than that postulated by Reaction 2. This indicates that the alumina phase does not completely react with cadmium to form a cadmium aluminate. This could be due to the alumina phase having low porosity thereby preventing complete sorbent utilization. Also, formation of a higher volume product on the outer surface could inhibit further capture of cadmium.

From the mechanistic study it is clear that the sorption process under present experimental conditions is influenced by diffusional resistances. In practical systems, the rate of sorption can be increased by optimization of the sorbent pore structure to facilitate intraphase transport of cadmium vapors into the sorbent. The interphase mass transport limitations can also be reduced to increase the overall sorption rate. For in-situ applications, the efficiency of the sorbent can be increased by decreasing the particle size.

Removal of Other Metal Compounds

Results from previous studies (1,2) indicate that these sorbents can also be used to remove alkali and lead compounds from simulated flue gases. For removal of sodium chloride vapors, kaolinite, bauxite and emathlite were found suitable. However, kaolinite and bauxite were found more suitable than emathlite at temperatures higher than 1000°C since the final reaction products of NaCl adsorption on bauxite and kaolinite (nephelite and carnegiete) have melting points higher than 1500°C. The adsorption of alkali chloride on kaolinite and emathlite is irreversible with the release of chlorine back to the gas phase as HCl vapor. The adsorption on bauxite is partially reversible and part of the chlorine is retained.

Lime was not found to be a good sorbent for lead capture. Although bauxite and kaolinite captured approximately the same amount of lead chloride, kaolinite has a higher ultimate capacity for lead capture and a smaller water soluble fraction of captured lead. The mechanism of sorption for both alkali and lead vapors was found to be a combination of adsorption and chemical reaction.

Figure 3.   Cadmium concentration map on kaolinite flake surface.

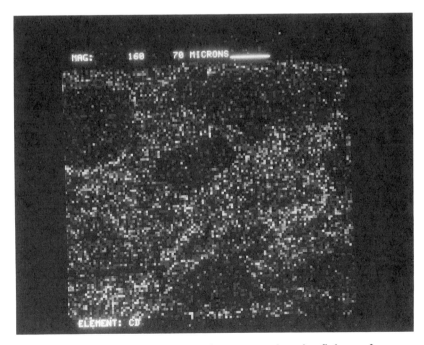

Figure 4.   Cadmium concentration map on bauxite flake surface.

Conclusions

1. Aluminosilicate type material are effective sorbents for removal of metal compounds in combustors and incinerators.

2. The sorption mechanism and the nature of final products formed have important implications in the way each sorbent may be used.

3. The overall sorption process is a complex combination of adsorption, condensation, diffusion and chemical reactions.

Literature Cited

(1) Uberoi, M.; Shadman, F. AIChE J., **1990**, 36(2), 307.
(2) Uberoi, M.; Punjak, W.A.; Shadman, F. Progress in Energy and Combustion Science **1990**, 16, 205.
(3) Lee, S.H.D.; Johnson, I.J. J. Eng. Power **1980**, 102, 397.
(4) Bachovchin, D.M.; Alvin, M.A.; Dezubay, E.A.; Mulik, P.R. A Study of High Temperature Removal of Alkali in a Pressurized Gasification System, DOE-MC-20050-2226, Westinghouse Research and Development Center, Pittsburgh, PA. **1986**.
(5) Palmer, S.A.K.; Breton, M.A.; Nunno, T.J.; Sullivan, D.M.; Suprenant, N.F. Metal/Cyanide Containing Wastes: Treatment Technologies **1988**.
(6) Pacnya, J.M.; Jaworski, J. Lead, Mercury, Cadmium and Arsenic in the Environment **1987**.
(7) Barton, R.G.; Maly, P.M.; Clark, W.D.; Seeker, R. Proceedings of 1988 National Waste Processing Conference, **1988**, 279.
(8) Holland, A.E.; Segnit, E.R.; Gelb, T. Journal of The Australian Ceramic Society **1976**, 10, 1.

RECEIVED June 12, 1992

Chapter 18

Analysis of Organic Emissions from Combustion of Quicklime Binder-Enhanced Densified Refuse-Derived Fuel–Coal Mixtures

Russell Hill[1], Baushu Zhao[1], Kenneth E. Daugherty[1], Matthew Poslusny[2], and Paul Moore[3]

[1]Department of Chemistry, University of North Texas, Denton, TX 76203
[2]Department of Chemistry, Marist College, Poughkeepsie, NY 12601
[3]Erling-Riis Research Laboratory, International Paper, Mobile, AL 36652

Emission testing for PCBs and PCDDs has been performed on the combustion gases produced by cofiring a quicklime binder enhanced densified refuse derived fuel/coal mixture. Analysis found that the PCBs were reduced in the presence of the $Ca(OH)_2$ binder and PCDD concentrations were below detection limits. Samples were obtained from the 567 ton, full scale, cofiring of bdRDF at various concentrations with high sulfur coal at Argonne National Laboratory, 1987. An EPA Modified Method 5 sampling train was used to iso-kinetically obtain samples prior to and after pollution control equipment. After sample clean up, analysis was completed using low and high resolution GC/MS. In an attempt to correlate PCB and PCDD reductions with binder concentrations, calcium and chloride analysis were performed on feedstock, fly ash and bottom ash samples.

Two of the nations current major concerns are municipal solid waste and energy. The combustion of refuse derived fuel (RDF) is increasingly being recognized as an attractive alternative to both problems.

On average 5 pounds of municipal solid waste (MSW) per day is being produced by each U.S. citizen. This waste cumulates to over 200 million tons per year, the majority of which is disposed of in landfills (1). Up until the formation of the Environmental Protection Agency (EPA) in 1970, most of the landfills were simple open dumping grounds with little control measures in place. In the mid 1970's under authority of the Resource Conservation Recovery Act, the EPA began shutting down open dumping and promoting sanitary landfills (2). Sanitary landfills are typically huge depressions lined with clay to minimize

0097–6156/93/0515–0223$06.00/0
© 1993 American Chemical Society

possible leaching into the area groundwater. These
landfills have guidelines detailing packing, covering, and
monitoring procedures (3). Industry has recently begun to
pay attention to this overwhelming problem and new products
for containment and monitoring are constantly being
developed. However, a slow start has postponed industrial
solutions to the distant future, while at the same time,
an increasingly concerned public is demanding immediate
solutions. Between 1982 and 1987 approximately 3000
landfills were shut down without replacement (4). The
Office of Technology Assessment (OTA) predicts that 80% of
idle landfills will be closed within twenty years (5).

The "Not in my Backyard Syndrome" (NIMBY), has
significantly contributed to the decrease in development
of many new landfill sites. A classic example of public
opposition is the 1988 Marbro barge incident. The barge
carried 3000 tons of garbage 6000 miles down the eastern
U.S. coastline looking for a landfill willing to accept
its cargo (6).

This new found public consciousness has forced
landfill sites to be placed farther from population
centers. The result has been higher transportation costs.
The northeastern states, which have traditionally exported
garbage, are finding it increasingly difficult and costly
to find anyone willing to accept the waste. Even states
with the potential to locate many new sites are being
cautious in response to public opinion. Texas, with its
open space, awarded 250 permits per year in the mid 1970s,
but was down to less than 50 per year by 1988 (7).
Stricter regulations proposed by the EPA in 1991 are
expected to eventually lead to a drastic cost increase of
approximately 800 million dollars per year (8). These
facts, coupled with the potential for large liability
costs, have forced the MSW industry to look for other
prospective solutions.

An obvious alternative to landfilling MSW is
incineration. Mass burn incineration, where refuse is fed
into a furnace with moving grates at temperatures of
$2400°F$, was a popular solution in the 1970s. Increased air
pollution, as well as ground water contamination due to
leaching of ash residue, has currently placed mass burning
in a state of disfavor. Mass burn incineration also tends
to be prohibitively expensive (up to $400 million per
facility) (7,9).

Pollution produced by the combustion of MSW is
certainly a justifiable concern. However, if a means for
controlling emissions to reasonable levels is found, then
another combustion product could be put to good use -
ENERGY.

The energy potential of the 200 million tons of MSW
produced annually in the U.S. equals approximately 326
million barrels of oil (9). This renewable resource can
only be tapped if an environmentally acceptable alternative

to mass incineration is found. A promising alternative is
that of resource recovery followed by incineration. This
process involves removing valuable recyclables such as
cardboard, plastics, glass, metals, and compost materials,
all from a central receiving station. The chief by-
product of this process, consisting primarily of paper, is
known as refuse derived fuel (RDF).

RDF has a energy value of about 7500 Btu/lb, similar
to a high grade of lignite coal. It can be cofired with
coal using cement kilns or other coal burning facilities.
The two most common forms are RDF-3 or "fluff" and RDF-5
or "densified".

Fluff RDF typically varies in particle size from a
few inches for spread stoker boilers, down to 0.75 inches
for suspension fired boilers. It is either burned in
dedicated boilers or cofired with coal. There are a number
of problems with fluff RDF.

1.   It tends to compress under its own weight
     limiting storage time.
2.   Its bulk density (2-3 lb/cubic foot)
     makes it difficult to transport.
3.   If wet it can decay rapidly.
4.   It can be difficult to deliver to a furnace
     in a controlled manner.

Many of these difficulties can be overcome by
extruding fluff RDF to create 2-3 inch long by 1/4 inch
in diameter pellets of densified RDF (dRDF). Densified
RDF has a bulk density on the average of 35 lbs/cubic foot.
Transportation, storage, degradation, and processing
problems are all reduced by densification.

Typically dRDF processing begins on the plant
tipping floor. Large items are removed from the waste and
the remainder is loaded onto conveyor belts. The material
first passes through a rotary drum sieve to remove the
fines (mostly organic material) for composting. The
material is then hand picked from the conveyor belt for
recyclables. Hammermills or grinders reduce the material
in size. Next, ferrous metals are removed magnetically.
Air classifiers are used to separate the fine,
predominantly paper material from small pieces of glass,
rock, etc. which must be landfilled. The fraction of
material requiring landfilling in an efficient plant will
only amount to about 10-15% of the total original volume.
The paper and plastic RDF is now dried to the proper
moisture level for densification. Binders are occasionally
added prior to densification to increase physical
durability and combustion performance.

In 1985 the University of North Texas (UNT), under
contract with Argonne National Laboratory, investigated
over 150 materials as possible binders for dRDF. A binder
was sought that would improve pellet integrity for
transportation and storage purposes. A sample of the

materials used exemplifies the diversity of materials investigated:

glues	carbon black
natural starches	urea resins
urea formaldehyde	epoxy resins
sulfite waste liquors	weed additive
beeswax	gelatins
fly ashes	asphalt
animal adhesive	roof tar sludges
limestone	paint sludges
cement kiln dust	cotton burrs
slags	portland cement

Initially, agents were selected based on expected availability, cost, environmental acceptance, and potential binding ability. Because of cost, it was felt that the binder used with RDF should show cementitious bonding as opposed to bonding necessitating a high temperature operation. Generally, cementitious bonding involves inorganic adhesive bonds holding together an aggregate, such as a RDF particle. The adhesion results primarily from hydrogen bonding. This type of cementing provides pellets that possess limited physical strength but are durable for transportation and long term storage. They are also resistant to water attack and biodegradation. Advantages of cementitious bonding include the facts that the cements can be poured or gunned directly into place. They cause little or no dimensional changes during binding, and they are relatively inexpensive.

The densification technique will remove the pores between the RDF particles (accompanied by shrinkage of the components), as a result of grain fusion and strong interactions between adjacent particles. Before densification of RDF can occur, the following criteria must be met. A mechanism for material transport must be present and a source of energy to activate and sustain this material must exist.

The two primary mechanisms for heat transfer are diffusion and viscous flow. Heat is the primary source of energy, in conjunction with energy gradients due to particle-particle contact and surface tension.

The difference in free energy or chemical potential between the free surfaces of RDF particles, is crucial to the densification process. Diffusion properties, temperature, and particle size are also important.

The 150 binding agents were first screened based on cost and environmental effects. Cost was weighted at 60%, environmental effects at 40%. Environmental acceptability, was further broken into the 3 categories of: toxicity 10%, odor 10%, and emissions 20%. The list was reduced to 70 potential binders in this manner. These binding agents were analyzed in a laboratory study broken into 11 characteristics:

binder Btu content   binder dispersibility
binder ability to wet   binder ash content
pellet water sorbability   pellet caking
pellet ignition temperature   pellet moisture content
pellet durability   pellet aerobic stability
pellet weatherability ($110/32°F$)

The tests were grouped into two classes, the first four characteristics being binder properties and the last seven characteristics being classified as pellet properties. The binder properties were weighted at 7% per characteristic, and the pellet properties were weighted as 4% per characteristic. Laboratory test results produce the 13 top binders for further processing. The highest ranked binders were:

calcium oxide (CaO)   bituminous fly ash
calcium hydroxide ($Ca(OH)_2$)   lignite fly ash
dolomite   western coal fines
$Ca(OH)_2$/dolomite   Iowa coal fines
cement kiln dust $Ca(OH)_2$/lignite fly ash
portland cement   calcium lignosulfonate
CaO/dolomite/Portland Cement

Some observations that can be made on the most effective binders are (a) the binder must have large surface areas, (b) the binders are basic. The basic binders (those using $Ca(OH)_2$) are believed to be effective due to:

(1)   the calcium binding agent breaks the RDF substrate down producing a by-product with acidic groups such as a carboxylic acid group

$$\text{cellulose} \xrightarrow{\text{oxidizes}} N[R-\overset{O}{\overset{\|}{C}}-OH] \xrightarrow{\text{CaO}} R-\overset{O}{\overset{\|}{C}}-O-\underset{\underset{O}{\overset{\|}{R-C-O}}}{Ca} + H_2O$$

(2)   the calcium from the basic binder reacts in a complex or chelate formation mode with the acidic groups of RDF

(3)   the calcium hydroxide produced by the basic groups reacting with moisture in the RDF, or present initially in the binder, gradually carbonates

$$Ca(OH)_2 + CO_2 \dashrightarrow CaCO_3 + H_2O$$

These final binding agents were tested in a pilot plant operation performed at the Jacksonville Naval Air

Station, Jacksonville, Florida in the summer of 1985. The test involved over 7000 tons of RDF, and pilot plant experiments. Mixing processes and testing procedures were previously published (10,11). Bulk density was used as an estimate of pellet durability during on-site testing. $Ca(OH)_2$ was found to increase bulk density over a variety of pellet moisture contents and allowed pressing of pellets at moisture levels normally too high for processing.

UNT suspected the binder's basic nature might help reduce the emission of acid gases. Sulfur dioxide is reduced by burning dRDF simply because of dRDF's lower sulfur content (0.1-0.2%) as compared to western coal's (2.5-3.5%). Additional reductions can be realized through the well known reaction of sulfur oxide, a lewis acid, with the base function of the calcium oxide's surface in ionic form. Hydrolysis of the oxide is known to increase the reaction kinetics. The reaction rate is believed to be determined by the concentration of hydroxide sites. The following equations depict this process. (12).

$$H_2O + CaO \; ---> \; Ca(OH)_2$$

$$Ca(OH)_2 + SO_2 \; ---> \; CaSO_3 + H_2O$$

In the combustion center the binder should be predominantly in the oxide form. After transport to cooler areas within the system the previous listed equations become applicable. The result is a reduction of SO_x emissions. Furthermore, lower combustion temperatures obtained when cofiring dRDF tend to favor the gas solid reaction of SO_2 with CaO. Lower combustion temperatures are also known to lead to decreases in nitrogen oxide emissions.

Hydrochloric gas emissions will tend to increase due to increased chlorine concentrations from plastics. The increase in HCl can be expected to be minimal compared to reduction in SO_x/NO_x emissions. It was further proposed that the binder might physically reduce the production of polychlorinated biphenyls and dioxins by adsorbing halogens prior to their formation or by absorbing the species directly as typically demonstrated in spray dryer absorber systems. (10,11)

These hypotheses were tested at a pilot plant operation at Argonne National Laboratory in 1987. The six week program combusted over five hundred tons of binder enhanced dRDF (bdRDF) blended with Kentucky coal at heat contents of 10, 20, 30 and 50 percent. The $Ca(OH)_2$ binder content ranged from 0 to 8 percent by weight of dRDF. Emission samples were taken both before and after pollution control equipment (multicyclone and spray dryer absorber). All samples were taken to UNT for analysis.

A discussion of the significant reductions in sulfur and nitrogen oxides has been previously published (13).

The following discussion reports results for analysis of polychlorinated biphenyls, polyaromatics, polychlorinated dioxins, and polychlorinated furans.

Experimental

Isokinetic samples were taken for the analysis of polychlorinated biphenyls (PCB's), polyaromatics (PAH's), polychlorinated dioxins (PCDD's) and polychlorinated furans (PCDF's). The following sampling sites were investigated.

Site 1 combustion zone (2000°F); sample site (1200°F)
Site 2 prior to pollution control equipment (300°F)
Site 3 after pollution control equipment (170°F)

Sample collection was completed using an EPA modified Method 5 sampling train, and a XAD-2 resin for trapping the majority of organics. Various samples were spiked with isotopically labeled standards prior to soxhlet extraction for 24 hours with benzene. The percent recovery of these standards provided a means to determine extraction efficiency and detection limits. Possible interferants were removed by sample elution through a minimum of two acid and base modified silica gel columns followed by an alumina column. Standard mixtures of native and isotopically labeled analytes were prepared for calculation of relative response factors and calibration of the gas chromatography mass spectrometer (GC/MS). The GC/MS analysis was performed with a Hewlett-Packard Model 5992B. (14)

Results and Discussion

The EPA's sixteen most hazardous PAH's (Table V) and all congener groups of PCB's were tested. The results are reported as total PCB and PAH concentrations found at a particular site because it is felt that the overall production of PCB's/PAH's is the parameter of primary concern. No specific compound was produced at inordinately high levels. The results of site 2 and site 3 sampling areas are found in Tables I and II. Figures 1 and 2 clearly depict a reduction in PAH and PCB emission as the binder concentration is increased. Data are not available on all compositions due to the mixing and sampling methods used. Tables III and IV show the calcium and chloride contents of fly ash from the multicyclone for a series of specific compositions. The calcium is used as an indicator of the binder present in the boiler system during a particular sampling period. The increase in calcium occurring between the first coal "blank" run and the second blank occurring 3 weeks later, suggest a build-up of residual binder throughout the boiler configuration. It is noteworthy that the increase in water soluble chloride

Table I

Polyaromatic Hydrocarbons (PAH's);
Polychlorinated Biphenyls (PCB's) at Site 2

Run#/Sample#	Site	mg PAH's cubic meter of gas sampled	mg PCB's cubic meter of gas sampled
Coal	2	1.7×10^{-2}	6.2×10^{-3}
Coal/10%RDF/0%B	2	1.0×10^{-3}	1.3×10^{-2}
Coal/10%RDF/0%B *	2	7.6×10^{-2}	2.7×10^{-1}
Coal/10%RDF/4%B	2	1.6×10^{-2}	1.4×10^{-2}
Coal/10%RDF/8%B	2	4.0×10^{-3}	7.6×10^{-3}
Coal/10%RDF/8%B	2	8.1×10^{-3}	7.7×10^{-3}
Coal/20%RDF/0%B	2	3.5×10^{-2}	9.7×10^{-3}
Coal/20%RDF/0%B	2	4.6×10^{-2}	7.7×10^{-3}
Coal/20%RDF/4%B	2	2.2×10^{-1}	2.0×10^{-3}
Coal/20%RDF/4%B	2	3.5×10^{-1}	2.9×10^{-1}
Coal/20%RDF/8%B	2	2.4×10^{-1}	1.3×10^{-2}
Coal/20%RDF/8%B	2	3.0×10^{-1}	3.4×10^{-3}
Coal	1	3.4×10^{-1}	5.4×10^{-3}
Coal	2	1.3×10^{-1}	3.9×10^{-4}

* This sample was lighter in color than all the rest

B = Binder

Table II

**Polyaromatic Hydrocarbons (PAH's);
Polychlorinated Biphenyls (PCB's) at Site 3**

Run#/Sample#	Site	mg PAH's cubic meter of gas sampled	mg PCB's cubic meter of gas sampled
Coal	3	4.6×10^{-3}	5.3×10^{-4}
Coal/10%RDF/0%B	3	6.3×10^{-3}	1.2×10^{-3}
Coal/10%RDF/0%B	3	1.5×10^{-2}	*
Coal/10%RDF/0%B	3	8.1×10^{-3}	1.6×10^{-3}
Coal/10%RDF/4%B	3	7.3×10^{-3}	9.1×10^{-3}n
Coal/10%RDF/8%B	3	7.3×10^{-3}	1.1×10^{-4}
Coal/10%RDF/8%B	3	3.1×10^{-3}	3.1×10^{-3}
Coal/20%RDF/0%B	3	3.6×10^{-4}	2.8×10^{-4}
Coal/20%RDF/0%B	3	4.0×10^{-3}	1.2×10^{-3}
Coal/20%RDF/4%B	2	7.9×10^{-2}	4.2×10^{-2}
Coal/20%RDF/4%B	3	4.9×10^{-2}	6.5×10^{-3}
Coal/20%RDF/8%B	3	1.0×10^{-3}	2.4×10^{-3}
Coal/20%RDF/8%B	3	8.1×10^{-3}	8.5×10^{-4}
Coal	3	7.0×10^{-2}	4.0×10^{-3}
Coal	3	1.4×10^{-3}	4.3×10^{-4}

* Interference made it impossible to determine the
quantity of PCB's in this run

B = Binder

Table III

Calcium Levels in Fly Ash

Fuel	ppm of Calcium
First coal blank	3,000
Coal - 10% dRDF (0% binder)	6,700
Coal - 10% dRDF (4% binder)	10,600
Coal - 10% dRDF (8% binder)	15,000
Second coal blank	4,200

Table IV

Soluable Chloride Levels in Fly Ash

Fuel	ppm of Chloride
First coal blank	100
Coal - 10% dRDF (0% binder)	190
Coal - 10% dRDF (4% binder)	280
Coal - 10% dRDF (8% binder)	320
Second coal blank	280

Table V

EPA Priority PAH's

Naphthalene	Benzo-a-anthracene
Acenapthylene	Chrysene
Acenapthene	Benzo-b-fluoranthene
Fluorene	Benzo-k-fluoranthene
Phenanthrene	Benzo-a-pyrene
Anthracene	Dibenzo-a,h-anthracene
Fluoranthene	Benzo-g,h,i-perylene
Pyrene	Idendo-1,2,3,-g,d-pyrene

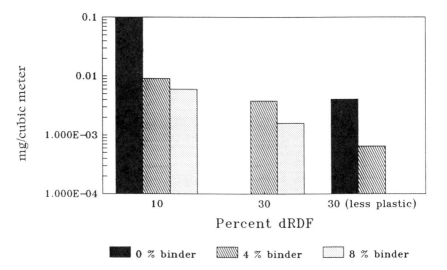

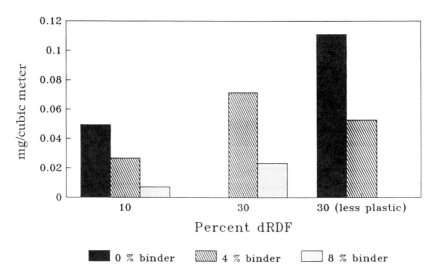

Figure 1. PAH Concentrations Detected

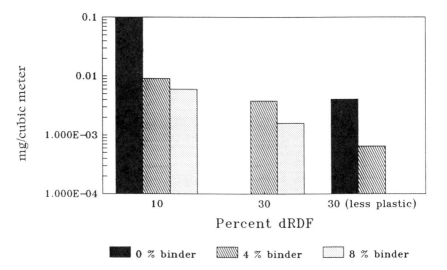

Figure 2. PCB Concentrations Detected

in the fly ash for the 10% RDF samples is on the same order as the reduction of PCB's seen in Figure 2. This is presumably due to the lime's ability to bind the chloride in the combustion area. The much higher chloride content of the second coal blank ash relative to the first blank can be explained by assuming a longer contact time for chloride adsorption on the residual binder as the binder became saturated throughout the boiler system. (15)

The dioxin and furan analysis initially concentrated on the tetra chlorinated species present after the pollution control equipment at sample site 3. Priority was given to detecting the 2,3,7,8 congeners because of their higher toxicity. Sample site 3 was chosen because of the importance of determining what concentrations of the analyte, if any, were reaching the environment. Table VI shows that no dioxins or furans were found at the listed detection limits.

In order to discover if the analyte was simply depositing on fly ash, samples of composited fly ash were also analyzed for adsorbed PCDD's and PCDF's. The five fly ash samples chosen to represent the full spectrum of test burn compositions are listed in Table VII. Analysis results repesented by Tables VIII through XII again demonstrate no detectable trace of dioxins or furans. Additional analyses were subsequently performed for penta, hexa, hepta, and octa congeners at sample site 3. In all cases, no dioxins or furans were detected.

These somewhat surprising negative results led to a search for a means to improve detection limits. Lower detection limits were thought to be obtainable through the use of a high resolution mass spectrometer (HRMS). Triangle Laboratories at Research Triangle Park, NC, was contracted for use of their equipment and expertise in this area. Eight representative samples were chosen for HRMS analysis. Three of the previously analyzed samples from site 3 were chosen as verification on the accuracy of UNT's results. The remaining samples were of various burn compositions from site 2 prior to pollution control equipment. These samples were selected because UNT's results suggested that no PCDD's or PCDF's were present after pollution control.

Triangle Laboratories generally improved results only on the order of one magnitude, and confirmed UNT's findings with negative results for dioxins and furans at site 3. Only one of the site 2 samples had any detectable analyte. Emissions of 0.41 ng/m^3 total HxCDD and 0.66 ng/m^3 total OCDD were found for the sample from a 20% dRDF/ 0% binder/ coal burn mixture. It should be noted that the toxicity equivalent factor (TEF) for HxCDD is 0.0004 leading to a emission rate of 0.16 pg/m^3 while the TEF for OCDD is 0. Emission rates for both are certainly much lower than any current standards. Results of the same sample at site 3 found no trace of dioxin or furan.

Table VI

Tetra-Chlorinated Dioxins and Tetra-Chlorinated Furans at Site 3

Run#/Sample#	Site	Tetra-Chlorinated Dioxin Level	Tetra-Chlorinated Furan Level	Detection Limit ng/m^3
Coal	3	BDL	BDL	0.72
Coal/10%RDF/0%B	3	BDL	BDL	1.99
Coal/10%RDF/0%B	3	BDL	BDL	4.07
Coal/10%RDF/0%B	3	BDL	BDL	5.24
Coal/10%RDF/4%B	3	BDL	BDL	4.80
Coal/10%RDF/8%B	3	BDL	BDL	4.27
Coal/10%RDF/8%B	3	BDL	BDL	4.27
Coal/20%RDF/0%B	3	BDL	BDL	0.49
Coal/20%RDF/0%B	3	BDL	BDL	0.47
Coal/20%RDF/4%B	3	BDL	BDL	4.16
Coal/20%RDF/4%B	3	BDL	BDL	4.10
Coal/20%RDF/8%B	3	BDL	BDL	4.78
Coal/20%RDF/8%B	3	BDL	BDL	4.78
Coal	3	BDL	BDL	3.85
Coal	3	BDL	BDL	4.85

ng/m^3 = nanograms per cubic meter

BDL = Below Detection Limits

B = Binder

Table VII

Analyzed Fly Ashes Compositions

Date	Composite Sample #	Composition
June 21-22, 1987	1	20% Btu Content Reuter RDF and 70% Btu Content Coal - 0% Binder
July 7, 1987	2	30% Btu Content Reuter RDF and 70% Btu Content Coal - 0% Binder (no plastic)
June 15-16, 1987	3	Coal Only
June 23-24, 1987	4	30% Btu Content Reuter RDF and 70% Btu Content Coal - 8% Binder
July 4-5, 1987	5	50% Btu Content Reuter RDF and 50% Btu Content Coal - 4% Binder

Table VIII

GC/MS Analysis of Ash Sample 5 for Polychlorinated Dioxins and Furans

Mass	Analytes	Amount Found ng/g sample)	Detection Limit (ng/g sample)
304/306	TCDF	BDL	0.9966
320/322	TCDD	BDL	0.9966
340/342	PeCDF	BDL	2.4916
356/358	PeCDD	BDL	2.4916
374/376	HxCDF	BDL	2.4916
390/392	HxCDD	BDL	2.4916
408/410	HpCDF	BDL	2.4916
424/426	HpCDD	BDL	2.4916
442/444	OCDF	BDL	4.9832
458/460	OCDD	BDL	2.4916

% Extraction Efficiency = 57.9

BDL = Below Detection Limit

Table IX

GC/MS Analysis of Ash Sample 4 for Polychlorinated Dioxins and Furans

Mass	Analytes	Amount Found (ng/g sample)	Detection Limit (ng/g sample)
304/306	TCDF	BDL	1.0001
320/322	TCDD	BDL	1.0001
340/342	PeCDF	BDL	2.5002
356/358	PeCDD	BDL	2.5002
374/376	HxCDF	BDL	2.5002
390/392	HxCDD	BDL	2.5002
408/410	HpCDF	BDL	2.5002
424/426	HpCDD	BDL	2.5002
442/444	OCDF	BDL	5.0005
458/460	OCDD	BDL	2.5002

% Extraction Efficiency = 9.2

BDL = Below Detection Limit

Table X

GC/MS Analysis of Ash Sample 3 for Polychlorinated Dioxins and Furans

Mass	Analytes	Amount Found (ng/g sample)	Detection Limit (ng/g sample)
304/306	TCDF	BDL	1.0022
320/322	TCDD	BDL	1.0022
340/342	PeCDF	BDL	2.5055
356/358	PeCDD	BDL	2.5055
374/376	HxCDF	BDL	2.5055
390/392	HxCDD	BDL	2.5055
408/410	HpCDF	BDL	2.5055
424/426	HpCDD	BDL	2.5055
442/444	OCDF	BDL	5.0111
458/460	OCDD	BDL	2.5055

% Extraction Efficiency = 53.1

BDL = Below Detection Limit

Table XI

GC/MS Analysis of Ash Sample 2 for Polychlorinated Dioxins and Furans

Mass	Analytes	Amount Found (ng/g sample)	Detection Limit (ng/g sample)
304/306	TCDF	BDL	0.9965
320/322	TCDD	BDL	0.9965
340/342	PeCDF	BDL	2.4913
356/358	PeCDD	BDL	2.4913
374/376	HxCDF	BDL	2.4913
390/392	HxCDD	BDL	2.4913
408/410	HpCDF	BDL	2.4913
424/426	HpCDD	BDL	2.4913
442/444	OCDF	BDL	4.9825
458/460	OCDD	BDL	2.4913

% Extraction Efficiency = 81.2

BDL = Below Detection Limit

Table XII

GC/MS Analysis of Ash Sample 1 for Polychlorinated Dioxins and Furans

Mass	Analytes	Amount Found (ng/g sample)	Detection Limit (ng/g sample)
304/306	TCDF	BDL	0.9992
320/322	TCDD	BDL	0.9992
340/342	PeCDF	BDL	2.4980
356/358	PeCDD	BDL	2.4980
374/376	HxCDF	BDL	2.4980
390/392	HxCDD	BDL	2.4980
408/410	HpCDF	BDL	2.4980
424/426	HpCDD	BDL	2.4980
442/444	OCDF	BDL	4.9825
458/460	OCDD	BDL	2.4980

% Extraction Efficiency = 133.6

BDL - Below Detection Limit

Thus the pollution control equipment is apparently efficient in removing the analyte at these low concentrations. (16)

There is a natural disinclination to report results such as found in Table VI and Tables VIII through XII. No concrete values for dioxin and furan concentration produced during the combustion of the RDF are presented. Instead, one finds detection limits for which PCDD and PCDF levels were found to be below.

When attempting to investigate the environmental impact of a developing industry, it is important to contribute all technically responsible information for the development of a reliable database. This becomes particularly imperative when dealing with sensitive topics, such as the production of dioxins and furans during the combustion of refuse derived fuel.

The presented data provides baseline values as a means for prediciting potential pollution levels for future industrial demonstrations and for the development of adequate detection techniques.

Conclusion

Results of the pilot plant program indicates that the binder enhanced densified refuse derived fuel can be cofired with coal, at the levels tested, without producing detectable amounts of dioxins or furans. PCB's and PAH's are aypparently reduced as a function of the quicklime binder content.

Literature Cited

1.   Ohlsson, O., et. al., Proceedings of the American Association of Energy Engineers, **1986**, p. 1.
2.   Gershman, H.W., Brickner, R.H., Brutton, J.J., Small-Scale Municipal Solid Waste Energy Recovery Systems; VNR, NY, **1986**.
3.   Smith, F.A., Waste Age, April **1976**, p. 102.
4.   Rice, F., Fortune, **1988**, 117, pp. 96-100.
5.   Office of Technology Assessment, OTA Report No. 052-003-01168-9, **1989**.
6.   Vogel, S., Discover, **1988**, 42, No. 4, p. 76.
7.   Rathje, W.L., The Atlantic Monthly, December **1989**, 264, pp. 99-109.
8.   Johnson, P., McGill, K.T., USA Today, August 26, **1988**.
9.   Carpenter, B., Windows, **1988**, p. 9.
10.   Daugherty, K.E., et. al., Proceedings of the American Association of Energy Engineers, **1986**, p. 930-935.
11.   Daugherty, K.E., et. al., Phase 1, Final Report, U.S. Department of Energy, ANL/CNSV-TM-194, **1986**.

12.   Mobley, J.D., Lim, J.K., In Handbook of Pollution Technology, Calvert, S., Englund, H.M.; John Wiley & Sons, N.Y., **1984** pp. 210-211.
13.   Jen, J.F., Ph.D. Dissertation, Analysis of Acid Gas Emissions in the Combustion of the Binder Enhanced Densified Refuse Derived Fuel by Ion-Chromatography, University of North Texas, TX, **1988**.
14.   Poslusny, M., Moore, P., Daugherty, K., Ohlsson, O., Venables, B., American Institute of Chemical Engineers Symposium Series, **1988**, 84, No. 265, pp. 94-106.
15.   Poslusny, M. Ph.D. Dissertation, Analysis of PAH and PCB Emissions From the Combustion of dRDF and the Nondestructive Analysis of Stamp Adhesives, University of North Texas, TX, **1989**.
16.   Moore, P. Ph.D. Dissertation, The Analysis of PCDD and PCDF Emissions From the Cofiring of Densified Refuse Derived Fuel and Coal, University of North Texas, TX, **1990**.

RECEIVED April 6, 1992

Ash and Slag Utilization

Chapter 19

Ash Utilization and Disposal

Carl A. Holley

Ferro-Tech, Inc., Wyandotte, MI 48192

There are many changes taking place in the utilization and
disposal of ash and related materials and each day it seems
that there are more restrictions on the disposal of ash.
In one project, mass burn ash has been screened to remove
the plus one inch pieces and then combined with the fly ash
and portland cement to produce a pellet that is very slow-
leaching with all of the heavy metals "fixed".

In the northeastern United States, the ash from circulating
fluidized bed combustors needs to be pelletized so it can
be transported back to the mine for easy disposal. The ash
can also be blended with digested municipal sewage sludge
to form a soil additive which contains lime and nutrients.

The new clean air standards are making it necessary for
utilities to install sulfur dioxide scrubbers which produce
gypsum. This gypsum filter cake can be pelletized so that
the pellet produced can be utilized in the final grind of
portland cement or can be disposed of in a very slow-
leaching pile.

We have taken the lead in producing these agglomeration
techniques and many more processes which we will describe
and show the detailed process flow diagram for each of
these methods.

The utilization and disposal of fly ash in a legal and
environmentally safe manner is becoming a major engineering challenge
because of rapidly changing government laws and regulations.   In
addition, the ongoing search for lower cost power and alternative
fuels and combustion methods is continually changing the fly ash.
Government regulations limit the amount of sulfur released during
combustion of coal. To limit sulfur emissions from a pulverized coal
combustor we are now adding a dry sulfur dioxide scrubber in which a
lime [Ca(OH)$_2$] solution is sprayed into the exhaust gas stream where

0097–6156/93/0515–0242$06.00/0

some of the lime reacts with the sulfur dioxide to form calcium sulfate ($CaSO_4$) or calcium sulfide ($CaSO_3$) which is a solid and can be removed from the gas stream with the fly ash. Another new type fly ash is that from a circulating fluidized bed combustor where the sulfur is reacted with calcined limestone in the bed to form calcium sulfate ($CaSO_4$) or calcium sulfide ($CaSO_3$) which again is removed with the fly ash.

There are many changes taking place in ash management based on the fact that pellets can be produced which are very slow-leaching. Pellets can be moved with conventional bulk material handling equipment without producing a dusty environment. The pellets can be disposed of in almost any type of state permitted site and do not need to be placed in a lined landfill. In addition, slight modifications to the pelletizing process can produce usable products such as lightweight aggregate or aggregate to be utilized in asphalt.

We see that each situation or plant needs to be treated as a separate problem and a process developed to solve that particular problem. We had been trying to pelletize fly ash for twenty years with only limited success until we discovered that the real requirement to allow us to produce consistent pellets was to first condition the fly ash in an intense mixer so that each particle becomes coated with water. The two key pieces of equipment required to pelletize fly ash are the Ferro-Tech-Turbulator (Figure 1) and the Ferro-Tech Disc Pelletizer (Figure 2). The Ferro-Tech-Turbulator is a proprietary, intense, highly efficient, agitative agglomeration device covered by Patent No. 4,881,887. The unit has high efficiency because it operates by fluidizing the material and atomizing the injected liquid which may be water or a liquid binder. Each whirling dust particle is uniformly coated with a very thin layer of the liquid binder. Ideally, the liquid binder coating is one molecule thick, however, this thickness is not practical so the layer is a few molecules in thickness. This coating of the individual particles with a liquid is called "conditioning". These coated dust particles, moving at a high velocity in suspension, collide and impact with great force with the other coated, spinning particles within the turbulent wake created behind the pins, forming a very uniformly sized and dense particulate or microgranule which is held or pulled together by the surface tension of the liquid layer. The thick, resilient polymer liner in the Turbulator combined with critical close pin tip tolerance causes the pins to fully sweep or wipe the liner, eliminating product build-up on the inner casing of the Turbulator body. Because of the intensity of the Turbulator, it can efficiently pre-blend, de-dust, blend, "condition", densify, hydrate and micropelletize all types of fine powders, dust, fume and hard-to-wet particles. The fly ash particles are uniformly coated with water in the Turbulator before they are discharged to be pelletized in the disc pelletizer. The conditioning step in the Ferro-Tech-Turbulator compensates for the variation in composition, particle size and surface area which is a characteristic of fly ash from coal combustion due to changes in combustor operations and variation in fuel.

Figure 1. Ferro-Tech-Turbulator

Figure 2. Ferro-Tech Disc Pelletizer

Circulating Fluidized Bed Combustor Ash Agglomeration

Most of the new cogeneration plants are utilizing a circulating fluidized bed combustor principle which burns coal having reasonable levels of sulfur (up to 3%). The ash from these combustors consists of two fractions, bed drain or bottom ash and fly ash. The bed drain ash is approximately 25 to 30% of the total ash. The best ash systems keep the two ash fractions separated so they can be blended uniformly for agglomeration. The fly ash can be agglomerated alone, but the bed drain ash must be blended with at least 50% fly ash before quality pellets can be produced.

The basic process flow diagram for producing a 1/4" x 1/2" pellet is shown in Figure 3. This system includes two (2) surge bins with feeders feeding the ash into a Ferro-Tech-Turbulator where approximately 2/3 of the 18 to 20% (wet basis) of the water required for pelletizing is added. This ash should be near ambient temperature, however, it may be as high as 200°F when discharged from the feeder. The conditioned ash from the Turbulator discharges directly into the disc pelletizer where pellets are formed. This disc pelletizer is designed for this specific agglomeration application. The pellets or green balls from the disc pelletizer have a modest amount of strength so that they can withstand the treatment of the material handling system to the pellet curing area. The curing of these pellets is very similar to the curing of concrete or any cementitious blend. In this case the curing can take place in a ground pile where temperatures can be up to 150°F. Heat is gained as the hydration reaction of the curing takes place. Normally the pellets have gained enough strength to be handled in 24 to 48 hours. Strength enough to be handled is a crush strength of 10 to 15 pounds for a 1/2 inch diameter pellet. Figure 4 shows the effect of curing temperature on the strength of the pellets. The 150°F can easily be reached when the pellets cure in a pile or bin. After extended curing (30 to 60 days), a typical 1/2 inch diameter pellet may require a force of 200 to 300 pounds to crush it. These pellets can easily be utilized as road bed material or as aggregate in concrete.

High Sulfur Coal. When a high sulfur coal (above 3%) is burned in a circulating fluidized bed combustor, this fly ash acts very differently. When water is added to the fly ash in a Ferro-Tech-Turbulator, the ash very quickly (5 to 10 minutes) gains temperature. A typical temperature curve is shown in Figure 5 where the temperature rose to 290° F in seven minutes. This heating is a result of the hydration of the lime (CaO) by the water to form hydrated lime [Ca(OH)$_2$]. Since this fly ash can contain 30 to 40% lime (CaO), we can see that there is much heat generated. If enough water has been added to produce pellets, that is, 18 to 20% (wet basis) of water, the pellets will quickly begin to hydrate, heat up and expand and will completely disintegrate becoming an even finer dust than it was in the beginning.

The process shown in flow diagram Figure 6 is necessary if a pellet is to be produced which will hold together and gain strength. This process is covered by Patent No. 5,008,055. The fly ash and bed drain ash are added at a uniform rate into Turbulator No. 1 where 8 to 10% water (wet basis) is added. The discharge from Turbulator No.

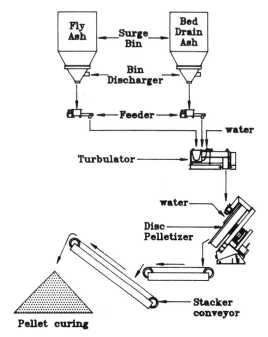

Figure 3. CFB Ash Pelletizing System

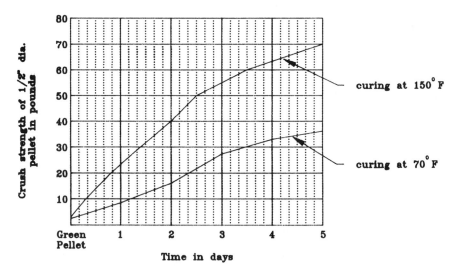

Figure 4. CFB Ash Pellets

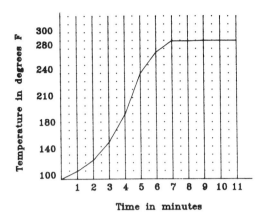

Figure 5.   Hydration Temperature Curve

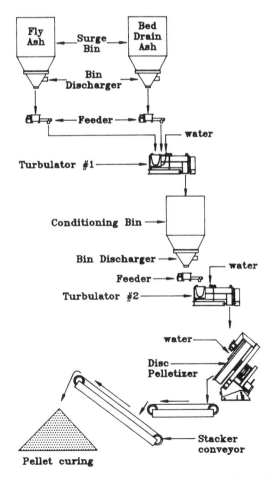

Figure 6.   CFB Ash Pelletizing System for High Sulfur Coal

1 drops directly into a conditioning bin where the temperature, due
to the hydration reaction, is as much as 300°F. The material is
retained in the conditioning bin until all of the free water is
chemically reacted and a hydrate is formed. The conditioning bin is
carefully designed to be a mass flow bin so that the ash moves
uniformly toward the discharge. The partially hydrated ash is
discharged from the bottom of the conditioning bin through a screw
feeder into Turbulator No. 2 where 10 to 12% (wet basis) of water is
added to the ash. The uniformly wet ash is discharged from
Turbulator No. 2 directly into a disc pelletizer where 8 to 10% (wet
basis) of additional water is added and pellets are formed. As
before, the pellets are placed in a curing area for 24 to 48 hours or
until they gain satisfactory strength.

Fly Ash Lightweight Aggregate

One of the better uses for Class F and Class C fly ash from
pulverized coal power plants is to produce lightweight aggregate.
This aggregate can be utilized to produce lightweight concrete block
and other masonry forms. The basic process flow diagram for the
process is shown in Figure 7. The aggregate seems to have many
advantages over other lightweight aggregates including its spherical
shape. The aggregate produced by this process bonds to the mortar
with both a mechanical bond and a chemical bond instead of just a
mechanical bond as is true of other non-reactive aggregates.

The bulk density of the aggregate is approximately 45 to 50
lbs./cu.ft. The typical sieve analysis (ASTM C136) of the aggregate
required to produce concrete for the manufacture of concrete blocks
is:

Sieve Size	Product Percent Passing	Specification Percent Passing
1/2"	100	100
3/8"	99	90 - 100
4 mesh	83	65 - 90
8 mesh	49	35 - 65
16 mesh	31	---
50 mesh	19	10 - 25
100 mesh	15	5 - 15

Synthetic Gypsum & Fly Ash Pelletizing

One of the major problems at pulverized coal power plants burning
high sulfur coal which have a wet sulfur dioxide lime scrubber is how
to dispose of the synthetic gypsum slurry. One very simple solution
is to produce a pellet by blending dry fly ash with the gypsum filter
cake or even with the gypsum slurry. Figure 8 shows the basic
process flow diagram for the required system. In this system, the
filter cake is metered into the Ferro-Tech-Turbulator at a constant
rate. The feed rate of the fly ash is controlled by a moisture
control system on the disc pelletizer, Patent No. 5,033,953. This
system adjusts the feed rate to maintain a constant surface moisture
of 18 to 20% (wet basis) on the pellets which are about to discharge

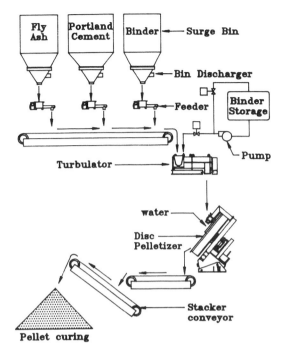

Figure 7.   Fly Ash Lightweight Aggregate Pelletizing System

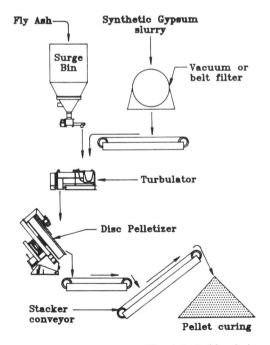

Figure 8.   Synthetic Gypsum & Fly Ash Pelletizing System

from the disc pelletizer. The moisture control system utilizes an
infrared emitter and sensor to optically determine the surface
moisture of the pellets. The pellets from the disc pelletizer are
placed in a curing pile where they gain strength before being placed
in a landfill. Figure 9 shows a typical curing curve for these
pellets.

Mass Burn Ash Agglomeration

The processing of mass burn ash is in a state of change since there
is obviously a need to produce a product that has all of the heavy
metals "fixed" so they do not leach from a landfill. The requirement
is to process both the bottom ash and the fly ash. Processing the
bottom ash alone is very difficult because of the extreme variation
in the size of the pieces and in the high moisture [20 to 25% (wet
basis)] of the material. A finger screen can be utilized to separate
the plus one inch particles from the minus one inch pieces even
though the moisture may be as high.

As shown in Figure 10, the plus one inch material in the bottom
ash is removed and the minus one inch bottom ash is blended with the
available fly ash (normally 50% of bottom ash) along with portland
cement. The portland cement is normally 15 to 20% (wet basis) of the
bottom ash. The three of these materials are all fed directly into
a disc pelletizer which is equipped with a reroll ring. Moisture is
added to the disc pelletizer through spray nozzles until pellets are
produced. It normally requires a total of 16 to 20% (wet basis)
moisture to produce pellets.

The completed pellets are discharged from the disc pelletizer
pan into the reroll ring where 1 to 3% (wet basis) portland cement is
added to coat the pellets. The coating of the pellets with portland
cement serves two purposes. The first purpose is to seal the surface
of the pellets which will assist in minimizing any leaching, the
second purpose is to keep the pellets from sticking together as they
cure in the tote bin.

After approximately seven days of curing, the heavy metals in
the ash are fixed to the point that they will pass the T.C.L.P.
leaching test. The pellets will have strength enough to be easily
handled and should not crush under normal handling.

Orimulsion Fly Ash Agglomeration

With the world searching for lower cost fuels we now have an interest
in the combustion of Venezuelan heavy bitumen. Due to its extremely
high viscosity at ambient temperature, the raw bitumen is unsuitable
for conventional transport to utility stations for subsequent
combustion. As a result, a bitumen-in-water emulsion (hereinafter
called Orimulsion) has been developed to ease handling of the bitumen
by reducing its viscosity to a range similar to heavy fuel oils.

The fly ash from combustion of Orimulsion is very unusual with
98% (by weight) of the particles being less than 10 microns in size.
This leads to an aerated bulk density of 5 lbs./cu.ft. and a
deaerated bulk density of 10 to 12 lbs./cu.ft. The fly ash contains
10 to 15% vanadium which makes it valuable as a concentrate for
vanadium production.

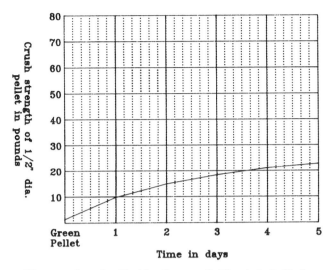

Figure 9. Synthetic Gypsum & Fly Ash Pellets

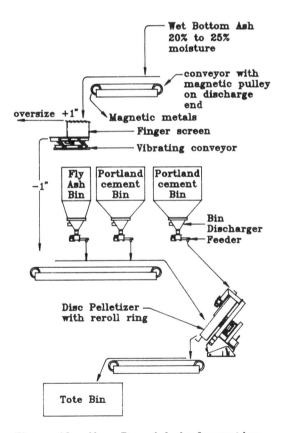

Figure 10. Mass Burn Ash Agglomeration

The fly ash is very fine, reactive and very difficult to wet; therefore, the wetting of the fly ash with water is very critical. It has been found that the only effective method of wetting this fly ash is the Ferro-Tech-Turbulator. Once the fly ash is wet, a hydration reaction takes place. The reacting material quickly heats and all of the water is either chemically bound or is evaporated. The partially hydrated fly ash is again introduced into a Ferro-Tech-Turbulator where more water ia added. The wet fly ash is discharged onto a disc pelletizer where pellets are formed. This process is covered by Patent No. 5,008,055.   The bulk density of the pellets will be increased to 60 to 70 lbs./cu.ft. from the initial 5 to 10 lbs./cu.ft.   The pellets will cure in a few minutes and become hard enough for normal material handling; that is, a 1/2 inch diameter pellet will require 40 to 50 pounds force to crush it.

Conclusions

Essentially any fly ash can be agglomerated into non-dusting and very slow-leaching agglomerates. The requirements for the agglomerates keep changing and become more restrictive as the regulators try to make the environment safer. Our goal is to assist the customer in developing the best process to fulfill the regulators present and future demands. Many of the processes appear to be simple, but there has been much experience involved in the development. We feel that the simplest process is the best process if it meets all of the customer´s goals.   We know that each situation is unique and a process must be adapted to the special ash from a specific plant. It is advisable to cooperate with a someone who has the most experience so you can have confidence in their recommendations.

RECEIVED April 6, 1992

Chapter 20

Utilization of Coal Gasification Slag
Overview

Vas Choudhry and Steven R. Hadley

Praxis Engineers, Inc., 852 North Hillview Drive, Milpitas, CA 95035

Coal gasification generates solid waste materials in relatively large quantities, and their disposal can represent a significant expense. For example, a 100-MW power plant based on IGCC technology using 1000 tons of 10% ash coal per day may generate over 110 tons/day of solid waste or slag, consisting of vitrified mineral matter and unburned carbon. As coal gasification technologies, considered clean and efficient methods of utilizing coal, find increasing applications for power generation or chemical feedstock production, it becomes imperative that slag utilization methods be developed, tested, and commercialized in order to address the costly problems associated with its disposal as solid waste. This chapter presents an overview of the experimental work that has been conducted to characterize samples of slag from various gasifiers and to identify and test a number of commercial applications for their utilization, and discusses various issues with regard to slag utilization. In the course of examining various utilization applications for a number of coal gasification slags that parallel those developed for fly ash, a better understanding of slag as a construction material has been achieved. The applications tested include the use of slag as an aggregate for road construction, cement concrete and asphalt concrete, and production of lightweight aggregate from slag.

In the past decade, fly ash, bottom ash, and boiler slag have increasingly been utilized in construction and other applications. A number of scientific studies have been conducted to develop utilization manuals applicable for general use. Examples include: *Fly Ash as a Construction Material for Highways: A Manual*, in 1976,(*1*) and *Fly Ash Structural Fill Handbook* in 1979.(2)   In addition, recent applications for fly ash were summarized at the Eighth International Ash Utilization Symposium in 1987.(*3*)

0097–6156/93/0515–0253$06.00/0

A comprehensive survey of coal combustion by-products was undertaken by the Electric Power Research Institute (EPRI) in 1983 in the preparation of the *Coal Combustion By-Products Utilization Manual*.(*4*)   Volume 2 of this manual comprises an annotated bibliography.(*5*) As indicated in the manual, in 1984, 66.4 million tons of boiler ash (comprising fly ash, bottom ash, and boiler slag) was generated nationwide, of which 12.4 million tons (or 18.7%) was utilized in a variety of applications including blasting grit, fill material, roofing granules, roadway construction, de-icing grit, and cement additives.   By comparison, in Europe and Japan, combustion solid wastes are utilized to a greater extent; this is attributed to the demand for construction aggregate and fill materials, the shortage of space for waste disposal, and environmental and economic factors.

The search for utilization applications for coal gasification slag parallels that of fly ash, which has been tested successfully for a variety of applications including aggregate stabilization in airport, highway, and dam construction, engineered backfill, soil amendment, cement additive, and lightweight aggregate production. A number of similar applications for gasification slag have been studied by Praxis Engineers, Inc. under a series of contracts funded primarily by the Electric Power Research Institute (EPRI) with additional support from Texaco, Inc. and Southern California Edison.

Development of Slag Utilization Technology

Using fly ash utilization as a model, Praxis started work in 1986 to develop the utilization of gasification slag. The steps involved in this approach can be summarized as follows:

- Measurement of physical and chemical properties expected to affect the utilization of slag,

- Screening of conventional industrial and construction materials and products for potential slag utilization applications,

- Comparison of the properties of gasification slag with the specifications established for other materials to identify possible slag substitution,

- Testing of promising applications at the bench scale,

- Enhancement of the properties deemed significant from the utilization viewpoint by simple, low-cost preparation techniques, and

- Selection of successful applications for further testing at the pilot or demonstration level.

Initial testing was undertaken using slag samples from the Cool Water Demonstration Plant (CWDP). The Cool Water Demonstration Program was based on a Texaco gasification process and used a number of major U.S. coals, including a western bituminous coal and an Illinois Basin coal. Details of the program are provided in a 1985 EPRI report.(*6*)

Sixteen potential applications for the utilization of gasification slag were initially identified based on a comparison of the preliminary characteristics of a CWDP slag generated from a western bituminous coal feed with those of ash by-products. The most promising of these applications included use of slag as a soil conditioner, abrasive grit, roofing granules, ingredient in cement and concrete manufacture, road construction aggregate, and lightweight aggregate. The results of this work were summarized in a paper(*7*) and presented in an EPRI report.(*8*)

Once successful utilization concepts had been identified for this slag, samples of another slag generated at CWDP from a different coal feedstock and three other slags generated from different gasifiers were also evaluated. The three additional gasifier technologies were the Shell Coal Gasification Process, the British Gas Corporation/Lurgi slagging gasifier, and the Dow Chemical entrained-flow gasification process, using single-stage operation.

In a parallel study, use of slag as a feed material for the production of synthetic lightweight aggregate was investigated. The findings of this study were presented in an EPRI report.(*9*) In a follow-on project, the production of lightweight aggregate from slag was successfully advanced to the pilot scale.

Coal Gasification Slag Properties

Among the early work on the identification of coal gasification slag properties is a 1983 chemical characterization study comparing coal gasification slag samples obtained from pilot projects with coal combustion solid wastes, conducted by Oak Ridge National Laboratory.(*10*) It was determined that, though there are some differences between various slags, in general gasification slag was nearly as inert chemically as slag from cyclone furnaces.

In characterization testing conducted by Praxis Engineers in 1987-88, focusing on slag utilization, the physical and chemical properties of five coal gasification slags were found to be related to the composition of the coal feedstock, the method of recovering the molten ash from the gasifier, and the proportion of devolatilized carbon particles (char) discharged with the slag ("Characterization of Coal Gasification Slags," EPRI Contract No. RP2708-3. Final report in publication). The rapid water-quench method of cooling the molten slag inhibits recrystallization, and results in the formation of a granular, amorphous material. Some of the differences in the properties of the slag samples that were characterized may be attributed to the specific design and operating conditions prevailing in the gasifiers. For instance, the British Gas/Lurgi gasifier produced a slag with a distinct iron-rich phase in addition to the silicate phase, and the Texaco gasifier generated slag containing a higher proportion of discrete char particles.

In general, slag is nominally in the 5-mm x 0.3-mm size range, which is equivalent to the classification for fine aggregates used in cement concrete and asphalt concrete. The apparent specific gravity of slag ranges between 2.64 and 2.81, and its dry compacted unit weight is between 70.1 and 104.9 lb/ft^3. The water absorption capacity of slag varies from 2 to 16% and increases with its char content. The leachability of the slags is generally very low. Both EP toxicity and ASTM extraction tests reported nonhazardous levels of the eight metals regulated under these tests. These findings confirm the work done on slag by the developers of several gasifier technologies such as Texaco, Shell, and Dow.

The elemental composition of the slag samples with respect to both major and trace elements is similar to that of the gasifier feed coal ash, as shown in Table I. The major constituents of most coal ashes are silica, alumina, calcium, and iron. Slag fluxing agents, when used to control molten ash viscosity inside the gasifier, can result in an enrichment of calcium in the slag.

The Cool Water slag was classified as nonhazardous under the RCRA regulations. EP toxicity and ASTM extraction tests were run on a number of slags to evaluate their leachability. The slags appear to be nonleachable with respect to RCRA-listed metals. Tests for eight RCRA heavy metal anions were run, with only sulfate anions being detected at significant concentrations (25 to 200 mg/l).

Table I. Comparative Composition of Typical Blast-Furnace Slag, Cool Water Feed (Utah) Coal Ash, and Cool Water Slag

Mineral	Blast Furnace Slag	Cool Water (Utah) Coal Ash	Cool Water Slag
SiO_2	32-42	48.0	40-55
Al_2O_3	7-16	11.5	10-15
CaO	32-45	25.0	10-15
MgO	5-15	4.0	2-5
Fe_2O_3	0.1-1.5	7.0	5-10
MnO	0.2-1.0	NA	NA
S	1.0-2.0	NA	<1

Evaluation of Potential Utilization Concepts

Selection of applications to utilize gasification slag must take into account the fact that it is in competition with conventionally used materials whose acceptability has been established over long periods. In this effort, the emphasis was placed on evaluating the functional requirements of various applications (such as compressive strength in the case of cement concrete) in order that existing specifications--written for natural materials--do not rule out slag utilization. Ultimately, if slag is found to satisfy the functional requirements of an application, suitable standards can be established for its use in particular cases. A precedent for this procedure is the creation of a standard such as ASTM C 989-87a which was adopted for utilization of ground blast-furnace slag as cement.

Selection of the specific utilization concepts was guided by the following criteria:

- Similarity between the properties of slag and those of the material it replaces, and

- Achievement of comparable final products meeting the necessary functional requirements.

Based on these criteria, a number of utilization concepts were identified. These include:

Agriculture:	Soil conditioner, lime substitute, low analysis fertilizer, carrier for insecticides
Industrial material:	Abrasive grit, catalyst and adsorbent, roofing granules, industrial filler, mineral (slag) wool production, filter media
Cement and concrete:	Concrete aggregate, mortar/grouting material, pozzolanic admixture, raw material for portland cement production, masonry unit production
Road construction and maintenance:	De-icing grit, fine aggregate for bituminous pavement, base aggregate, subbase aggregate, seal-coat aggregate
Synthetic aggregate:	Lightweight construction aggregate, landscaping material, sand substitute
Landfill and soil stabilization:	Soil conditioner to improve stability, structural fill, embankment material
Resource recovery:	Source of carbon, magnetite, iron, aluminum, and other metals

Of these, a number of high-volume applications were tested at the laboratory scale and found to be suitable. For example, the potential for using slag as a fine aggregate for base, subbase, and backfill applications is suggested by the slag size gradation. Shear strength, permeability, and compaction test data also indicate that slag would perform well as an aggregate fill material.

While these applications would consume large quantities of slag they provide few economic incentives to the construction industry to replace cheap and abundant conventional materials with slag at this stage. However, as concern about the environment increases and recycling of waste products becomes a priority, this situation could change rapidly.

Use of Slag in Road Construction

Slag produced by the Cool Water Demonstration Plant was studied in road construction applications by testing various asphalt mix designs incorporating slag. By itself, the slag was not found to be suitable for surface pavement applications due to the lack of coarse particles and the tendency to degrade when abraded. However, its use as a subbase and base material in road construction is quite feasible as it meets the resistance value requirements for the California Department of Transportation standards for Class 1, Class 2, and Class 3 subbases, and Class 2 aggregate base. In order to offset the high proportion of fines in the slag, it may need to be mixed with coarser conventional materials prior to use as a specification base material and as an asphalt concrete aggregate.

Asphalt concrete hot mixes containing varying concentrations (6-10%) of asphalt and 30-50% slag by weight as the fine aggregate were tested for their strength (S-values) in a laboratory. A mix in which 30% slag was used with 6% asphalt yielded an S-value of 50, which is much higher than the minimum value of 30-37 required for various grades of asphalt concrete. This mix, which had good workability, compares favorably with the control test mix containing 5% asphalt, which has an S-value of 58.

Use of Slag in Cement and Concrete Applications

The composition of the CWDP slag and its natural pozzolanic properties are similar to the raw material used to make portland cement clinker. In this application, the slag carbon (char) content may be beneficial and may provide some of the fuel needed to make the clinker. The slag could also be added to cement clinker and ground with it.

The char content of some of the slags is far higher than the 1% limit placed on aggregate. This makes it necessary to recover the unconverted carbon from the slag, both in order to meet the standard for aggregate and to improve the process economics. Char removal was accomplished by means of simple specific gravity devices. The recovered char is a usable by-product.

Several batches of concrete were prepared using slag to replace varying quantities of the sand in the mix. Specimens in which 50% and 75% of the sand was replaced by slag had compression strengths of 2790 and 2480 psi respectively, over a 28-day curing period. This compared well with the control sample containing no slag, which had a compression strength of 3410 psi. These results indicate that slag could be used to replace a large proportion of the fine aggregate in making nonstructural concrete.

Tests to replace some of the fine aggregate used to make concrete with slag were performed by substituting 50% of the sand by slag. The test specimens achieved satisfactory results, with compressive strengths ranging from 3000 to 3500 psi, compared with a control strength of 3900 psi at the same cement content. These results satisfy typical compressive strength requirements of 2000 psi for concrete pads for sidewalks, driveways, and similar applications. Based on these tests, a large batch of concrete was prepared using 50% slag to replace the fine aggregate in the cement concrete mix and used to cast a concrete pad at the Cool Water Demonstration Plant site. In subsequent monitoring, its performance was found to be comparable to that of standard cement concrete.

A similar study was conducted by Shell Development Company to test the utilization of slag for replacing part of the fine aggregate in cement concrete.(*11*) In a mix in which nearly all of the fine aggregate was replaced by slag, a 28-day compressive strength of 3000 psi was obtained. Slag was subsequently routinely used on site in the preparation of cement concrete mixes.

In a series of tests conducted by Praxis Engineers, slag ground to a fine powder was used as a cement replacement ("Characterization of Coal Gasification Slags," EPRI Contract No. RP2708-3. Final report in publication). Cement additive requirements have been established for blended cements in ASTM C 595, which covers five classes of blended hydraulic cements made from conventional materials for both general and specific applications. Following initial exploratory tests, it was concluded that it was necessary to process the slag samples to remove potentially deleterious substances. The lighter char fraction was removed from one of the slags by density separation, and an iron phase was recovered from another slag by magnetic separation. The percentages of cement replaced by slag in these tests were 15% and 25% respectively. The use of processed slags resulted in a more successful replacement of cement by slag. All of the 15% slag-cement blend samples exceeded the 3-, 7-, and 28-day strength requirements of 1800, 2800, and 3500 psi respectively, and one of the four slags tested exceeded these requirements at the 25% replacement level. The other three 25% replacement level slag samples achieved the required 28-day strength but did not satisfy the 3- and 7-day requirements. The average 28-day strength for the 15% blend was 5600 psi, and that of the 25% blend was 4900 psi.

The success of the prepared slag-cement blends in achieving long-term compressive strength suggested that ground slag would also qualify as a pozzolanic mineral admixture. A pozzolan is a finely ground siliceous material which can react with calcium ions, in the presence of water and at room temperature, to form strength-producing calcium silicate minerals in a manner similar to cement reactions. A 35% replacement of cement by slag was evaluated in accordance with the procedures outlined in ASTM C 311. The success of a pozzolanic test is measured by the Pozzolanic Index, which indicates the ratio of the sample's compressive strength to that of an ordinary portland cement control sample. All of the concrete samples thus produced exceeded the

Pozzolanic Index requirement of 75%, with index values ranging between 90 and 118%.

Slag Lightweight Aggregate

Lightweight aggregates (LWA) have unit weights that are approximately 40-60% those of standard aggregates. Annual consumption of LWA in the United States for various applications is approximately 15 million tons. Major applications of LWA are in the production of lightweight structural concrete used in highrise buildings and lightweight precast products such as roofing tiles, masonry blocks, utility vaults, cement concrete pipes, etc. Conventional LWAs are produced by pyroprocessing of naturally occurring expansible shales or clays at temperatures ranging between 1880 and 2200°F, after pulverizing, working into a paste, and extruding them to produce pellets of the desired size. The strength requirements for lightweight concretes made from LWA are given in Table II.

Slag-based lightweight aggregates (SLA) were produced by Praxis Engineers (see ref. 9) by duplicating the processing methods used for commercial LWA manufacture. These steps included grinding the slag, mixing it with a clay binder and water, and extruding it to form long strands that were cut to the desired sizes. These wet green pellets were then dried and fired in a laboratory muffle furnace at 1800°F for 4 minutes. A unit weight of 45 lb/ft^3 was measured for the SLA, which is below the minimum coarse LWA specification of 55 lb/ft^3. Concrete made from the SLA had a 28-day compressive strength of 3100 psi and a unit weight of 105 psi, which exceeds the ASTM requirements shown in Table II.

Table II. Lightweight Aggregate Unit Weight, Minimum Compressive Strength, and Tensile Strength (28-Day Requirements for Structural Concrete, ASTM C 330)

100% LWA Mix		Sand/LWA Mix	
Unit Weight lb/ft^3	Compressive Strength, psi	Unit Weight lb/ft^3	Compressive Strength, psi
115	--	115	4000
110	4000	110	3000
105	3000	105	2500
100	2500	100	--

SOURCE: Reprinted with permission from *1987 Annual Book of ASTM Standards, Section 4: Construction. Volume 04.02: Concrete and Aggregate,* Designation C 330-87, p. 248. Copyright ASTM.

Further tests have confirmed that the density of the SLA can be controlled as a function of the firing temperature, as shown in Figure 1. This indicates that SLA products can be produced to meet specific density requirements such as those for cement concrete LWA, lightweight concrete masonry units, or ultra-lightweight material used in insulating concrete.

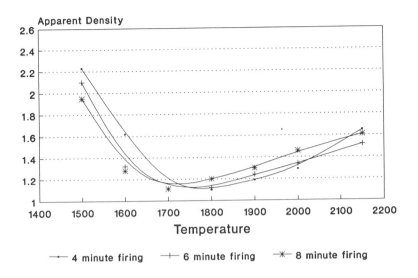

Figure 1. Time/Temperature/Density Relationship for Expansion
of Slag Pellets

Tests on discrete 2-mm particles of each slag showed that they also expand to form a lightweight material when fired at 1600-1900°F. Further tests on all particles larger than 0.3 mm, without pelletization, confirmed this phenomenon. The materials resulting from these tests had unit weight values of 15-25 lb/ft^3. The concrete produced from one of the expanded slag samples had a unit weight of 33 lb/ft^3, which qualifies it to be classified as an insulating concrete. However, it had a compressive strength of only 125 psi, which is somewhat lower than the strength of commercially available insulating concretes at 200-250 psi. It is expected that the strength can be considerably increased with minor adjustments to aggregate gradation and the cement proportions used to formulate these test samples.

The experimental work on slag utilization has been developed to the continuous pilot scale for the production of SLA from slag. The results from this test program have been very encouraging and have confirmed the bench-scale test results. During the tests, engineering information on energy requirements, scale-up information, and off-gas analysis was obtained, and the mechanism of slag expansion was investigated. The energy requirements for SLA production are considerably lower than those for conventional LWA production due to the lower kiln temperatures required for the expansion of slag. Samples of SLA with different unit weights produced during the pilot tests are undergoing extensive testing ("Production and Testing of a Large Batch of Slag Lightweight Aggregate," Final Report for EPRI Project No. RP1459-26, in publication.)

In 1989, Ube Ammonia Industry Company, Ltd. of Japan reported having expanded the slag produced from their Texaco-type gasification process.(12) A flash furnace was used to expand a selected portion of prepared slag to make a lightweight product equivalent to perlite for the local market.

Conclusions

Gasification slag has been determined to be an environmentally nonhazardous material, whose unique properties may be attributed to the composition of the mineral matter in the coal feedstock and the method of quench-cooling applied in the gasifier. Bench-scale test data have shown that there are a number of promising applications for the utilization of gasification slags. In particular, the utilization of slag in applications such as road, cement concrete, and construction aggregates, cement additives, and lightweight aggregates has been demonstrated. Production of slag-based lightweight aggregates (or SLA) is feasible and should be established as a priority. The high unit price of lightweight aggregates will permit SLA to be transported for greater distances while remaining economically competitive, thus rendering slag utilization less sensitive to the location of the gasifier.

Future of Slag Utilization

Currently, in most utilization scenarios, gasification slag would be used as a replacement for materials that have a relatively low unit cost, such as cement concrete and road aggregates. Unless potential commercial users of slag are provided with extensive characterization and utilization data, the economic incentives alone are unlikely to be sufficient to cause them to incorporate slag into their mix designs and production scenarios. The initial resistance to the use of new materials that may be encountered in the construction materials manufacturing industry can be addressed in two ways. First, the slag producer can supply complete engineering data on slag utilization to the prospective end user, who would then be responsible for any slag processing steps that might be required for a particular application. An alternative, more comprehensive approach would be for the slag producer to deliver physically prepared slag to the end user in a form that meets the user's specifications; these specifications could vary depending on the market demand in the vicinity of the gasifier. If recent legislation in California can be used to gain an insight into coming regulatory trends, at least 50% of the slag produced in the state will be required to be utilized in the coming decade, thereby creating additional incentives for producers and prospective end users to work together to realize its utilization potential.

Literature Cited

1.   *Fly Ash as a Construction Material for Highways. A Manual.* Federal Highway Administration, Washington, D.C., May 1976. PB 259 302.
2.   *Fly Ash Structural Fill Handbook.* Electric Power Research Institute, Palo Alto, Calif., December 1979, EPRI EA-1281.
3.   *Proceedings: Eighth International Ash Utilization Symposium,* Volumes 1 and 2. Prepared by the American Coal Ash Association and published by the Electric Power Research Institute, Palo Alto, Calif., October 1987. EPRI CS-5362.
4.   *Coal Combustion By-Products Utilization Manual.* Electric Power Research Institute, Palo Alto, Calif., Volume 1, February 1984. EPRI CS-3122, V1.
5.   *Coal Combustion By-Products Utilization Manual.* Electric Power Research Institute, Palo Alto, Calif., Volume 2: Annotated Bibliography, July 1983. EPRI CS-3122, V2.
6.   *Cool Water Coal Gasification Program: Third Annual Progress Report,* Electric Power Research Institute, Palo Alto, Calif., January 1985. EPRI AP-3876.
7.   Choudhry, V.; Deason, D.; Sehgal, R., "Evaluation of Gasification Slag Utilization Potential." Presented at the 6th EPRI Coal Gasification Contractors' Conference, Palo Alto, California, October 1986.
8.   *Potential Uses for the Slag from the Cool Water Demonstration Plant,* Electric Power Research Institute, Palo Alto, Calif., February 1987. EPRI AP-5048.
9.   *Synthetic Lightweight Aggregate from Cool Water Slag: Bench-Scale Confirmation Tests,* Electric Power Research Institute, Palo Alto, Calif., May 1991. EPRI GS-6833.
10. *Comparison of Solid Wastes from Coal Combustion and Pilot Coal Gasification Plants,* Electric Power Research Institute, Palo Alto, Calif., February 1983. EPRI EA-2867.
11. Salter, J. A. et al, "Shell Coal Gasification Process, By-Product Utilization." Presented at the Ninth International Coal Ash Utilization Symposium in Orlando, Florida, January 21-25, 1991.
12. Sueyama, T.; and Katagiri, K., "Four-Year Operating Experience with Texaco Coal Gasification Process in Ube Ammonia." Electric Power Research Institute, Palo Alto, Calif. In *Proceedings of the Eighth Annual EPRI Conference on Coal Gasification,* August 1989. EPRI GS-6485.

RECEIVED June 15, 1992

Chapter 21

High-Strength Portland Cement Concrete Containing Municipal Solid Waste Incinerator Ash

James T. Cobb, Jr., C. P. Mangelsdorf, Jean R. Blachere, Kunal Banerjee, Daniel Reed, Clayton Crouch, Coby Miller, Jingqi Li, and Jeanette Trauth

School of Engineering, University of Pittsburgh, Pittsburgh, PA 15261

The commercial use of solid wastes from energy-producing units, such as coal-fired boilers and oil shale combustors, has been practiced for several decades in the United States and in Europe. Recently, work by numerous organizations has begun on a variety of methods to render hazardous solid residues non-hazardous and to create beneficial uses for ash from municipal solid waste incinerators. One method for both purposes is the replacement of a portion of the fine aggregate in Portland cement concrete. The strength of the concrete drops significantly as the portion replaced increases, even with normal additives. This chapter presents the greatly improved strengths obtained with ash, which has been exposed to a new additive. These results show that up to 35% of the concrete can be made up of ash, while still obtaining compressive strengths of over 5000 psi (34.5 MPa). Micrographs of the original ash, ash and additive, concrete with ash but without additive, and concrete with ash and additive indicate the role of the additive. TCLP extractions of this novel new concrete and evaluations of its engineering properties have yet to be conducted. The economics, commercialization and extension of the development to other situations are discussed.

Power generation facilities and the construction industry many years ago recognized a remarkable opportunity for an exchange of great mutual profitability. With considerable development work, they transformed fly ash from coal-fired power plants, a waste requiring ever more expensive landfill for disposal, into a useful substance for inclusion in materials for building highways and other structures. The operators of municipal solid waste (MSW) incinerators are investigating a parallel opportunity for the beneficial use of the ash from their facilities. This chapter begins with a brief survey of the extent of coal ash utilization and what was required to institutionalize it. Next, the article will survey the current state of development of beneficial use of MSW ash and the effort that is yet required for its institutionalization. The article will conclude with a report on the status of one program for developing the beneficial use of MSW ash in precast concrete.

Commercial Use of Solid Wastes from Coal-Fired Boilers

Coal ash is formed by an extensive, complex processing of the mineral matter in coal, through a sequence of high temperatures and both reducing and oxidizing environments. As the composition of the mineral matter and the nature of the processing conditions vary, so will the characteristics of the resulting ash. However, the principal constituents of mineral matter are always silica and alumina, which are two of the principal ingredients of manufactured Portland cement (the third ingredient of which is lime). The third principal constituent of coal mineral matter is usually iron sulfide (bituminous coals) or limestone (subbituminous coals and lignites), although recently certain sulfur removal technologies provide lime (or limestone) as a cofired ingredient, particularly with bituminous coal.

A comparison between the processing of mineral matter in pulverized coal (PC) boilers and the manufacturing of Portland cement in rotary kilns is instructive. In a PC boiler the processed mineral matter experiences temperatures up to 1900°K, which are well above the melting point of the ash (usually between 1300 and 1500°K). The resulting fly ash is composed largely of very small particles of alumina-silica glass, in the form of mixtures of glasses and crystalline mullite ($3Al_2O_3 \cdot 2SiO_2$), quartz (SiO_2) and metal silicates (*1*). In a cement kiln, producing a glassy Portland cement clinker, a mixture of typically 60 percent lime, 25 percent silica, 5 percent alumina and 10 percent iron oxide and gypsum are exposed to temperatures between 1700 and 1900°K. The cooled clinker is ground to a fine powder and marketted as cement.

It is not surprising, then, to find that fly ash from a PC boiler is a pozzolan, having good cementitious characteristics when mixed with lime and water (*1*). It is also not surprising, considering the large amounts of electricity produced by PC boilers in this country, to find that electric utilities and concrete manufacturers years ago began to develop methods for directly using fly ash from PC boilers as an admixture with Portland cement in concrete production. This development has been very successful. In 1986, for example, of 49 million short tons of fly ash produced in the United States, 9 million short tons (or 18%) were in beneficial use, half of this in concrete (*2*). This represents a major business activity, which has seen the establishment of its own "niche" trade organization, the American Coal Ash Association.

It may be noted in passing at this point that ash from the combustion of oil shale is also an excellent pozzolanic admixture for Portland cement. While this material has not yet been commercialized in the United States (in the absence of the commercial use of oil shale itself as an energy source), it has found acceptance in other countries, notably in China and certain of the republics of the former USSR (3).

The first use of coal fly ash with Portland cement was to repair a tunnel spillway at the Hoover Dam in 1942. The first relevant specifications for fly ash in concrete were promulgated by the American Society for Testing and Materials (ASTM) over twenty years later in the mid-1960's. The current version of these standards is the very comprehensive ASTM C 618. The amount of fly ash, having properties within the range allowed by ASTM C 618, in typical concrete applications, for example, will vary from 15 to 25 percent by weight with amounts up to 70 percent for massive walls and girders (2).

As noted above, however, less than 10 percent of the fly ash produced by PC boilers in the United States is used in concrete. The reason for this limited utilization is that only one-quarter of the fly ash produced in this country meets the ASTM C 618 standard. Thus, to be used beneficially the remaining three-quarters of the fly ash, as well as all of the bottom ash and boiler slag produced -- a total of nearly 60 million short tons per year, must find its way into products with less restrictive standards. About 10 million short tons per year have done so, going into such uses as concrete products (as aggregate), structural fills, road base, asphalt fillers, snow and ice control, blasting grit and roofing granules, grouting and mine back-filling (2).

The electric power industry continues to seek ways in which to expand the beneficial use of coal ash. In 1984 the Electric Power Research Institute (EPRI) established an Ash in Cement/Concrete Products research project (RP 2422). Its purpose is to better define coal ash characteristics as they relate to ultimate material properties of the wide range of products, mentioned in the previous paragraph. The project continues at this writing. Ultimately, EPRI will make recommendations for broadening the standards for these products, which will allow expanded utilization of coal ash and further reduce the need of the U.S. electric utility sector for landfills (4).

EPRI has also established an Advanced SO_2 Control Management research project (RP 2708) to define the characteristics of a new solid residual from coal-fired power plants -- by-products of SO_2 removal technologies, including:

- Atmospheric Fluidized Bed Combustion
- Calcium Spray Drying
- Limestone Furnace Injection
- Sodium Sorbent Duct Injection
- Calcium Sorbent Duct Injection

These materials all contain significant amounts of gypsum and other sulfates, along with the coal ash as previously generated. A variety of markets and products are being explored for these by-product materials, similar to the markets, listed above, which are already enjoyed by coal ash itself. Work on this project continues at this writing also. Preliminary research suggests that proper by-product conditioning and overall product mix design modification may result in acceptable product performance in "concrete-type" applications (5).

Beneficial Use of MSW Incinerator Ash

As landfill space becomes more limited, the operators of MSW incinerators are beginning to follow the same path toward possible beneficial use of ash from their plants, that electric power plant operators created for themselves over the past few decades. Ash from MSW combustors can have less than 10% of the original volume entering the facility. It should be noted, however, that for the public, which bears the ultimate ownership for MSW incinerators, there are two other approaches to reducing the amount of waste needing landfilling. Recycling, of course, is one of these, while waste minimization is the other. But no matter how much these other two approaches are implemented, there will always remain a residual whose volume can be effectively reduced in a waste-to-energy process, such as an MSW incinerator, and that waste-to-energy process will ultimately yield an ash.

Because of the very low quality of incinerator ash, it would appear rather inconceivable to use it as an admixture in Portland cement. The most commonly considered beneficial use for it, therefore, is as an aggregate in either bituminous or Portland cement concrete. This method has the further advantage, just as does the use as aggregate of bottom ash and boiler slag from coal-fired plants, of displacing sand and gravel, which must be mined from sometimes environmentally sensitive locations. Another short-term advantage of this method is its ability to bind the toxic metals, entering with the MSW (6), into the concreted mass. Controversy over the long-term implication of the presence of these metals in the concrete has arisen, however. The Environmental Defense Fund has expressed vigorous concern for the long-term release of these metals as the concrete eventually degrades or is reduced to rubble (7).

Utilizing MSW incinerator ash in Portland cement concrete has been investigated by several organizations. At a number of locations, bottom ash (slag) from high-temperature MSW combustors is utilized as coarse aggregate in regular concrete (8,9). Utilization of low-strength concretes containing either fly ash or mixed fly and bottom ash (combined ash) are moving toward commercial use in three directions.

One of these directions is to make block secondary products, such as artificial reef blocks, construction blocks, and shore protection devices (10,11). Artificial reef blocks, prepared at the State University of New York, Stony Brook, consist of 85% combined ash and 15% Portland (Type II) cement, and have compressive strengths of about 1000 psi (6.9 MPa) (10). By comparison, standard precast Portland cement concrete contains 40% coarse aggregate, 40% fine aggregate and 20% Portland cement and has a compressive strength of over 3500 psi (24.2 MPa).

Construction blocks, prepared at SUNY Stony Brook, consist of 35-60% combined ash, 25-50% sand, 15% Portland (Type IP) cement, and sufficient Acme-Hardesty superplasicizer to allow the mix to flow easily (10). Compressive strengths vary from 1600 psi (11.0 MPa) to 2600 psi (17.9 MPa). A boathouse, built recently at SUNY Stony Brook from 14,000 such construction blocks, is being evaluated for structural and environmental acceptability (10,12). Finally, shore protection devices require the use of a patented admixture, Chloranan (manufactured by Hazcon, Inc., of Brookshire, Texas) at a ratio of cement to admixture of 10:1. Using cement percentages between 17 and 33, compressive strengths up to 4200 psi (29.0 MPa) are reported (10).

An extension of the work at SUNY Stony Brook on the manufacturing of construction blocks containing incinerator ash is being conducted at SUNY Buffalo (11). There, concrete masonry units (CMU's or construction blocks) have been prepared using a modified ASTM C 109 standard. A wide range of ASTM standards have been used to characterize the ash which is used in the CMU's. In manufacturing the CMU's a variety of techniques, known to improve the durability, strength and sulfate resistance of Portland cement concrete, have been used to improve the commercial properties of the ash-containing mixes. Compressive strengths in excess of 1500 psi (10.3 MPa), satisfactory for commercial CMU's, have been obtained from mixes utilizing Portland (Type V) cement.

A second approach to MSW ash utilization is the accretion of combined ash with Portland cement (8-14%) into coarse aggregate for use in roadbeds and concrete (12,13). This level of concretization provides aggregate with strengths of about 1200 psi (8.3 MPa), similar to the material described in Reference 10.

Finally, a third approach is the stabilization of combined ash with Portland cement (6-10%) to create landfill covers (12,14). In all of these approaches, the ability of Portland cement concrete to reduce the leachability of trace metals from the concreted mass to meet TCLP standards has been a key element of their development.

The U.S. Environmental Protection Agency's Risk Reduction Engineering Laboratory (Municipal Waste Technology Section) is currently completing an evaluation of the effectiveness of five different solidification/stabilization processes for treating MSW residues (12,15). These processes include Portland cement only, Portland cement with polymeric additives, Portland cement with soluable silicates, waste pozzolans, and phosphate addition treatments. A major portion of the work is being conducted at the Waterways Experiment Station of the U.S. Army Corps of Engineers in Vicksburg, Mississippi. There, physical properties and durability testing include unconfined compressive strength (UCS), UCS after immersion, and wet/dry and freeze/thaw testing. Leaching characterization include TCLP, serial distilled water extraction, availability tests and monolith leaching tests (16). Results of this work are expected to be published in late 1991 (12).

A major area which these projects do not appear to address is the precast concrete market. For entrance into this arena, compressive strengths between 3500 psi (24.2 MPa) and 5500 psi (38.0 MPa) must be achieved. In addition, extensive physical testing must be applied and minumum standards met. These tests include tensile strength, freeze-thaw, deicing, and abrasion. As the preceding brief review shows, obtaining the compressive strengths required by precasters has proven impossible with significant amounts of combined ash, even with standard additives. Only those concretes made with the addition of large amounts of Chloranan enter this range.

The School of Engineering at the University of Pittsburgh has recently discovered an inexpensive method which permits high-strength concrete to be produced, containing large amounts of combined ash from a MSW combustor. The method, which has just been disclosed to the University as the first step in the patent process, will now be described as thoroughly as possible within the limits imposed by that process.

Portland Cement Concrete Manufacture

In October 1988 the University of Pittsburgh began to study the utilization of MSW combustor ash in Portland cement concrete. Four 750-lb (340-kg) samples of ash have been obtained during the past two years from the MSW combustor at Poughkeepsie, New York, operated for the Dutchess County Resource Recovery Agency by Dutchess Resource Energy, a subsidiary of the Resource Energy Systems Division of Westinghouse Electric Corporation. The primary combustor in this facility is an O'Conner water-wall rotary kiln. The bottom ash drops into a water-filled pit, from which it is reclaimed past a grizzly to take out large particles, and a magnet for iron removal. The hot gases from the primary combustor pass through a secondary combustor, a boiler, and a dry scrubber. The fly ash and spent limestone (injected in the scrubber for acid gas removal) are added to the bottom ash just before the combined ash is loaded into trailers for hauling to a land disposal site.

The first two samples of ash were composited from 5-pound (2.3-kg) sub-samples of combined ash, collected six times daily for 25 days. Ash Sample #1, combined ash collected from mid-July to early September 1989, contained 8-21% moisture, 15.1% calcium oxide, 2.21% sulfur trioxide and 7.2% iron oxide and exhibited a 7.26% loss of ignition. Ash Sample #2 was bottom ash collected during January 1990. Ash Samples #3 and #4 were bottom ash collected in essentially one quick draw on May 24-25, 1990, and March 1-2, 1991, respectively. Ash Sample #3 contained 11.9% calcium oxide, 2.58% sulfur trioxide and 9.8% iron oxide and exhibited a 8.04% loss of ignition.

Six batches of concrete were made with Ash Sample #1 and fifteen batches with Ash Sample #2. The basic recipe was 17% coarse aggregate, 40% ash and 43% Portland cement. An air entrainment additiive, a water reduction additive and a silica fume additive were all tried individually with certain of these batches. The slump of each batch was held as close as possible to 1.75 inches (4.45 cm) by varying the water content. Concrete batches were produced in a small commercial mixer in the Concrete Laboratory of the Civil Engineering Department, following standard procedures. A number of cylinders (3 inches (7.6 cm) in diameter and 6 inches (15.2 cm) long) were formed from each batch. Cylinders were stored in an environmentally controlled room. Compressive strengths were measured on a Baldwin hydraulic compression tester, also located in the Concrete Laboratory of the Civil Engineering Department. Sets of four cylinders were cracked, and the load at breaking were averaged to obtain reported values of compressive strength.

Compressive strengths were generally in the range of 1000 psi (6.9 MPa) to 2300 psi (15.9 MPa). The ash in the first six batches of concrete were subjected to the Standard Extraction Procedure Method (the EP TOX procedure). The extractions were performed in the Environmental Laboratory of the Civil Engineering Department and leachates were sent to a commercial laboratory for analysis. Neither the ashes nor the concretes exceeded the EP TOX limits for any constituent. However, the ash and one of the six concrete batches (which happened to contain a lower amount of Portland cement and a higher amount of coarse aggregate than according to the usual recipe, and which also contained some sand) exceeded the NYCRR limits for cadmium and lead. The presence of fly ash in Ash Sample #1 appeared to cause a number of "popouts". A white crystalline material, identified as a physical assemblage of aluminum chloride and calcium oxide crystals, was found at the focal point of all the popouts examined.

As Ash Sample #2 was running out, it was decided to attempt to modify the chemical and/or physical makeup of the surface of the ash particles to increase the strength of the bond between the ash and the cement. Two different commonly available chemicals were tried as additives in the last two batches of concrete made from Ash Sample #2. One of these novel additives did, in fact, yield compressive strengths of nearly 4000 psi (36.0 MPa). Therefore, a set of batches with varying amounts of this effective, inexpensive, novel additive were prepared with ash from Ash Sample #3. The results are shown in the Table I. The amount of additive is given as a percentage of the weight of cement present in each batch. A second batch of concrete, Batch Number 44, was made with the last portion of Ash Sample #3, using the same recipe as Batch Number 32.

Table I. Variation of Compressive Strengths with Additive

Batch Number	Additive (%)	90-Day Compressive Strength, Psi (MPa)
29	3.47	0 (0)*
30	1.73	4940 (34.1)
32	0.87	6200 (42.8)
36	0.65	3750 (25.9)
37	0.43	3210 (22.1)
34	0.17	2260 (15.6)
42	0.07	1560 (10.8)**

*   All cylinders of this batch broke apart by 90 days.
**   14-day compressive strength.

Samples of ash and concrete were examined in the Scanning Electron Microscope (SEM) in the secondary electron imaging mode. The phases present were also compared using X-ray microanalysis in the SEM. Figures 1 and 2 are examples of micrographs for Batches Number 42 and 44, respectively, in the secondary electron imaging mode. It appears from an examination of these micrographs that, even with a 12-fold increase in additive from Batch Number 42 to Batch Number 44, the same basic cement structure is present in both concretes.

Figures 3 and 4 are SEM micrographs of ash. Figure 3 is for Ash Sample #4 as received. Figure 4 is for Ash Sample #3, which has been mixed with an aqueous solution of the chemical additive, using the same amounts of ash and additive as was used in preparing Batch Number 32. It may be observed from an examination of these micrographs that the surface of the as-received ash is heavily contaminated with fine particles, which are not present on the treated ash.

It is hypothesized from the observations reported above that the elimination of the fine particles in the concrete formulations with the chemical additive renders the surface of the ash particles (serving as fine aggregate) more amenable to strong bonding with the hydrated Portland cement, yielding high-strength concrete as a product.

A preliminary cost analysis of one ton (908 kg) of concrete (dry basis), composed of

- 420 pounds (191 kg) coarse aggregate
- 750 pounds (340 kg) of ash
- 830 pounds (377 kg) of Portland cement
- chemical additive at the level of Batch Number 32

shows that purchase of the materials for its manufacture would require $23.50. The cost of materials for a 2:2:1 precast concrete is $10.00. The ash-containing concrete with the formulation of Batch Number 32 would be $13.50 more expensive than the standard concrete. Since the former contains 750 pounds (340 kg) of ash, the ash would have to be forced into this beneficial use with a tipping fee of $36.00 per ton ($79.30 per Mg), which is less than one-third of the current tipping fee at landfills.

Future Work

One final technical step in developing concrete with the same recipe as Batch Number 32 needs to be taken. Durability tests (tension strength, expansion, freeze/thaw, deicing and abrasion) and a TCLP extraction need to be carried out. When these results are available, precasters in the vicinity of the MSW combustor, operated by the Dutchess County Resource Recovery Agency, can make an informed technical decision on the use of the ash from this combustor in their products. Regulatory permission for this use of the MSW combustor ash would then have to be sought from the State of New York.

Several other aspects of the beneficial use of MSW combustor ash in Portland cement concrete should be examined. First, the current 0.5:0.9:1 recipe may not be the optimal one for commercial use. A range of formulations should be explored. Second, additives similar to the one used in these first experiments could be tried, especially if durability and TCLP tests uncover any problems. Third, ash from other types of combustors could be tested for the applicability of

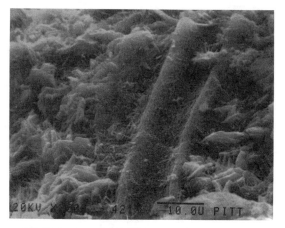

Figure 1.   SEM micrograph of the fractured surface of Batch Number 42,
made with 0.07% additive; the bar is 10 micrometers.

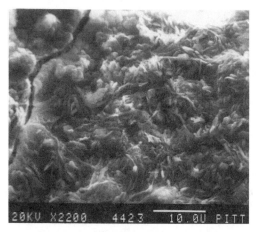

Figure 2.   SEM micrograph of fracture surface of Batch Number 44, made
with 0.87% additive; the bar is 10 micrometers.

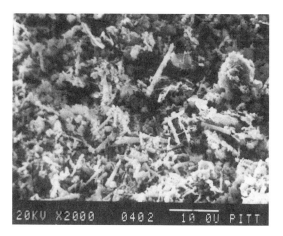

Figure 3.   SEM micrograph of Ash Sample #4 as received; the bar is 10 micrometers.

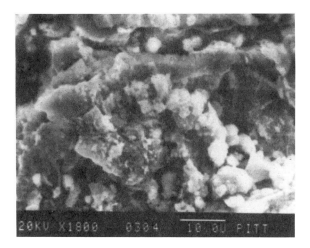

Figure 4.   SEM micrograph of Ash Sample #3, mixed with additive; the bar is 10 micrometers.

the new additive . These combustors would include other MSW combustors with different methods for ash removal from the unit. They could also include industrial and hazardous waste combustors.

Acknowledgements

The financial and technical support of the Resource Energy Systems Division of Westinghouse Electric Corporation (W-RESD) for this work is gratefully acknowledged. Dr. Suh Lee and Mr. Patrick Gallagher of W-RESD have given substantial advise and counsel. Mr. Richard Melville of Dutchess Resource Energy was most helpful in obtaining the ash samples.

References

(1)   Burnet, G. and Murtha, M. J. "Application of Current Utilization Technologies to Coal Solid Wastes of the Future," Proceedings of the Second Annual International Pittsburgh Coal Conference, September 16-20, 1985, pp 543-557.
(2) Jablonski, G. J. and Tyson, S. S. "Overview of Coal Combustion By-Product Utilization," Proceedings of the Fifth Annual International Pittsburgh Coal Conference, September 12-16, 1988, pp 15-22.
(3) Smadi, M., Yeginobali A. and Khedaywi, T., "Potential Uses of Jordanian Spent Oil Shale Ash as a Cementive Material," *Magazine of Concrete Research*, Vol. *41*, No. 148, pp 183-190 (**1989**).
(4)   Golden, D. M., "EPRI Coal Combustion By-Product Utilization R&D," Proceedings of the Fifth Annual International Pittsburgh Coal Conference, September 12-16, 1988, pp 32-46.
(5) Perri, J. S. and Golden, D. M. "Assessing Technical and Market Factors for Advanced SO_2 Control By-Product Utilization," Proceedings of the Sixth Annual International Pittsburgh Coal Conference, September 25-29, 1989, pp 915-924.
(6) Franklin, M. A., "Sources of Heavy Metals in Municipal Solid Waste in the United States, 1970 to 2000," Proceedings of the Third International Conference on Municipal Solid Waste Combustor Ash Utilization, Arlington, Virginia, November 13-14, 1990.
(7) Denison, R. A., "The Hazards of Municipal Incinerator Ash and Fundamental Objectives of Ash Management," presented at "New Developments in Incinerator Ash Disposal," Northwest Center for Professional Education, April 26, 1988.
(8)   Hartlen, J.," Incinerator Ash Utilization in Some Countries in Europe," Proceedings of the First International Conference on Municipal Solid Waste Combustor Ash Utilization, Philadelphia, Pennsylvania, October 13-14, 1988.
(9) Mahoney, P. F. and Mullen, J. F. "Use of Ash Products from Combustion of Shredded Solid Waste," Proceedings of the First International Conference on Municipal Solid Waste Combustor Ash Utilization, Philadelphia, Pennsylvania, October 13-14, 1988.

(10) Roethel, F. J. and Breslin, V. T. "Stony Brook's MSW Combustor Ash Demonstration Programs," Proceedings of the Third International Conference on Municipal Solid Waste Combustor Ash Utilization, Arlington, Virginia, November 13-14, 1990.

(11) Berg, E., "Utilization of Municipal Waste Incinerator Residuals as an Ingredient in Concrete Masonry Units," Paper No. 41d, 1991 Summer National Meeting of the American Institute of Chemical Engineers, August 18-21, 1991, Pittsburgh, Pennsylvania.

(12) "Treatment and Use: The Future of Ash Management?" *Solid Waste & Power*, October **1991**, pp 20-28.

(13) Huitric, KR. L., Korn, J. L. and Wong, M. M. "Characterization and Treatment of Ash Residue at the Commerce Refuse-to-Energy Facility," Proceedings of the Third International Conference on Municipal Solid Waste Combustor Ash Utilization, Arlington, Virginia, November 13-14, 1990.

(14) Chesner, W. H., "The Long Island Ash Utilization Program -- Preliminary Engineering, Economics and Environmental Findings, and Guidelines for MSW Combustor Ash Utilization," Proceedings of the Third International Conference on Municipal Solid Waste Combustor Ash Utilization, Arlington, Virginia, November 13-14, 1990.

(15) Wiles, C. C., "U.S. EPA's Research Program for Treatment and Utilization of Municipal Waste Combustion Residues," Paper No. 41b, 1991 Summer National Meeting of the American Institute of Chemical Engineers, August 18-21, 1991, Pittsburgh, Pennsylvania.

(16) Holmes, T., "Durability and Leaching Characteristics of Municipal Waste Combustor Residues Treated by Solidification/Stabilization during the USEPA Municipal Innovative Technology Evaluation Program," Paper No. 41c, 1991 Summer National Meeting of the American Institute of Chemical Engineers, August 18-21, 1991, Pittsburgh, Pennsylvania.

RECEIVED May 1, 1992

TECHNOLOGY DEVELOPMENT

Chapter 22

Assessing the Feasibility of Developing and Transferring New Energy Technology to the Marketplace

A Methodology

H. M. Kosstrin[1] and B. E. Levie[2]

[1]W. Beck and Associates, 51 Sawyer Road, Waltham, MA 02154
[2]W. Beck and Associates, 1125 17th Street, Suite 1900,
Denver, CO 80202

Any new process intended to produce clean energy from waste should be characterized for its ultimate feasibility of becoming commercially successful. A four-phase strategy to analyze the process and plan for scale-up is presented. First, the new technology is assessed in terms of its market potential based on laboratory, bench scale or pilot data. A comparison with competing commercial technology is performed to compare the technology with its competition by estimations of factors such as life cycle cost, public acceptance, and adaptability to changing conditions and fuels. Second, the current status is reviewed with respect to theory, laboratory or pilot scale results, and available cost data. Third, the path to commercialization is outlined. The stages of scale-up and data required to prove the concept and remove risks of commercialization are identified. Finally, the financing needs for the various stages of scale-up and for a commercial unit are determined.

Development of any new technology has traditionally been a controversial subject due to high expectations shared by proponents and results which many times fall short of these expectations. Solid and liquid waste management has seen both success and failure in the implementation of new technology. For example, promises to commercially produce liquid or gaseous fuels and/or chemicals from municipal solid waste (MSW) or refuse derived fuel (RDF) have so far been unfulfilled after several attempts at demonstrating various technologies. These failures encourage us to examine new and undeveloped technology in a more sophisticated and step-wise manner than has been previously done. By learning from past failures and taking a methodical and proactive approach to scaling-up suitable technology, we can better direct development so that realistic expectations can be made and met. The approach discussed here will increase chances for successful development of new waste management technologies.

The following phases outline the approach to be presented:

I.   Determine if a technology at its current state of development, either conceptual, bench, or pilot scale, can be potentially competitive with commercial technologies today.

II. Establish the current status of the technology and determine what needs to be better understood before progressing.

III. Establish the path which would most logically be taken to result in commercializing the technology.

IV. Identify the requirements of different financing options necessary to commercialize the technology.

These phases follow a progression in which the results of each builds on the results from the preceding phases. This review can be started at any time in the development process and should be updated to account for new data on the technology, the competition, or the market as they become available.

Review and analysis of new technology can be biased according to the perspective of the reviewer. The investigators, developers, and sponsors all have vested interests in the technology which may prevent a balanced view of the technology, its development, or its commercialization. Investors and lenders typically look for independent reviews of the technology prior to committing large amounts of capital. This can best be accomplished by persons without conflicts of interest and with an adequate background reviewing development of the technology.

Review of the Concept

In this initial stage of analysis, the technology is looked at objectively to assess its niche in current markets. The first step is to identify the market or markets where the technology would most likely be competing and to broadly establish a range of competitive pricing for the service provided or product produced. It is important to consider all areas where the technology could potentially compete, including those outside the primary field of interest. High value chemicals, resins, and plastics, for instance, may be more economically feasible to produce than fuels from certain feedstocks.

Questions to resolve before proceeding are those which would be important to an investor. These generally will establish if the market is potentially strong and lasting. In the area of solid waste management, the following questions can be used as guidelines to ascertain the market's potential. Similar questions can be developed for any particular field.

Is the market for the products expanding? A market analysis should be made to determine the value of the primary products at the location of the initial commercial facility and subsequent facilities. While the products may have a high value to the end user, transportation and broker costs should be accounted.

- <u>Is the market monopolistic or controlled by a few companies?</u>   Some commodity markets may be hard to break into due to certain restrictions. Long-term supply contracts may be in place with the end users. The end user may be unwilling to bear the risk of a new supplier who is unknown in terms of quality and interruptions in service. Significant discounts on product price may need to be made to obtain contracts which will lead to a track record of quality and supply.

- <u>What are the minimum or maximum requirements for waste needed to be processed?</u> There will be physical constraints on the process in terms of maximum throughput. Economic and/or physical constraints will dictate the minimum throughput possible. The local waste "market" and need for waste disposal must be examined to see if the technology makes sense for a given location.

- <u>Is the waste composition changing due to recycling, composting, changes in consumption, etc.?</u> The waste composition changes by locality, season, consumers spending habits, product packaging, and programs or technology put into place to manage waste. Source separation and recycling, yardwaste composting, automated mixed waste separation with recycling, waste reduction, and reuse can make significant changes to the remaining waste's composition.   The trends for waste composition changing should be determined to the excess possible.

- <u>Do long-term contracts for feedstocks already exist which would interfere with this technology?</u>   The feedstock (waste) must be available before a commercial facility can be introduced. While a market for products may be strong in a certain location, the ability to secure long-term waste system contracts may be poor.

- <u>What are the standards for the product(s) produced?</u> Product quality and consistency must be met to insure long-term markets. Product specifications should be identified and the process evaluated as to its ability to meet these specifications.   Examples of products are steam, electricity, fuel, and chemicals and recovered and/or processed materials.

- <u>Can environmental permits be obtained?</u> Environmental issues can become a barrier to the commercialization of a technology in a specific locale. Non-attainment areas will require additional analysis of options. State and local regulations may also increase the cost of compliance.

- <u>What is the public perception of the technology?</u> Public perception will vary from place to place. This is an intangible property which can prevent a technology from being developed.

-   <u>What are the characteristics of markets for byproducts of the process?</u> The byproducts of a process can make or break the economic feasibility of the technology. A byproduct here is something which is produced in a relatively low quantity, or something which has relatively little value, or no value at the location where it is produced. For instance a process which produces a high value chemical, may also make a gas with a reasonable BTU value. But in some cases, as with waste, the byproducts could be a concentrated mix of hazardous materials which could cost more to dispose of than the original MSW. Hazardous byproducts will also increase public resistance, as in the case of incineration ash from MSW waste to energy plants.

The next step in reviewing a new technology is to compare the technology with those commercial technologies currently in the identified market(s). This comparison can be as brief or extensive as is desired, depending on if we are considering a revolutionary change or just an evolutionary advance in the market. At a minimum, cost and environmental comparison should be made between the new technology and what is available in the market. The cost should be assessed on a life-cycle basis, accounting for capital, operation and maintenance, disposal of residue costs, and revenues from tipping fees, the primary product and any byproducts. The general environmental assessment could include a number of considerations including impacts on air, water, workers, flora and fauna.

There are many other considerations in performing this initial assessment. The feedstock must be compatible with the technology, and the product(s) compatible with the existing markets. Flexibility can be quite advantageous in the waste management industry, as the quantities and composition of waste is rarely fixed. While some technologies might only be competitive for a certain type and quantity of waste, others could take many types of waste, in a range of quantities. Effects on other related technologies should be assessed, as today municipalities and other organizations are interested in integrated solid waste management. Generally no one technology can solve the waste problems for a given location. Thus, technologies which can work effectively together may be more desirable than those which prevent other technologies and strategies from being employed successfully.

In order to compare the new technology with existing ones, it must be emphasized that the new technology should be judged on a realistic basis. A conservative estimate for costs, revenues and efficiency of the new process should be used for comparison purposes. Often a new technology assessment underestimates commercialization costs and greatly over-estimates potential revenues from products.

Establishing Current Status of the Technology

The second phase of this review is to establish the current status of the new technology, providing a baseline or framework from which further development can be compared. The initial limited economic feasibility developed in Phase I, can be updated with new information gained in this phase. Technical and economic gaps in knowledge should be identified in this phase and either resolved now or targeted for later development work and/or analysis.

Existing Data Review

This stage of review is many times performed by the researchers in order to propose further expenditures or justify previous funding. Therefore, some data may already exist for this analysis.

The first part of establishing the current status is to verify that the process proposed is physically possible and practically attainable. This will require checking previous assumptions, reviewing theory and obtaining correct parameter values for thermodynamic, kinetic and mass and heat transfer. Mass and energy balances should be done to check process feasibility. A second law analysis could be performed on the process to identify inefficiencies. This second analysis can identify the appropriateness of the technology with respect to energy conversion. It can also help to point to process steps which could be improved to yield more efficient conversion rates.

Once the theory has been reviewed, operational data from the lab, bench, and/or pilot facility should be assessed to determine the deviation from theory. This will allow a better estimate of expected yields as the process is scaled-up further. It will also serve to highlight areas where the process can be improved or is not performing as well as expected. In some cases, it will point to the fact that the data is inaccurate or insufficient for reasonable analysis and that additional and more accurate data must be obtained before further progress can be made. It is important in this review that sufficient data be available to determine the precision of the data. This review should determine if enough replicates have been run to achieve the desired confidence in the results. QA/QC protocol used to perform experiments and acquire data should be reviewed and verification of its use may be established. This may be a weakness in many developing technologies, as sufficient independent testing to produce adequate statistical data can be costly.

There should be an adequate review of the instrumentation and data acquisition system to determine any measurement biases which exist. For example, biases occur in high temperature measurements, and when measurements are made close to the detection limit of the instrumentation. There may be a need to perform an uncertainty analysis. This allows for a couple of benefits. First, we can see how all the measurement errors are propagated into the errors of calculated quantities such as conversion efficiency. Secondly, the analysis can show which measurement errors are the most critical to measure accurately. For example, a flow measurement can be 100 times as important as a temperature measurement in closing a mass balance.

Once a thorough review of the available data has been accomplished, we need to update our original economic model. Existing cost data should be reviewed to better establish costs of the technology at its present state of development. These costs should be segregated as much as possible into standard technologies and developmental technologies to identify which areas need more accurate estimates as development proceeds. If possible, costs for each piece of equipment or unit operation should be tabulated.

Many costs will not be available based on pilot plant data, such as upstream and downstream equipment which may not be implemented at this stage of development. But this equipment can be estimated if standard technology is used. Equipment in this category may include material waste recovery systems, gas cleaning, liquid cleaning, heat recovery equipment, and emissions, effluent and residue treatment systems.

Costs for operation and maintenance (O&M) are difficult to determine as pilot scale or smaller equipment will rarely run for long continuous periods of time. Some costs may be determined such as on energy requirements, energy losses, and other requirements of the process such as gases, water, or other utilities. Costs associated with running the process for long periods of time will generally not be available. But preliminary estimates can be made, and ranges input to the economic model to establish a current economic status.

Technical and Economic Questions

The review of the current status will raise various questions on both a technical and cost basis. Technical questions which are easily resolved with current equipment should be addressed as soon as possible prior to going on to this Phase.

It is generally far less expensive to acquire data at the initial stages of development than later on, and this data can provide many benefits. The additional data taken may indicate unusual phenomena occurring which need to be understood for successful scale-up. Extra information may point to flaws in the technology such as larger heats of reaction than calculated, poor kinetic rates, or poor catalytic activity. Such results might be indicated using extra thermocouples, calorimetry, pressure transducer, etc. These may be economical to measure at this stage of development, but not once the technology is scaled up. Instruments used for pilot plant scale equipment can be less durable due to less demanding use and shorter operations requirements than in a commercial plant. Pilot plant work allows the fixing of varying operational parameters widely without the risk of interfacing with revenues or ruining costly equipment.

Development is a time to understand the processes as well as possible in order to be able ultimately to optimize and control it. This can be even more important with a variable feedstock like waste, since we need to be able to predict performance of a technology for changes in the waste stream. Modeling of the process is an important tool to use to take our understanding of the process and mathematically describe it. The model can eventually be used as a tool for scaling up the process. Discrepancies between actual operating data and theoretical projections may indicate poorly understood phenomena, inaccurate data, or invalid assumptions in the theory. These technical data gaps may need to be filled before further progress should be attempted. Otherwise, scale up may bring with it some unforeseen problems.

Technical questions which may be unanswerable include environmental impacts, reliability of equipment over time, labor necessary to run and maintain operations full time, and degradation of the process over time due to unknown phenomena. These questions will need to be revisited in later phases of development, and should be noted to trigger later activity.

Economic questions which may not yet be answered should be identified at this point and noted for later resolution. These may include questions of costs for upgrading the products and byproducts for sale, prices for the products and byproducts, and disposal costs for residues and effluents. Some of these questions are best left for later stages in development, when more representative products and residues will be produced. By initially establishing costs of upgrading or treating products or residues, it may be revealed that further consideration needs to be given to different methods of treatment. This may need to be worked on before, or concurrent to scaling-up.

Establishing a Path to Commercialization

Now that the current status of the emerging technology has been established and we have updated the economic model with new information, which still projects a competitive technology, we can establish how to proceed. This third phase of development can consume fairly large amounts of capital, so a critical assessment should be conducted to establish a deliberate agenda so that an investor may be convinced to fund this phase.

The initial task of this phase is a risk assessment to identify any technical flaws in the concept, and establish a plan to address and overcome any obstacles. As an example, the process data from bench scale operations has confirmed the kinetic viability of the process but has left unanswered certain mechanical questions. For instance, we know the reactor works but we have assumed in our model a feeder that can use unprocessed feedstock. The problem identified is, how do we introduce the solid feedstock in a uniform, continuous manner without excessive preprocessing. This risk assessment, which should include all components, is intended to identify components of the process that either require further development prior to proceeding to the first scale-up or to find an acceptable alternative.

For solid waste, feeding and residue removal have been historical non-trivial unit operations. There are still problems observed in feeding and ash removal in state-of-the-art incinerators which operate at pressures slightly below atmosphere to prevent refugee emissions. With reactors demanding air tight seals for control of certain processes, solids feeding and removal takes on a more important role and is naturally more complex.

The final piece of the risk assessment is to critically look at the question of scale-up. One may ask the question, how far can we proceed, in this initial step from bench scale? But the right question is, what is the maximum scale-up possible from the final development unit to the commercial demonstration? Answering this question is a key to determining the total path to commercialization. We can then decide on how many scale-up steps to take and when critical components should be scaled-up. These steps may include any or all of the following: an integrated pilot plant, a semi-works to prove out critical components or a complete demonstration system.

After planning the global technical approach, and the required component development has been identified, we need to feed any revisions to our overall cost model to reconfirm feasibility. The next stage is to determine the additional technical data, whether mechanical or process, that is required.

Typical questions which help identify such data include:

- Does each component work as intended? The importance of each component to a commercial facility's success cannot be over-emphasized. The failure or chronic problems associated with a single component can mean an inability to meet availability guarantees, which will have a direct financial consequence.

- Does the system as a whole work together in a safe manner? Safety is a key element of commercial operation. Without the ability to 1) build passive and active safety protection and 2) maintain a strong safety record, the process will never garner interest of financial institutions due to the risk and liability associated with it.

- Does proper selection of materials take into consideration "corrosive and erosive" elements in the process? Solid waste, even when put through separation equipment such as trommels, magnets, and air classifiers, contains metal and glass. These can cause wear on surfaces with which it contacts, and at higher temperatures will lead to fouling and corrosion of surfaces. Proper design of equipment in handling these effects is essential for long-term and cost effective operation.

In addition, the duration of acquiring the answers to these questions should be established. Typical goals of this first scale-up may be 5,000 hours of total test time with perhaps 1,000 hours of continuous operation under design conditions. The purpose of this scale-up is to work through the operational and process problems, confirm yields and product quality, and obtain an indication of reliability. The "other" objective is to be able to again refine the economic model with the data obtained from this first scale-up for both capital and operations costs. We will need this information, since we are approaching the time that additional capital will be needed for the next scale-up or for a continuously operational demonstration facility.

Once we have established technically what type of data and scaled-up system the technology requires, we need to establish a cost of this phase of the work and raise additional capital. At this stage, it is important to consider if any revenue can be derived from the operation of the development unit to offset the operational costs. This may not be realistic, but an investor typically likes to see some "pay as you go" operation while development is progressing.

After the development unit has completed the technical data acquisition, it is again time to refine our feasibility model with new cost data, operational data and reliability data. The level of success of the development unit will at times determine the type of financing that the process developer can consider. The final section discusses the various options.

The First Commercial Unit - How Can It Be Financed

As we complete the previous phase, the additional data collected from the scaled-up operation is again fed back into the economic model to reconfirm feasibility. A positive result will now enable the project to proceed to raise the capital necessary to build a full scale facility which, by definition, when successful will be the first commercial unit.

Financing can be obtained from a variety of sources, ranging from total equity, where the investor assumes all the risk, to non-recourse project financing where the risk of failure is divided between the lender and the equity participant. Technologies concerned with the disposal or processing of solid waste are currently eligible to obtain tax exempt bonding authority. The lower cost of capital by using tax exempt bonds is a commonly used method to enhance the overall economics of a project.

As we proceed to raising the required funds to build the first commercial project, it is important to understand the risk posture of the various sources of capital that are available to build the first commercial plant. Two general types of capital are available, loans and equity.Lenders, either banks, insurance companies or the public bond market, have one basic concept in common. Their upside potential, in any project, is a return of capital with interest at the agreed upon rate. While the downside risk, if the project totally fails, is a total loss of capital. Equity participants, although they have the same potential loss as a lender, have a much larger upside potential if the project is successful. Due to this basic difference between risk and reward, a lender typically assumes a much more conservative view of a project that is going to be a stand alone business entity.

As a direct consequence of this risk/reward posture of various sources of capital, the cost of this capital, to a technology developer, differs. Although equity capital does not carry a defined interest rate, suppliers of this type of capital desire a substantially higher return on that capital than does a lender.

Funding a new technology using non-recourse project financing, typically requires either some level of equity participation or a guarantee to pay back the debt, or some combination of the two. The level of equity participation or debt guarantee depends on the characteristics of the project and the projected economics as determined by an Independent Engineering Review. This Independent Engineering Review is critical to both lenders and equity participants, since it is intended to confirm both the technical and economic viability of the technology.

From a lender's point of view, the typical characteristics of a strong project include some or all of the following:

o   A turnkey construction contract including a fixed price, fixed completion date, detailed performance test and penalties for non-performance.

o   An operations and maintenance contract with a fixed price and incentives for positive performance and penalties for poor performance.

o   Independent projections based on the technology and contract structure which show adequate cash flow to cover all expenses and debt service.   These projections should be done for both the expected operational scenarios and in cases where potential problems may arise that are either technical or economical in nature.

In the ideal world of a supplier of capital only second of a kind facilities would be financed, thereby minimizing any technology risk. Today we are seeing more first of a kind facilities looking for funding.   The more detailed the process of technology development, producing data that can be evaluated in a positive light by an independent engineer, should allow the financing of a first commercial facility with a maximum of debt capital.

Raising the capital for a new technology can be as challenging as completing the technical development. However, this job is easier when the proper groundwork has been laid by following the methodology presented here.

RECEIVED June 15, 1992

INDEXES

Author Index

Affiliation Index

Subject Index

Bestsellers from ACS Books

The ACS Style Guide: A Manual for Authors and Editors
Edited by Janet S. Dodd
264 pp; clothbound, ISBN 0–8412–0917–0; paperback, ISBN 0–8412–0943–X

Chemical Activities and Chemical Activities: Teacher Edition
By Christie L. Borgford and Lee R. Summerlin
330 pp; spiralbound, ISBN 0–8412–1417–4; teacher ed. ISBN 0–8412–1416–6

Chemical Demonstrations: A Sourcebook for Teachers,
Volumes 1 and 2, Second Edition
Volume 1 by Lee R. Summerlin and James L. Ealy, Jr.;
Vol. 1, 198 pp; spiralbound, ISBN 0–8412–1481–6;
Volume 2 by Lee R. Summerlin, Christie L. Borgford, and Julie B. Ealy
Vol. 2, 234 pp; spiralbound, ISBN 0–8412–1535–9

Writing the Laboratory Notebook
By Howard M. Kanare
145 pp; clothbound, ISBN 0–8412–0906–5; paperback, ISBN 0–8412–0933–2

Developing a Chemical Hygiene Plan
By Jay A. Young, Warren K. Kingsley, and George H. Wahl, Jr.
paperback, ISBN 0–8412–1876–5

Introduction to Microwave Sample Preparation: Theory and Practice
Edited by H. M. Kingston and Lois B. Jassie
263 pp; clothbound, ISBN 0–8412–1450–6

Principles of Environmental Sampling
Edited by Lawrence H. Keith
ACS Professional Reference Book; 458 pp;
clothbound; ISBN 0–8412–1173–6; paperback, ISBN 0–8412–1437–9

Biotechnology and Materials Science: Chemistry for the Future
Edited by Mary L. Good (Jacqueline K. Barton, Associate Editor)
135 pp; clothbound, ISBN 0–8412–1472–7; paperback, ISBN 0–8412–1473–5

Personal Computers for Scientists: A Byte at a Time
By Glenn I. Ouchi
276 pp; clothbound, ISBN 0–8412–1000–4; paperback, ISBN 0–8412–1001–2

Polymers in Aqueous Media: Performance Through Association
Edited by J. Edward Glass
Advances in Chemistry Series 223; 575 pp;
clothbound, ISBN 0–8412–1548–0

For further information and a free catalog of ACS books, contact:
American Chemical Society
Distribution Office, Department 225
1155 16th Street, NW, Washington, DC 20036
Telephone 800–227–5558